THE OFF-GRID LIVING BLUEPRINT

Complete & Proven DIY Solutions to Achieve 100% Self-Sufficiency. Secure all essential resources, fortify your home & safeguard your family's future through any crisis

by

Jim K. Rockwell

Table of Content

Welcome to This Journey 5

Module A. Establishing Your Off-Grid Home Base 6
    1. Foundational Principles for Constructing an Off-Grid Residence 6
    2. Deep Dive on Location Factors: Choosing the Ideal Spot for Your Off-Grid Home 7
    3. Strategic Budgeting and Resource Management 9
    4. Disaster-Proofing Structures: Building for Extreme Weather 12
    5. Long-Term Sustainability and Resource Management 13
    6. Off Grid Foundational Projects 14

Module B. Water Self-Sufficiency 19
    7. Water Sourcing and Collection 19
    8. Water Storage Solutions 25
    9. Purification Techniques: Ensuring Safe Water Off the Grid 29
    10. Water Distribution 37
    11. Backup and Emergency Storage Strategies 42
    12. Water Self-Sufficiency Module Projects 44

Module C. Energy & Power Systems for Total Independence 63
    13. Generating Power: Effective Off-Grid Energy Production 63
    14. Storing Energy: Power Reserves for Low-Output Times 65
    15. Distributing Power Safely & Efficiently 68
    16. Optimizing Energy Usage for Maximum Efficiency 71
    17. Energy and Power Independence Projects 75

Module D. Waste & Recycling Solutions 90
    18. Composting Toilets: Eco-Friendly Off-Grid Sanitation 90
    19. Greywater Recycling & Wastewater Treatment 93
    20. Biogas Production: Converting Organic Waste into Usable Energy 95
    21. Hazardous Waste Disposal: Handling Batteries, Chemicals, and Electronics Safely 98
    22. Household Waste Management & Off-Grid Recycling 100
    23. Waste & Recycling Solutions Projects 101

Module E. Security & Property Defense 107
    24. Fortifying Your Property: Keeping External Threats at Bay 107
    25. Personal & Home Defense Strategies 110
    26. Security & Property Defense Projects 115

Module F. Sustainable Food & Self-Sufficiency 121
    27. Growing Food: Off-Grid Farming & Gardening Techniques 121
    28. Fundamentals for Raising Livestock Off the Grid 123

29. Hunting, Fishing, & Foraging 127

30. Food Preservation and Storage 129

31. Hydroponics & Aquaponics: Advanced Off-Grid Food Production 140

32. Sustainable Food & Self-Sufficiency Projects 143

Module G. First Aid & Medical Emergency Preparedness 161

33. Emergency Medical Preparedness & First Aid 161

34. Trauma Care & Managing Medical Emergencies Off-Grid 163

35. First Aid & Medical Emergency Preparedness Projects 166

Module H. Cooking, Heating, Cooling & Lighting 169

36. Cooking Without Power: Alternative Cooking Methods 169

37. Heating Systems: Warming Your Space Efficiently 170

38. Natural Cooling: Keeping Your Home Comfortable 172

39. Off-Grid Lighting Solutions 174

40. Cooking, Heating, Cooling & Lighting Projects 175

Module I. Essential Skills, Tools & Communication 182

41. General Repair Works, Basic Carpentry, and Tool Use 182

42. Communication Methods: Staying Connected Beyond Traditional Grid Services 183

43. Basics of Vehicle Maintenance 184

44. Essential Skills, Tools & Communication Projects 186

Module J. Community Building & Financial Preparation 189

45. Starting an Off-Grid Community 189

46. Financial Planning: Budgeting for Sustainable Independence 191

47. Income Opportunities: Monetizing Skills and Resources Off the Grid 192

48. Barter Systems: Trading Skills and Goods 194

49. Community Building & Financial Preparation Projects 195

Module K. Off-Grid Healthy Living, Mental & Emotional Preparedness & Medical Preparedness 198

50. Building a Resilient Body 198

51. Mental & Emotional Preparedness for Off-Grid Living 200

52. Managing Medications & Chronic Illness Off the Grid 200

53. Sustainable Hygiene & Natural Sanitation 202

54. Off-Grid Healthy Living, Mental & Emotional Preparedness & Medical Preparedness Projects 203

Module L. Essential Outdoor Survival & Disaster Readiness 207

55. Essentials of Bug-Out Planning & Bug-Out Location 207

56. Shelter Building: Staying Protected in Any Environment 209

57. Wilderness Water Solutions: Finding, Purifying & Storing Water 213

58. Navigation Without GPS: Finding Your Way in Any Terrain 215

59. .Emergency Power Solutions 217

60. Fire-Starting: Reliable Methods for Any Condition	220
61. Handy Knots Everyone Should Learn	223
62. Essential Outdoor Survival & Disaster Readiness Projects	225

Thank You 235

Access Your Extra Content 236

References 237

Disclaimer and Legal Notice

This publication is provided strictly for educational and entertainment purposes. Every reasonable effort has been made to ensure that the information contained herein is accurate, current, and reliable. However, the author and publisher make no guarantees or warranties, either express or implied, regarding the completeness, accuracy, or suitability of the content for any specific purpose.

Nothing in this publication should be interpreted as legal, financial, medical, or professional advice. The strategies, methods, and suggestions presented are general in nature and may not be appropriate for your individual circumstances. Before relying on or applying any information from this material, you are strongly advised to consult with a qualified professional in the relevant field.

By accessing and using this publication, you expressly acknowledge and agree that the author and publisher accept no responsibility or liability for any loss, damage, or harm—whether direct, indirect, incidental, or consequential—that may result from the use or misuse of the information provided. This includes, but is not limited to, any errors, omissions, inaccuracies, or misinterpretations.

Copyright © 2025. All Rights Reserved.

No portion of this publication may be copied, reproduced, stored in a retrieval system, or transmitted in any form or by any means—electronic, mechanical, photocopying, recording, or otherwise—without the prior written consent of the publisher. Exceptions are granted only for brief excerpts used in reviews, academic works, or other legally permissible non-commercial purposes as allowed under applicable copyright law.

Welcome to This Journey

Bringing this guide to life has been a mission fueled by purpose and passion. This guide was created for people like you—those ready to take control of their lives, secure their future, and build true independence in an uncertain world. Every strategy, and solution inside these pages is designed to give you practical, proven tools to thrive no matter what challenges come your way.

Knowing that this knowledge is now in your hands gives this work meaning. Together, we are part of a growing movement—one that values self-sufficiency, resilience, and the freedom that comes from living in harmony with the land and on your own terms. Thank you for stepping into this way of life and for becoming part of a community that's redefining what it means to be prepared.

Before you roll up your sleeves and dive into the projects, plans, and techniques that will help you achieve 100% self-reliance, there are two things you can do to make sure you get the most out of this blueprint:

1) Unlock Your Extra Materials

We've pulled together a collection of exclusive resources to support your off-grid lifestyle. These extras go beyond what's in the book—offering checklists, templates, and detailed project guides to make your journey smoother and your results stronger. To access them, flip to the section titled "**Access Your Extra Content**" at the back of this guide.

2) Share Your Excitement

Your voice has the power to inspire others to take charge of their future. Whether you're starting fresh or already deep into your off-grid transformation, we'd love to hear what this journey means to you. There are a couple of easy ways to share:

Option I: Create a short video sharing what excites you most about building your self-sufficient life.

Option II: Snap a photo of the book in action—on your workbench, in your garden, or wherever it's fueling your progress or just share a few written thoughts about how exited you are to start this journey.

By sharing, you're helping others find the courage to begin—and together we're strengthening a community of resourceful, capable people who are ready for anything.

Scan the QR code below to share your excitement!

Module A. Establishing Your Off-Grid Home Base

1. Foundational Principles for Constructing an Off-Grid Residence

A house connected to the grid is simple. Electricity, water, and waste management are handled by massive infrastructures that you rarely think about. When you flip a switch, power flows. When you turn on a faucet, water appears. But when you remove the grid, every resource becomes something you must actively create, store, and manage. That changes everything.

Building an off-grid home isn't just about putting up walls and a roof—it's about designing an entire self-sustaining system that provides for your needs, whether that means generating electricity, collecting water, or regulating indoor temperatures. Every decision, from site selection to materials to orientation, influences how efficiently your home will function without external support.

Understanding Self-Sufficiency in Off-Grid Home Design

A conventional home can afford inefficiencies. A poorly insulated house will still stay warm if enough electricity is pumped into the heating system. A high-water-use household won't notice the waste as long as municipal supply is constant. An off-grid home doesn't have these luxuries. It must be self-sufficient, energy-efficient, and climate-responsive from the very beginning.

At its core, an off-grid home functions on a few fundamental principles:

• It harvests its own resources—power, water, heat, and sometimes even food.
• It minimizes energy loss through smart design, natural insulation, and passive heating and cooling.
• It reduces reliance on mechanical systems by using architecture itself as part of the solution rather than relying entirely on generators, pumps, and filters.
• It is designed for resilience, capable of operating efficiently even when weather, seasons, or unforeseen events challenge normal function.

Rather than forcing modern comforts into a space that fights against nature, the key is to design a home that works with the environment rather than against it.

Site Selection: The Foundation of Efficiency

The location of an off-grid home is the single most critical factor influencing its performance. Land determines water availability, energy potential, soil quality for food production, exposure to natural hazards, and even how much insulation the structure will require.

A well-chosen site naturally provides many of the essential components of survival, reducing the complexity of off-grid systems and making the entire project more cost-effective.

Some factors to assess before building include:

• Sun exposure. In colder regions, maximizing southern exposure (in the Northern Hemisphere) increases passive solar heating. In hotter climates, shading and airflow take priority.
• Wind patterns. Natural windbreaks from trees, hills, or constructed barriers can prevent heat loss in winter, while openings can encourage cooling breezes in the summer.
• Water access. Proximity to natural water sources (springs, underground aquifers, rain catchment potential) reduces the need for energy-intensive water transport.
• Soil quality. Even if gardening isn't a priority at first, soil quality will influence future food production and even building materials if using techniques like adobe or rammed earth.
• Elevation and drainage. Proper elevation prevents flooding, and good drainage ensures stable foundations and healthy water storage conditions.

A well-positioned home will require less artificial heating, cooling, and energy input, making it more sustainable in the long run.

Designing for Passive Heating and Cooling

Heating and cooling account for the majority of energy use in a home. When a house is grid-connected, the easy solution is to install a heater or air conditioning unit and forget about it. Off-grid living demands a smarter approach.

Passive solar design uses the natural movement of the sun to regulate indoor temperatures. A house that takes advantage of passive solar heating can stay warm even in cold winters without relying on artificial heat sources.

Key elements of passive solar design include:

- South-facing windows (Northern Hemisphere) to maximize solar gain in winter.
- Thermal mass walls and floors (stone, concrete, adobe) to absorb heat during the day and release it at night.
- Roof overhangs designed to block high summer sun while allowing winter sunlight to penetrate.
- Cross-ventilation windows placed to encourage airflow, reducing summer heat buildup.
- Earth-sheltered construction (partially built into a hillside) for natural insulation.

A well-designed home can reduce heating and cooling energy needs by as much as 80 percent, making it much easier to maintain comfort without large energy inputs.

Choosing the Right Building Materials

An off-grid home must be structurally resilient, well-insulated, and made from materials that complement its environment. The right materials can provide natural insulation, reduce long-term maintenance, and even be sourced from the land itself.

Some of the most effective materials for off-grid construction include:

- Adobe and cob – Naturally insulating, fire-resistant, and made from local soil.
- Rammed earth – Extremely durable, with excellent thermal mass.
- Straw bale construction – High insulation value, ideal for cold climates.
- Earthbag construction – Affordable, strong, and adaptable to various climates.
- Recycled or reclaimed materials – Shipping containers, salvaged wood, and repurposed building materials reduce costs and environmental impact.

The choice of materials should be dictated by climate, availability, and the specific demands of the land.

Building for Water and Waste Management

Unlike a grid-connected home, where waste disappears down a drain and water flows endlessly from the tap, an off-grid house must handle every drop of water as a precious resource and every waste product as something to be managed responsibly.

A successful water system includes:

- Rainwater harvesting with proper filtration.
- Groundwater wells with backup hand pumps.
- Greywater recycling for irrigation.
- Composting toilets to avoid septic system dependence.

The more integrated the home is with its environment, the less energy and effort will be required to sustain it.

Resilience: Designing for the Unexpected

An off-grid home is not just a shelter—it's a survival system. That means designing for redundancy, adaptability, and self-reliance.

- Redundant energy sources ensure that power is available even if one system fails.
- Multiple water sources prevent shortages.
- Adaptable spaces allow for future expansions or changes in household needs.

The goal is not just to build a house but to create a living system that can sustain itself indefinitely.

2. Deep Dive on Location Factors: Choosing the Ideal Spot for Your Off-Grid Home

Here we will deep dive about the location choice. An off-grid home is only as strong as the land it stands on. No matter how well-built a house is, if the land does not support sustainability, the lifestyle will become a struggle. The choice of location determines access to water, the efficiency of energy production, natural insulation, food production potential, and overall security. It affects everything from how much effort is needed to maintain systems to how well the home withstands environmental changes.

Selecting the right spot is a decision that should never be rushed. A beautiful piece of land may seem ideal at first, but without considering long-term factors such as water access, climate adaptability, soil quality, and exposure to natural disasters, it can turn into an expensive and impractical challenge. The key is to analyze every potential location through a practical lens, ensuring that it supports self-sufficiency rather than making survival more difficult.

Evaluating Sunlight Exposure and Natural Energy Potential

Sunlight dictates how much solar energy can be harvested, how well plants grow, and how much passive heat a home can retain during cold months. Not all land is equal in terms of sun exposure. A property that is mostly shaded by tall trees or mountains will struggle to generate solar power, while one in an open, sunny location may experience overheating in the summer.

The first step in evaluating a site is observing how sunlight moves across it throughout the day. This will determine where to position solar panels, gardens, and the house itself. If a home is being built in a cold climate, maximizing southern exposure (in the Northern Hemisphere) or northern exposure (in the Southern Hemisphere) allows for natural heating. If the land is in a hotter region, shade from trees or terrain becomes an important cooling factor.

To further evaluate the land's potential, it helps to track the sun's movement over several days. This can be done by marking shadow lengths and observing how much direct sunlight reaches different areas at different times. If a home is placed incorrectly, it may receive too much shade in winter or too much heat in summer, forcing reliance on artificial heating and cooling.

Wind exposure is another crucial factor. A well-placed home takes advantage of natural wind patterns, using them for cooling while protecting against harsh winter winds. Observing wind movement on the land will help determine where to position windows for cross-ventilation and where windbreaks like trees or berms may be needed.

Water Availability and Drainage

Water is one of the most critical resources for off-grid living. Without reliable access to fresh water, survival becomes a constant struggle. A site must have either a natural water source, such as a well, spring, or river, or enough rainfall to support collection and storage.

When evaluating a property, the first priority is to determine where water will come from. Wells provide a long-term source, but not all locations have accessible groundwater. Rainwater collection is another solution, but it requires large storage capacity and proper seasonal planning. Surface water, such as a pond or stream, may seem like a good option, but it requires filtration and legal access rights must be verified.

Water drainage is just as important as water access. A property that floods after heavy rain can create problems with foundation stability, erosion, and water contamination. The best way to assess drainage is to visit the land after a storm and observe how water moves across it. Areas where water pools for extended periods may not be suitable for building.

A well-positioned home will be built on high ground, reducing the risk of flooding while allowing for effective water collection from roofs and land contours.

Soil Quality and Food Production Potential

Growing food on-site reduces reliance on external sources and provides long-term food security. However, not all land is suitable for cultivation. Soil must be tested for nutrient content, drainage capability, and contamination before assuming that it will support crops.

A simple test involves taking a handful of soil and observing its texture. Sandy soil drains too quickly and lacks nutrients, while heavy clay retains too much water and can suffocate plant roots. Loamy soil, which has a balanced mix of sand, silt, and clay, is ideal for gardening.

If the land does not have fertile soil, food production is still possible using raised garden beds, composting, or hydroponic and aquaponic systems. However, these methods require additional infrastructure, which must be accounted for in planning.

Terrain, Accessibility, and Safety

A property's terrain affects how difficult it will be to build and maintain an off-grid home. Steep land may offer good protection from wind but can make construction more complex and costly. Flat land is easier to develop but may be prone to flooding. Ideally, the land should have a natural balance, offering both elevation for drainage and flat areas for construction.

Accessibility is another factor that can be overlooked. A remote location offers privacy and security, but if it is too difficult to reach, transporting materials and supplies becomes expensive. Checking road conditions and seasonal access issues will help determine if the location is practical for long-term living.

The site should also be assessed for potential hazards. This includes checking for:

- Risk of wildfires in dry areas
- Earthquake-prone zones
- Landslide potential on steep slopes
- Dangerous wildlife that could threaten livestock or crops

A secure location should have a balance of isolation and accessibility, ensuring that it is self-sufficient but not impossible to reach in case of emergencies.

Legal Considerations and Zoning Restrictions

Even in remote areas, legal restrictions can impact off-grid living. Some locations have zoning laws that limit alternative building techniques, water rights that restrict collection, or land-use regulations that prevent full self-sufficiency.

Before committing to a property, research:

- Zoning laws and building codes for off-grid structures
- Water rights and restrictions on collection or well drilling
- Land use regulations regarding farming, energy production, and waste management
- Permitting requirements for alternative construction methods

In some regions, off-grid homes may need to meet specific codes for septic systems or energy generation. Ensuring that the chosen location allows for full independence prevents legal issues and fines in the future.

Evaluating the Community and Local Resources

A truly isolated location may seem appealing, but complete separation from society can create difficulties when it comes to supplies, medical care, and emergency situations. Understanding the local community and available resources helps determine if the location is practical for long-term living.

Some considerations include:

- Proximity to the nearest town for supplies and medical care
- Availability of local materials, such as lumber or stone, for building
- Community attitudes toward off-grid living and self-sufficiency

A supportive community can make off-grid living easier by providing trade opportunities, shared knowledge, and emergency assistance.

By carefully assessing all these factors, an ideal location can be chosen that supports long-term sustainability and independence.

3. Strategic Budgeting and Resource Management

Building an off-grid home is more than a construction project. It is an investment in self-sufficiency, long-term security, and sustainable living. Every dollar spent should bring lasting value, not just immediate convenience. Unlike a grid-connected home, where you can make changes later with minimal impact, an off-grid home demands careful financial planning from the start. Once built, every system must function efficiently, or unexpected costs will quickly become a burden.

Budgeting for off-grid living is not just about the upfront costs of land and materials. It also requires planning for ongoing maintenance, system expansions, and emergency contingencies. Proper resource management ensures that every aspect of the home, from energy to water to food production, is both cost-effective and sustainable.

Understanding the True Cost of Off-Grid Living

It is easy to underestimate how much it costs to go off-grid. Some assume that once they leave the grid behind, expenses will disappear. The reality is different. While off-grid living eliminates utility bills, it introduces new costs—water storage, energy systems, and long-term infrastructure maintenance. The key is to shift spending from recurring expenses to one-time investments that provide lasting benefits.

The total cost of going off-grid depends on several factors, including land price, climate, and lifestyle expectations. The fewer luxuries you require, the lower the cost. However, cutting corners in critical areas, such as insulation, energy storage, or water access, can lead to higher long-term expenses.

Some of the biggest costs to account for include:
- Land purchase and development
- Water sourcing, storage, and filtration
- Power generation and battery storage
- Building materials and labor
- Heating and cooling solutions
- Food production and storage
- Waste management systems
- Tools and equipment for maintenance and repairs

A well-planned off-grid budget is not about spending the least amount possible but about spending wisely to avoid future costs and inefficiencies.

Prioritizing Essential Investments

A strategic budget starts with defining what is essential and what can wait. Some expenses are non-negotiable, while others can be phased in over time. The goal is to start with the core systems that allow for basic survival and expand when resources allow.

The first step is to categorize expenses into three groups: immediate needs, short-term additions, and long-term upgrades.

Immediate needs are the foundation of an off-grid life. These are the systems that must be in place before moving to the land:

- A reliable water source and storage system
- A shelter with adequate insulation
- A basic power system (solar, wind, or generator)
- Sanitation solutions (composting toilet, septic system, or greywater recycling)

Short-term additions improve efficiency and comfort. These include:
- Additional solar panels or battery capacity for extended power use
- Expanded water filtration and storage to cover seasonal variations
- Food production systems, such as raised garden beds or a greenhouse
- More advanced heating and cooling solutions, including passive solar design enhancements

Long-term upgrades are investments that further improve resilience and independence. They include:
- Alternative energy sources, such as micro-hydro or biomass energy
- Larger-scale food production, such as aquaponics or small livestock
- Advanced water recycling systems for near-total independence from external sources
- Expanding storage and backup systems **to handle long-term emergencies**

A budget focused on these priorities prevents unnecessary spending on non-essential features in the early stages.

Finding Affordable Land and Reducing Development Costs

Land is often the most expensive part of an off-grid project. Choosing the right property can mean the difference between long-term sustainability and constant struggle. While a large, remote property may seem ideal, it comes with added costs for access roads, water transport, and energy infrastructure.

The most cost-effective land for off-grid living has:

- Natural water sources nearby, reducing the need for expensive well drilling
- A good mix of sunlight and shade to balance energy generation and cooling needs
- Fertile soil, if food production is a priority
- Existing structures or salvageable materials that can be repurposed

Some land purchases include restrictive zoning laws or building codes that make off-grid construction difficult. Researching local regulations before purchasing can save thousands in unexpected legal or compliance costs.

Clearing land, building roads, and preparing a homesite can add significant expenses. Cost-saving strategies include:

• Choosing a site that requires minimal land clearing
• Using existing natural features, such as hills for wind protection and tree cover for insulation
• Repurposing natural materials from the site, such as logs for construction or stone for retaining walls

Reducing land development costs frees up more of the budget for essential off-grid systems.

Managing Energy Costs and Avoiding Overinvestment

Many first-time off-grid builders make the mistake of either overspending or under-investing in their energy system. A large, expensive solar array may be unnecessary if the home is designed efficiently. Likewise, underestimating energy needs can lead to costly system upgrades later.

The first step is to reduce power consumption before building the energy system. This can be done by:

• Designing the home for passive heating and cooling
• Using energy-efficient appliances and lighting
• Implementing manual alternatives, such as hand pumps instead of electric pumps

Once energy demand is minimized, the next step is to determine how much power is actually needed. A small solar array with battery storage is often sufficient for basic lighting, refrigeration, and electronics. A larger system may be required for high-energy appliances, such as power tools or heating elements.

Backup power solutions, such as generators or propane appliances, add reliability without requiring an oversized primary energy system. This allows for a lower upfront cost while still ensuring energy security.

Water and Food Budgeting Strategies

A self-sufficient water system eliminates the need for expensive municipal connections, but it requires a combination of storage, filtration, and conservation techniques. The most cost-effective water systems rely on multiple sources, including rainwater collection, wells, and surface water.

A rainwater harvesting system with adequate storage can significantly reduce reliance on groundwater. Large storage tanks are expensive, but costs can be reduced by:

• Using multiple smaller tanks instead of one large system
• Repurposing used containers or building underground cisterns
• Implementing water-efficient fixtures to reduce consumption

Food production is another major expense, but a well-planned system pays for itself over time. Instead of immediately investing in large-scale farming, a staged approach allows for gradual expansion. This includes starting with high-yield, low-maintenance crops and expanding into permaculture systems over time.

Reducing Waste and Maximizing Material Use

Waste is a hidden cost in off-grid construction. Every discarded material represents lost money and resources. A waste-conscious approach to building and resource management includes:

• Salvaging and repurposing materials from demolition sites
• Using modular construction techniques that minimize offcuts and waste
• Choosing materials that serve multiple functions, such as thermal mass walls that provide both structure and insulation

A well-planned off-grid home requires fewer external resources, making it cheaper to build and maintain.

Emergency Funds and Long-Term Financial Planning

Even with careful budgeting, unexpected costs arise. Equipment failures, extreme weather, and medical emergencies can strain finances. Setting aside an emergency fund ensures that essential repairs and upgrades can be made without financial hardship.

A long-term financial strategy includes:

- Building a surplus of essential supplies, reducing the need for emergency purchases
- Learning DIY repair and maintenance skills to reduce reliance on outside labor
- Developing small-scale income streams to cover unexpected expenses

Strategic budgeting ensures that an off-grid home remains sustainable, not just in theory but in practice. Proper resource management makes it possible to build and maintain a resilient, self-sufficient home without excessive costs.

4. Disaster-Proofing Structures: Building for Extreme Weather

Every home faces environmental challenges, but an off-grid home must be built to withstand the harshest conditions with minimal reliance on external repair services. The structure itself must provide protection against storms, wildfires, heavy snowfall, earthquakes, and any other natural forces that the region is prone to. Designing with resilience in mind ensures that the home remains safe and functional no matter what nature throws at it.

Understanding Regional Threats

Before construction begins, the specific environmental risks of the location must be identified. A home built in a hurricane-prone area will need different reinforcements than one in an earthquake zone. Each climate and terrain type presents its own challenges, requiring strategic design choices to mitigate risks.

The most common environmental hazards include:

- High winds, tornadoes, and hurricanes
- Flooding and excessive rainfall
- Earthquakes and shifting ground
- Wildfires and extreme heat
- Heavy snow loads and freezing temperatures

Understanding these risks allows for the selection of materials, structural reinforcements, and positioning strategies that minimize exposure to potential disasters.

Reinforcing the Structure Against Wind and Storms

High winds can cause extensive damage to roofs, windows, and even entire structures. Wind-resistant design focuses on reducing uplift forces, reinforcing the building envelope, and using aerodynamically stable shapes.

A low, aerodynamic design reduces wind resistance, helping the home deflect rather than absorb strong gusts. The use of reinforced roof tie-downs and anchored walls prevents structural failure during storms.

Windows and doors are the most vulnerable points in a windstorm. Impact-resistant glass or storm shutters protect against flying debris. Entryways should be positioned away from prevailing winds whenever possible.

The foundation must be securely anchored to prevent the home from shifting. Concrete slab foundations or deeply embedded footings provide stability in high-wind regions.

Flood-Proofing and Water Management

Flooding is one of the most destructive forces a home can face. Water damage compromises foundations, weakens walls, and leads to mold and structural decay. Choosing an appropriate site is the first step in flood prevention. High ground is always preferable, reducing the need for artificial elevation.

For locations where flooding is unavoidable, building on stilts or raised foundations keeps living spaces above rising water levels. Grading the land to direct water away from the home prevents accumulation near the foundation.

Drainage systems such as French drains or gravel trenches guide excess water away from critical areas. A properly sloped roof with extended overhangs prevents water from pooling near walls and entrances.

Materials should be selected for their ability to withstand water exposure. Concrete, stone, and pressure-treated wood offer higher resistance to moisture than standard lumber.

Earthquake-Resistant Construction

Seismic activity places extreme stress on structures, requiring special reinforcements to prevent collapse. A flexible yet sturdy frame allows the home to absorb movement without breaking.

Wood and steel perform better in earthquakes than rigid materials like brick or concrete. If using masonry, reinforced concrete with embedded rebar prevents cracking under stress.

A symmetrical design distributes forces evenly, reducing the risk of structural failure. The foundation should be deeply anchored to bedrock or stable soil layers to prevent shifting.

Furniture and storage areas should be secured to walls to minimize falling hazards during tremors. Heavy objects should be placed low to the ground, and shelving should have built-in restraints to keep items from being thrown.

Fire-Resistant Home Design

For homes built in wildfire-prone areas, every material and design choice must prioritize fire resistance. A fire-resistant home relies on a combination of defensible space, non-combustible materials, and passive firebreaks.

Vegetation should be cleared in a buffer zone around the home, reducing the likelihood of flames reaching the structure. Fire-resistant exterior materials such as stucco, metal roofing, and treated wood reduce the risk of ignition.

Windows are vulnerable to heat exposure. Tempered glass or roll-down metal shutters provide protection during a fire. Roofs should be designed to prevent embers from lodging in gaps, with non-flammable coverings like metal or tile.

5. Long-Term Sustainability and Resource Management

Survival in an off-grid setting is not just about withstanding extreme weather—it is about ensuring that the home remains functional and self-sufficient for decades. Every resource must be carefully managed to prevent depletion, with systems in place for long-term sustainability.

Energy Independence and System Efficiency

An efficient off-grid energy system is designed with redundancy to prevent complete failure if one component malfunctions. Solar power is the most common solution, but it should be supplemented with wind, hydro, or biomass energy where possible.

Energy storage capacity determines how well the system handles fluctuations in weather. Battery banks must be properly sized to store excess energy for low-production periods.

Passive design features reduce overall energy consumption. A well-insulated home requires less heating and cooling, while strategically placed windows provide natural lighting and ventilation.

Energy usage should be monitored with a simple tracking system, allowing adjustments based on seasonal variations. Appliances should be selected for efficiency, prioritizing low-energy models.

Water Conservation and Reuse

Water availability fluctuates with climate and seasonal changes. A secure water system incorporates multiple collection and storage methods to prevent shortages.

Rainwater harvesting is a key component of off-grid water management. Storage tanks must be large enough to sustain long dry periods, with proper filtration to ensure safe use.

Greywater recycling extends the utility of water by repurposing it for irrigation. Simple filtration systems remove contaminants, allowing wastewater from sinks and showers to support plant growth.

Soil Health and Food Security

An off-grid homestead must produce its own food in a way that preserves soil health for future generations. Over-farming depletes nutrients, leading to poor crop yields and increased reliance on outside inputs.

Composting organic waste replenishes soil fertility naturally. Crop rotation and cover crops maintain nutrient balance, preventing degradation over time.

Food storage is equally important as food production. A well-ventilated root cellar extends the shelf life of fresh produce, reducing dependency on electricity for refrigeration.

Resource management requires long-term planning, ensuring that every aspect of the home functions efficiently without unnecessary waste. A well-maintained home built for resilience remains a secure and self-sufficient shelter regardless of external conditions.

6. Off Grid Foundational Projects

Project: Site Assessment and Sun Tracking Exercise

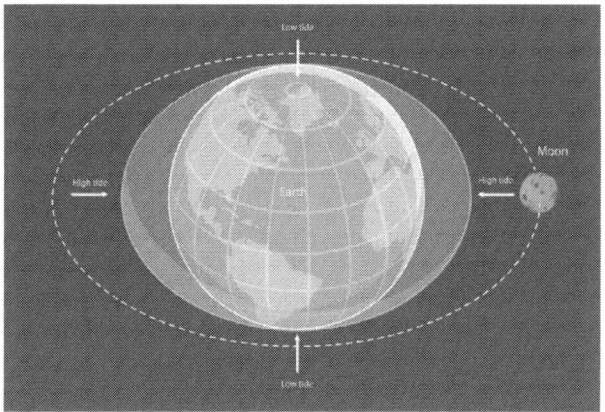

Building an off-grid home starts long before the first foundation stone is laid. The land you choose and how you position your home on it will determine its efficiency, sustainability, and long-term resilience. This project will guide you through assessing a potential site for your home, tracking sun exposure, evaluating wind patterns, and identifying natural resources like water access and insulation opportunities.

This exercise will give you firsthand knowledge of how your chosen location interacts with the elements, allowing you to design your home in a way that works with nature rather than against it.

Materials Needed
- Notebook or printed site assessment sheet
- Compass (or a smartphone with a compass app)
- Measuring tape (or a rough way to estimate distances)
- Colored markers or pencils
- A large piece of paper for sketching your site (or a digital mapping tool)
- A thermometer (optional but useful)
- Camera or phone to document findings

Step 1: Choose a Site to Assess

If you already own land, use this project to analyze your chosen home location. If you're still looking for land, select an area you're considering, or use a local park or open space to practice.

- Walk the site at different times of the day and get a general feel for the landscape.
- Identify key features like trees, hills, slopes, water sources, and any existing structures.
- Take note of areas that feel warmer, cooler, windier, or more protected.

If your land is large, choose a specific portion you think might be ideal for building. Mark it mentally or with small boundary markers.

Step 2: Track the Sun's Movement

Understanding how the sun moves across the land is essential for positioning your home in a way that maximizes natural light, warmth in winter, and shade in summer.

1. Find True South (or True North in the Southern Hemisphere)
- Use a compass or a smartphone app to determine where south is. This is where your largest windows should ideally face to capture sunlight in colder months.

2. Observe Shadows Throughout the Day
- At sunrise, mid-morning, noon, mid-afternoon, and sunset, stand in the center of your site and take note of the sun's position and the length of shadows.
- Pay special attention to how much direct sunlight reaches different areas of the land.

3. Document Seasonal Differences (If Possible)
- If you are in a location for multiple months, observe the shifting position of the sun at different times of the year.
- If you are short on time, use an online sun path calculator (such as suncalc.org) to estimate the seasonal movement of the sun at your location.

Step 3: Identify Wind Patterns

Wind direction plays a crucial role in heating, cooling, and overall comfort. A well-positioned home can use wind to its advantage while avoiding excessive heat loss in winter.

1. Look for Natural Windbreaks
- Identify hills, tree lines, or rock formations that could block cold winter winds.
- If your site is exposed, consider where you might plant trees or build fences to create wind protection.

2. Measure Air Movement
- Stand in different areas of the land at different times of the day and note when the wind is strongest.
- Use a light fabric, a ribbon, or even wet fingers to feel which direction the wind is coming from.
- If possible, observe how the wind shifts during storms or heavy weather patterns.

3. Mark Prevailing Winds on Your Map
- Strong winds from the north? You may need a windbreak.
- Gentle breezes from the south in summer? You may want to position windows to allow airflow.

Step 4: Identify Water Access and Drainage

Water is essential for any off-grid home. Even if you plan to rely on rainwater harvesting, understanding how water naturally flows across your land will help with placement of water storage, gardens, and home foundations.

1. Look for Natural Water Sources
- Are there springs, ponds, or wells nearby?
- Does the land collect rain in certain areas?

2. Observe Water Drainage After Rain
- If possible, visit your site after a rainstorm to see where water pools and how quickly it drains.
- If water collects in an area for too long, that may not be an ideal spot for a foundation.

3. Mark Potential Water Collection Points
- Identify high ground for a home site and lower areas for potential water retention ponds or rain collection.

Step 5: Sketch Your Findings

Once you have gathered information, create a rough sketch of your site. It does not need to be perfect—just a visual representation of what you observed.

1. Mark the sun's path, with notes on where sunlight is strongest and where shadows fall.
2. Indicate wind directions, noting strong or gentle breezes.
3. Draw existing features, such as trees, slopes, or water sources.
4. Mark potential home placements based on your findings.

Project: Off-Grid Expense Forecasting and Prioritization

Building an off-grid home requires careful financial planning. Every dollar spent should contribute to long-term self-sufficiency. This exercise will help create a structured budget, breaking down essential investments, future upgrades, and contingency plans. By the end, there will be a clear roadmap for where money needs to go and how to allocate resources efficiently.

Materials Needed
- Notebook or spreadsheet software
- Calculator
- Internet access for researching costs
- Printed or digital list of off-grid components (land, water systems, energy, materials, tools, etc.)
- A ruler or graph paper if creating a manual budget layout

Step 1: Identify Core Budget Categories

The first step is to list every major expense involved in setting up and maintaining an off-grid residence. Categories should cover both one-time purchases and ongoing costs.

Start with these essential groups:
- Land acquisition and site preparation
- Shelter construction and materials
- Water sourcing, storage, and filtration
- Energy systems (solar, wind, backup generators)
- Food production (gardening, livestock, preservation)

- Waste management (composting toilets, septic, greywater systems)
- Tools, equipment, and maintenance supplies
- Emergency and contingency funds

If additional categories apply, include them. It is better to overestimate expenses now than to be caught off guard later.

Step 2: Research and Estimate Costs

Each category now needs specific cost estimates. Some costs will vary depending on location, availability of materials, and personal preferences.

For each item within a category:

- Research average market prices
- Compare multiple suppliers for better deals
- Consider DIY options versus purchasing ready-made solutions
- Note any additional costs, such as installation fees or transport

A useful strategy is to create a price range instead of a fixed number. Some materials or equipment may fluctuate in price, and it is important to have a realistic upper and lower estimate.

Step 3: Categorize Expenses by Priority

Not all expenses are immediate. Some investments must be made upfront, while others can wait. Sorting expenses into three groups helps in structuring spending wisely.

Immediate Needs: These are non-negotiable. They include:

- Access to clean water
- Basic shelter with insulation
- A functional power system
- A sanitation setup (composting toilet or septic system)

Short-Term Investments: These will improve off-grid living but are not required for initial survival. They include:

- Additional solar panels or battery storage
- Garden expansion and food preservation tools
- More advanced water filtration
- Secondary structures like sheds or workshops

Long-Term Upgrades: These enhance efficiency and resilience over time. They include:

- Expanding renewable energy sources
- Installing automated energy management systems
- Building a root cellar or large-scale food storage
- Upgrading insulation for better climate control

Step 4: Create a Budget Framework

With expenses categorized, it is time to put everything into a structured budget.

- List the three priority categories in separate sections
- Assign estimated costs for each item, allowing room for price variations
- Identify areas where costs can be reduced by using salvaged materials, DIY solutions, or secondhand equipment
- Set a timeline for major purchases, ensuring that immediate needs are covered first

A table or spreadsheet works well for organizing this information. A simple layout should show:

- Expense category
- Item description
- Estimated cost (low and high)
- Priority level (Immediate, Short-Term, Long-Term)
- Notes on potential savings or alternatives

Step 5: Plan for Contingencies and Unexpected Costs

No budget is complete without accounting for unforeseen expenses. Equipment failures, medical needs, or extreme weather events can create sudden financial demands.

A realistic budget should include an emergency fund for:

- Unexpected repairs (solar panels, water pumps, heating systems)
- Medical and first aid supplies
- Backup food and water reserves
- Alternative energy sources if the main system fails

Setting aside a percentage of the total budget for contingencies ensures stability in unpredictable situations.

Step 6: Adjust and Optimize

Once the initial budget is complete, review it critically.

• Identify areas where spending can be reduced without compromising quality
• Look for opportunities to repurpose materials instead of buying new
• Adjust the timeline to align with available funds, ensuring that critical needs are covered first

Regularly revisiting and updating the budget ensures that spending remains efficient and aligned with long-term goals.

Project: Home Resilience Assessment and Reinforcement Plan

A well-built off-grid home must be prepared for extreme weather conditions, whether it's high winds, flooding, earthquakes, or wildfires. The goal of this project is to assess structural vulnerabilities and create a reinforcement plan to improve durability and safety. By systematically identifying weak points and applying cost-effective strengthening techniques, the home will become more resistant to environmental hazards.

Materials Needed

• Notebook and pen for documentation
• Measuring tape
• Camera or phone for taking reference photos
• Compass to determine wind direction
• Level for checking structural alignment
• Online access for researching reinforcement methods
• Basic construction tools (hammer, drill, screwdriver, wrench)
• Protective gear (gloves, safety glasses)

Step 1: Identify Key Environmental Risks

Before assessing the home's resilience, determine the primary weather threats in the area. The risks will vary depending on climate and geography. Some locations face frequent high winds, while others may deal with heavy snowfall, wildfires, or flooding.

1. Research local weather history and natural disaster patterns.
2. Observe the surrounding landscape for signs of past extreme weather damage, such as fallen trees or eroded land.
3. Use a compass to determine prevailing wind directions.
4. Identify whether the land is in a flood zone or has soil that is prone to shifting.
5. Check government hazard maps for earthquake fault lines or wildfire-prone regions.

Recording these risks will provide a reference for prioritizing reinforcement efforts.

Step 2: Structural Inspection for Wind and Storm Resilience

High winds can cause severe damage to homes that are not properly anchored. Wind-resistant construction focuses on roof stability, wall reinforcement, and securing openings.

1. Inspect the roof for weak or loose shingles. If any are missing or damaged, note them for repair.
2. Check the connection between the roof and walls. Roof tie-downs or hurricane straps should be in place to prevent uplift.
3. Examine windows and doors for gaps or signs of structural weakness. Reinforcement may be needed to withstand strong winds.
4. Inspect exterior walls for cracks or loose siding. Proper sealing and bracing will improve wind resistance.
5. Ensure that all outdoor structures, such as sheds and water tanks, are securely anchored to the ground.

If the home is located in a high-wind area, additional protection can be added, such as windbreaks or reinforced window coverings.

Step 3: Flood Risk Evaluation and Prevention

Flooding is one of the most common causes of structural damage. Water can weaken foundations, cause mold growth, and lead to long-term material deterioration.

1. Identify the lowest points around the home and observe where water collects after rainfall.
2. Check if the home is built on a natural slope. A higher elevation reduces flood risk.
3. Examine the foundation for cracks, gaps, or signs of water seepage. These areas need to be sealed with waterproof materials.
4. If the home is in a flood-prone area, consider raising outdoor electrical systems and using water-resistant construction materials.
5. Ensure that gutters and drainage systems are clear of debris to allow proper water flow.

For added flood protection, the landscape can be modified with trenches or French drains to redirect excess water away from the home.

Step 4: Earthquake-Resistant Construction Assessment

Seismic activity places intense pressure on a home's foundation and walls. Reinforcing structural elements helps prevent major damage in the event of an earthquake.

1. Check the foundation for visible cracks. Weak points in the foundation must be sealed to prevent structural failure.
2. Inspect support beams and columns. They should be securely fastened with steel brackets or seismic ties.
3. Examine wall connections to ensure they are properly braced and secured.
4. Check shelving, storage, and heavy furniture to ensure they are anchored to walls or floors to prevent falling during an earthquake.
5. Identify potential loose objects that could become hazards in a strong tremor and secure them with fasteners or brackets.

A reinforced foundation with flexible framing materials absorbs seismic movements better than rigid construction. Retrofitting older buildings with additional bracing improves earthquake resistance.

Step 5: Fireproofing for Wildfire-Prone Areas

Homes in wildfire-prone regions require additional protection to prevent ignition. Fire-resistant materials and proper landscaping can reduce the risk of flames reaching the structure.

1. Check the roofing material. Metal, tile, or asphalt shingles are more fire-resistant than untreated wood.
2. Clear dry vegetation within a buffer zone around the home. Maintaining a defensible space reduces the spread of fire.
3. Inspect exterior walls. Fire-resistant materials such as stucco, fiber cement, or stone provide better protection than traditional wood siding.
4. Ensure that all vents are covered with fine mesh to prevent embers from entering.
5. If there is an outdoor water source, confirm that it is accessible for emergency fire suppression.

Fire-resistant modifications should be combined with a fire escape plan to ensure safety in the event of an emergency.

Step 6: Create a Reinforcement Plan

Once vulnerabilities have been documented, the next step is to create a reinforcement plan. This plan should include:

1. A list of the most urgent improvements needed based on environmental risks.
2. Estimated costs and materials required for reinforcement.
3. A timeline for making improvements based on budget and urgency.
4. Identifying which reinforcements can be done as DIY projects and which require professional assistance.

Reinforcement efforts can be completed in stages, starting with the most critical upgrades. Each improvement adds to the overall durability of the home, ensuring that it remains a secure shelter under extreme conditions.

Module B. Water Self-Sufficiency

Water is the most fundamental resource for survival, yet many off-grid homesteads struggle with consistent access to clean, reliable water. Unlike urban areas where municipal supplies are available at the turn of a tap, living off-grid requires careful planning to secure water from natural sources, store it safely, purify it for drinking, and distribute it efficiently. A well-designed water system ensures that every drop is used wisely, minimizing waste while maximizing sustainability.

This module provides a complete framework for achieving long-term water independence. It begins with identifying and collecting water from viable sources, whether through wells, rainwater harvesting, or natural water bodies. Storage strategies are explored to ensure water remains safe and accessible throughout seasonal fluctuations. Effective purification methods are covered to eliminate contaminants and make water safe for drinking. The module also addresses efficient distribution systems, ensuring water reaches where it's needed with minimal effort and energy use. Finally, crisis preparedness strategies help safeguard against drought, contamination, or unexpected supply disruptions.

Mastering water self-sufficiency means eliminating dependence on external systems, reducing vulnerability to environmental changes, and ensuring a steady water supply for generations.

7. Water Sourcing and Collection

Identifying and Evaluating Water Sources

Water is more than just a necessity—it is the foundation upon which an off-grid life is built. Without reliable access to clean water, every other survival system falls apart. Unlike those who live in urban environments, where water flows effortlessly from a faucet, an off-grid homesteader must think ahead, ensuring that enough water is available every day, through every season, and in every unexpected situation.

Finding the right water source is not as simple as choosing the closest stream or digging a hole in the ground. It requires a deep understanding of local geography, climate, and sustainability, as well as careful planning to ensure the source remains viable for years to come. A river might seem convenient, but if it dries up in summer or floods in the spring, it's unreliable. A well may provide clear water today, but if the underground aquifer is depleted faster than it can refill, it won't last. Even rainwater, one of the most adaptable water sources, must be carefully calculated to ensure that storage capacity is sufficient during dry months.

Water scarcity is a challenge for many who seek self-sufficiency. The key to long-term water security is redundancy. No single water source is entirely foolproof. If a well pump fails, rainwater storage can step in. If a drought lasts longer than expected, surface water collection might become necessary. By understanding all the available water sources and how to make them work together, an off-grid home can ensure continuous and sustainable water independence.

Groundwater: The Reliability of Wells, Boreholes, and Springs

Groundwater is one of the most reliable water sources available, provided it can be accessed. It is naturally filtered as it moves through layers of rock and soil, reducing the likelihood of contamination compared to surface water. However, groundwater does not exist uniformly everywhere. In some locations, it is close to the surface, accessible with a simple hand-dug well. In other areas, it lies deep beneath layers of bedrock, requiring expensive drilling. Before deciding on groundwater as a primary source, an evaluation of depth, sustainability, and quality is essential.

The first step in groundwater assessment is determining how deep the water table is. This varies greatly by region. Some areas have groundwater within a few feet of the surface, while others require drilling through hundreds of feet of rock. The deeper the water table, the more expensive and difficult it is to access. A well dug too shallow risks drying out during hot months or after prolonged droughts, making it unreliable.

Equally important is the recharge rate of the aquifer. Just because a well produces water today doesn't mean it will continue indefinitely. If water is drawn faster than the underground reservoir can replenish itself, the well will eventually fail. Some aquifers recharge quickly, particularly those in areas with frequent rainfall and porous soil. Others, such as those beneath arid landscapes, replenish at an incredibly slow rate, sometimes taking decades or even centuries to recover lost water.

Quality is another critical factor. Groundwater is usually cleaner than surface water, but that doesn't mean it is free of contaminants. Certain regions have water naturally high in iron, sulfur, or other minerals, which may require filtration before use. Additionally, wells located near agricultural fields, industrial zones, or septic systems may be at risk of contamination from pesticides, fertilizers, or bacteria. Testing well water before committing to it as a primary source is essential for safety.

Springs offer a unique advantage over wells, as they naturally bring groundwater to the surface, eliminating the need for pumps or extensive drilling. A properly developed spring can be an effortless source of fresh, continuously flowing water. However, like wells, springs must be protected from contamination. If improperly sealed, surface runoff can pollute the water, introducing bacteria and debris. A spring box—a covered, sealed collection system—can keep the water clean while allowing for efficient collection.

Surface Water: Abundant but Challenging

Rivers, lakes, and ponds are among the most visibly abundant water sources, and their accessibility makes them tempting for off-grid water collection. However, surface water presents more challenges than any other water source. It is the most vulnerable to contamination, highly variable in availability, and often subject to strict legal restrictions.

Unlike groundwater, which is protected beneath layers of soil and rock, surface water is exposed to everything around it—rain runoff carrying pesticides, bacteria from wildlife, industrial pollutants, and natural contaminants like algae and silt. This means that even in areas where lakes and rivers seem abundant, they cannot be used without filtration and disinfection. Drinking untreated surface water is one of the most common causes of waterborne illness in self-sufficient living.

The location of the collection point also plays a role in determining water quality. Water drawn from the middle of a lake or river is typically cleaner than that near the shore, where stagnant water collects debris and organic material. The depth of collection matters as well—deeper water is usually cooler and less contaminated than surface water, where bacteria and algae thrive.

For those who rely on surface water, a proper collection and filtration system is essential. A settling pond can help remove debris before the water enters the main storage system. A layered filtration setup—including sand filters, charcoal filters, and biological filtration—can help improve water quality before final purification through boiling, UV treatment, or chlorination.

Rainwater Harvesting: A Sustainable and Adaptable Option

Among all water sources, rainwater harvesting is one of the most versatile and sustainable. It requires no drilling, no energy-intensive pumps, and can be collected directly where it is needed. However, its reliability depends entirely on local rainfall patterns.

A successful rainwater harvesting system begins with an understanding of how much rain actually falls in the area. A roof with a surface area of 1,000 square feet (93 square meters) can collect 620 gallons (2,350 liters) of water per inch (2.5 cm) of rainfall. That means that in a location with 20 inches (50 cm) of annual rainfall, over 12,400 gallons (47,000 liters) can be collected each year.

Storage is the limiting factor. Rain doesn't fall in perfectly timed amounts—some months might bring heavy downpours, while others may be completely dry. This means that a rainwater system must have enough storage capacity to bridge the gap between rainy and dry periods.

A well-designed system also includes filtration before storage. A first-flush diverter removes the initial runoff, which may contain dirt, dust, and contaminants. Screens and fine mesh filters help prevent debris from entering the storage tank, reducing sediment buildup and bacterial growth.

Atmospheric Water Collection: Extracting Water from the Air

In regions where traditional water sources are unreliable, alternative water generation methods such as atmospheric water generators (AWGs) and condensation traps can provide a steady water supply.

AWGs extract moisture from the air and convert it into liquid water. These systems work best in humid environments, but their energy consumption is significant, making them impractical for most off-grid situations unless paired with a dedicated renewable energy source.

A lower-tech alternative is fog nets, which capture moisture from mist and fog, directing it into collection containers. While their output is limited, they can be useful as a supplemental water source in coastal and mountainous regions.

By combining multiple water sources—groundwater, surface water, rainwater, and emergency alternatives—a self-sufficient home can ensure year-round water security, even in the face of changing environmental conditions.

Legal and Environmental Considerations

Water is the most critical resource for off-grid living, yet it is also one of the most regulated and environmentally sensitive. Unlike energy sources such as solar or wind, which are relatively unrestricted for personal use, water access is often subject to local laws, property-specific regulations, and sustainability concerns.

Failing to navigate these legal and environmental factors correctly can lead to unexpected fines, restrictions on water use, or even legal action. In some areas, collecting rainwater is limited, drilling a well requires permits, and drawing water from a stream could lead to conflicts with neighbors or authorities. Even beyond legality, ensuring that water use remains sustainable is essential—depleting a well too quickly or contaminating a natural water source could lead to long-term consequences that threaten an entire off-grid setup.

A well-planned water system accounts not only for availability and filtration but also for compliance with laws and long-term sustainability.

Water Rights and Local Regulations

Access to water is often assumed to be a basic right, but legal ownership of water is not always tied to land ownership. In many places, water is treated as a shared resource, meaning that how much you can collect, use, and store may be regulated. Before setting up any off-grid water system, it is essential to research the specific water laws and property regulations that apply in your area.

Understanding Water Rights

Water rights laws vary depending on location. In general, they fall under two main systems:

1. Riparian Rights – Common in Eastern US, UK, and some other regions, this system grants landowners adjacent to a natural water body the right to use water but not to divert or store it excessively. If a river, lake, or stream crosses a property, the owner cannot prevent downstream users from accessing it.
2. Prior Appropriation – More common in Western US and arid regions, this system operates on a "first in time, first in right" principle, meaning that earlier users of a water source have priority access. Even if a stream runs through a property, the landowner may not have rights to use it if someone else has an older claim.

Groundwater rights also differ depending on jurisdiction. In some locations, well water belongs to whoever owns the land above it, while in other regions, deep aquifers are considered public resources and require permits to access.

Permits for Well Drilling and Water Collection

Even if groundwater rights allow access to a well, permits may be required to drill, maintain, or expand a water source. Some regions impose limits on well depth, water extraction rates, or require environmental impact studies before drilling is approved.

Similarly, rainwater harvesting is regulated in some areas. While most places encourage rainwater collection, a few restrict the amount of water that can be stored, especially in areas where rainfall significantly contributes to local aquifers.

It is important to check with local authorities before making any modifications to water infrastructure, as failing to obtain proper permits could result in fines or even forced removal of a water system.

Avoiding Depletion and Contamination

Legal considerations are only part of the challenge. Even if water sources are legally accessible, using them irresponsibly can lead to depletion or contamination, reducing long-term viability. A sustainable water system must be designed with conservation in mind, ensuring that water use does not exceed natural replenishment rates or cause harm to the local environment.

Overdrawing from Groundwater

A well or borehole may seem like an unlimited source of water, but underground aquifers do not always replenish quickly. Some recharge within months or years, while others take decades or centuries to recover after excessive use.

Signs of groundwater overuse include:
- **Lowering of the water table**, requiring deeper drilling over time.
- **Seasonal drying of wells** that once provided year-round water.
- **Increased pumping costs** due to greater depth and effort required to extract water.

To avoid depleting groundwater, monitor well levels regularly and avoid excessive pumping during dry seasons. Alternative water sources, such as rainwater harvesting or surface water collection, can help reduce reliance on wells and preserve aquifer levels for long-term sustainability.

Managing Surface Water Use

Taking water from a river, lake, or pond may be legal, but improper collection methods can damage ecosystems and reduce the availability of clean water. If too much water is drawn from a stream, it may disrupt the natural flow, harming fish populations and plant life.

Erosion can also become a problem if water collection points are poorly designed. Removing vegetation near a water source can cause banks to collapse, leading to increased sediment in the water and potential contamination from runoff.

To minimize impact:
- **Use gravity-fed or low-impact pumping systems** that do not excessively alter natural flow.
- **Avoid clearing natural vegetation** around water sources, as plants help prevent erosion and filter contaminants.
- **Limit direct human and animal access** to surface water to reduce pollution risks.

Preventing Contamination

Contaminated water is a risk for both personal health and the wider ecosystem. While purification methods can remove harmful substances, the best approach is to prevent contamination in the first place.

Major sources of contamination include:
- **Septic system leaks** that introduce bacteria and waste into groundwater.
- **Agricultural runoff** containing fertilizers, pesticides, or animal waste.
- **Chemical spills** from fuel storage, paints, or industrial materials.
- **Poorly maintained wells** that allow surface contaminants to seep in.

Septic systems should be placed at least 100 feet (30 meters) away from wells or open water sources and checked regularly for leaks. If farming or livestock is part of the off-grid lifestyle, proper drainage systems should be used to redirect animal waste away from water collection points.

Natural filtration methods, such as constructed wetlands, can be used to treat greywater and surface runoff before it reaches the main water source. These systems allow plants and soil microbes to break down contaminants, improving water quality while maintaining sustainability.

Water Storage and Treatment Considerations

Storing water improperly can also lead to contamination. Rainwater, well water, or surface water stored for long periods must be protected from bacterial growth, algae, and sediment buildup.

Best practices for water storage include:

- Using **covered tanks** to prevent debris and algae growth.
- Keeping stored water **away from direct sunlight** to reduce bacterial growth.
- Filtering water before storage to **remove sediments and organic matter** that could promote contamination.

Even with good storage practices, water should still be tested periodically to ensure it remains safe for consumption.

Water Collection Methods

Water collection is the lifeblood of an off-grid home, and designing an efficient, sustainable system is at the heart of creating self-sufficiency. Without a reliable water source, the entire off-grid lifestyle can fall apart. It's essential to understand how to properly collect, store, and transport water while accounting for local conditions, seasonal variations, and potential challenges. A diverse, adaptable water collection strategy ensures your home is equipped for daily consumption, gardening, and emergencies, no matter the circumstances.

Each method has its strengths, and understanding these methods helps you decide the best approach based on your unique needs. From traditional well drilling to rainwater harvesting, surface water collection, and solar-powered pumps, the goal is to create a system that is efficient, easy to maintain, and environmentally sustainable.

Well Drilling: Manual vs. Machine-Drilled Wells

When it comes to groundwater, wells remain one of the most reliable sources for off-grid water. Unlike surface water that can fluctuate seasonally or be affected by contamination, groundwater often provides a stable, year-round supply. However, drilling a well requires a significant investment in time, effort, and sometimes money. Choosing between manual drilling and machine drilling depends on a variety of factors, including budget, depth to the water table, and available equipment.

Manual Well Drilling

Manual well drilling, while often viewed as a more primitive method, is cost-effective and feasible for shallow water tables. If you are fortunate enough to have a water table that is less than 50 feet (15 meters) below the surface, manual drilling may be the best option. This method requires basic tools such as a hand auger, a post-hole digger, or a simple jetting system, which can be rented or built yourself. The well itself is dug by hand or with a team of people, using a variety of traditional and modern techniques.

One of the biggest challenges of digging a well manually is that it is labor-intensive and time-consuming. It's important to have the right team and plan in place before starting. Additionally, manual wells have their limits. If the water table is deeper than the tools can reach, you may have to switch to machine drilling. The process generally includes:

1. Locating the well site – Ideally, the site should be close enough to your home for convenience but far enough to prevent contamination from human waste or chemicals.
2. Digging the well – Using the auger or post-hole digger, the hole is gradually deepened. Sometimes, a small team of people is needed to lower and raise the drilling tools.
3. Installing the casing – Once the well is dug, a PVC or steel casing is inserted to keep the well from collapsing and to prevent contamination from surface runoff.
4. Water retrieval system – Depending on your needs, a manual pump, bucket and rope system, or a simple hand-powered piston pump can be installed.

Manual well drilling can be a satisfying DIY project, but it is limited to specific conditions. For deeper wells, a different method is required.

Machine-Drilled Wells

Machine drilling is the most efficient and reliable method for accessing groundwater, especially when dealing with deep aquifers or hard rock layers. Unlike manual drilling, which is limited to shallow depths, machine drilling can reach depths of several hundred feet. This method is suitable for homes that require a higher volume of water or are located in areas where shallow groundwater is not available.

Machine drilling requires specialized equipment and expertise, and often the assistance of professional drillers. Here's how it works:

1. Site assessment – Before drilling, a geological survey is performed to determine the depth of the water table and the best drilling location. In some cases, professionals use a geophysical survey to map underground aquifers.
2. Drilling the borehole – The machine drills through the earth, using a rotary drill bit to break through soil and rock. Depending on the soil composition, this may take several hours to days to reach the desired depth.
3. Casing installation – As the borehole is drilled, a steel casing is inserted to prevent soil collapse and protect the water supply.
4. Pump installation – Once the well is drilled, a submersible pump or jet pump is installed to bring the water to the surface. Depending on the depth, the pump may need to be powered by electricity, solar energy, or even a windmill.

Machine-drilled wells can provide a reliable, high-volume water source for many years. However, they come with a high cost, ranging from several thousand dollars to tens of thousands, depending on the depth and complexity of the drilling.

Rainwater Harvesting: Designing an Efficient System

Rainwater harvesting is one of the most effective and eco-friendly methods of water collection. It is particularly useful in areas with moderate to high rainfall, and it allows you to capture and store water directly from the sky. However, even with the best roof area and rainfall, a properly designed system is essential for efficiency. The goal is to ensure that the collected water is free of contaminants, stored safely, and available year-round.

Calculating Harvestable Rainwater

Before setting up a rainwater harvesting system, it's important to calculate how much rainwater can be harvested. This depends on the size of the roof catchment area and the amount of annual rainfall.

To determine how much water you can expect to collect, use the following formula:

Total Collection (gallons) = Roof Area (sq. ft) × Rainfall (inches) × 0.623

For instance, if your home has a 1,000-square-foot roof and receives 20 inches of rainfall annually, you can collect:

$1{,}000 \times 20 \times 0.623 = 12{,}460$ gallons

In metric units, use this formula:

Total Collection (liters) = Roof Area (sq. meters) × Rainfall (mm) × 1

For a 93-square-meter roof in an area with 500 mm of rainfall, the total collection would be:

$93 \times 500 \times 1 = 46{,}500$ liters

This calculation helps you estimate how much water can be harvested and stored, ensuring you design the system accordingly.

Setting Up a Rainwater Harvesting System

A typical rainwater harvesting system includes several key components:

- **Roof catchment** – The roof surface must be **clean and non-toxic**, with materials like **metal or tiles** preferred over asphalt shingles, which can leach chemicals.
- **Gutters and downspouts** – Gutters should be properly installed to direct water into the collection system. Clean and maintain gutters regularly to avoid clogs.
- **First-flush diverter** – This ensures that the first few minutes of rainwater, which may contain debris, leaves, and contaminants, is diverted away from the storage tank.
- **Storage tank** – The collected water should be stored in **covered tanks** to prevent contamination and reduce algae growth. Tanks can be **above-ground** or **underground**, and they should be large enough to hold enough water for dry periods.
- **Filtration and purification** – Before consumption, the stored rainwater should be filtered. A basic **screen filter** can remove debris, and a **fine filter** or **UV treatment** can purify the water for drinking.

Direct Surface Water Collection: Filtering and Using Lake, River, or Stream Water Safely

Surface water, while abundant in many places, requires significant care when it comes to filtration and purification. Water from rivers, lakes, and streams can contain bacteria, parasites, and pollutants. Using this water safely requires understanding how to properly collect and purify it.

Collecting Surface Water

When collecting surface water, always aim to draw from deeper parts of rivers, lakes, or streams. Water at the surface is more likely to contain debris, algae, or contamination. If possible, avoid areas near roads, industrial sites, or agricultural runoff.

Water should be filtered immediately after collection to remove large particles and sediments. Simple pre-filtration can be done with cloth filters or mesh to remove visible debris. After that, further treatment is necessary to ensure the water is safe to drink.

Filtration and Purification

For Surface water is recommended to adopt a multi-stage filtration process:

1. **Sediment filters** – To remove visible particles and dirt.
2. **Activated carbon filters** – To remove odors, chlorine, and some chemicals.
3. **Biological filtration** – To remove bacteria and parasites.
4. **Chemical treatments** – Chlorine or iodine tablets can be used to kill pathogens.
5. **UV light** – For thorough disinfection.

With the right treatment methods, surface water can be made safe for drinking, cooking, and other uses. However, this process requires regular monitoring and proper maintenance of the filtration system.

8. Water Storage Solutions

Types of Water Storage Systems

Water storage is a critical aspect of off-grid living. It's not just about having access to water but about ensuring you have a reliable and sustainable system to store it for daily use and emergencies. The design and construction of your water storage system depend on several factors, including your local climate, the space available, and the materials you have access to. The two primary options for water storage are above-ground tanks and underground cisterns. Each has its advantages and disadvantages, and understanding these will help you make the best choice for your situation.

Above-Ground Tanks vs. Underground Cisterns: Choosing Based on Climate and Land Conditions

Both above-ground tanks and underground cisterns can provide ample storage for your off-grid water needs, but each type of system has characteristics that make it more suitable for certain climates and land conditions.

Above-Ground Tanks

Above-ground tanks are popular for their ease of installation and cost-effectiveness. These tanks are typically made from plastic, metal, or fiberglass, and they are placed on a flat, stable surface above the ground. They come in a variety of sizes and can be installed with minimal excavation, making them ideal for locations where digging a hole might be difficult or costly.

Pros of Above-Ground Tanks:

1. **Easy to Install**: Installation is relatively simple, requiring no excavation. You can set up a tank on a flat surface and start using it almost immediately.
2. **Low Maintenance**: Above-ground tanks are easier to inspect for damage or leaks. You can also clean them more easily since they are not buried in the ground.
3. **Cost-Effective**: These tanks are usually more affordable compared to underground cisterns, as they require fewer resources to install.
4. **Flexibility**: You can easily move an above-ground tank if necessary. This is useful if you want to change the location or expand your storage capacity in the future.

Cons of Above-Ground Tanks:

1. **Exposure to the Elements**: The biggest downside to above-ground tanks is their exposure to temperature fluctuations. In **hot climates**, they may be prone to **algae growth** if they are not properly shaded, and in **cold climates**, freezing can be an issue.

2. **Aesthetic Considerations**: Above-ground tanks can be visually intrusive, especially if you have limited space or want a more discreet setup.
3. **Vulnerability to Damage**: These tanks are exposed to the elements, and they can be damaged by external forces like falling debris or extreme weather conditions (e.g., strong winds or hail).

In areas with mild temperatures, above-ground tanks can be a reliable option. The water stored in them is typically more accessible for routine maintenance and monitoring. However, in extreme climates (either hot or cold), additional measures like insulation or UV-protective covers may be required.

Underground Cisterns

Underground cisterns are tanks that are buried beneath the surface of the earth. These tanks are typically concrete or fiberglass, though some homeowners may opt for metal or plastic cisterns. Cisterns are generally used for larger storage capacities and are a good solution when space is limited or when climate conditions make above-ground tanks impractical.

Pros of Underground Cisterns:

1. **Temperature Regulation**: One of the biggest advantages of underground cisterns is their **consistent temperature**. The earth naturally insulates the cistern, keeping the water **cool in summer** and preventing it from freezing in **winter**, making them ideal for **extreme climates**.
2. **Space-Saving**: Because the cistern is buried, it doesn't take up valuable surface space, making it a great option for **smaller properties** or areas where **space is limited**.
3. **Protection from Damage**: The tank is protected from physical damage, such as falling objects or extreme weather conditions, since it is buried underground.

Cons of Underground Cisterns:

1. **Installation Complexity**: Installing an underground cistern is **more labor-intensive** and expensive. Excavation is required to bury the tank, and you may need specialized equipment or professional assistance.
2. **Limited Access for Maintenance**: Unlike above-ground tanks, cisterns are harder to inspect, clean, or repair once they are buried.
3. **Higher Initial Cost**: The installation of underground cisterns tends to be more expensive, not only for the cistern itself but also for the labor involved in excavation and site preparation.

Underground cisterns are ideal in areas where temperature extremes are a concern, or where space is at a premium. They offer long-term durability and protection but come with higher installation costs and maintenance challenges.

Storage Materials: Concrete, Plastic, Metal, or Natural Reservoirs – Pros and Cons

The material you choose for your water storage system plays a significant role in the system's durability, maintenance needs, and safety. Each material offers unique benefits, and understanding the pros and cons of each can help you choose the best option for your needs.

Concrete

Concrete is a strong, durable material commonly used for underground cisterns. It can hold large volumes of water and withstand significant pressure, making it a preferred choice for large-scale water storage systems.

Pros of Concrete:

- **Durable**: Concrete cisterns can last for decades and are resistant to physical damage.
- **Ideal for Large Volumes**: Concrete can support much larger water storage systems than other materials.
- **Temperature Control**: Like underground tanks, concrete cisterns benefit from **natural temperature regulation** provided by the surrounding earth.

Cons of Concrete:

- **Expensive**: Concrete cisterns are costly to install, requiring significant resources and labor.
- **Susceptible to Cracking**: Over time, concrete can crack and degrade, which can lead to **water leaks**.

- **Heavy**: Concrete is heavy and may require heavy equipment to transport and install.

Plastic

Plastic water tanks, typically made from HDPE (High-Density Polyethylene), are a popular option for both above-ground and underground storage. These tanks are lightweight, cost-effective, and relatively easy to install.

Pros of Plastic:

- **Affordable**: Plastic tanks are generally less expensive than concrete or metal tanks.
- **Lightweight and Easy to Install**: Plastic tanks are easier to transport and install, requiring fewer tools and less labor.
- **Corrosion Resistant**: Plastic does not rust or corrode over time, making it suitable for long-term use.

Cons of Plastic:

- **Less Durable**: While durable, plastic tanks can crack under pressure, especially in extreme temperatures or if they are **overfilled**.
- **UV Damage**: Over time, exposure to the sun can weaken plastic tanks unless they are **UV-protected** or covered.
- **Limited Size**: While plastic tanks come in large sizes, they cannot match the scale of concrete cisterns.

Plastic is ideal for those who want an inexpensive, easy-to-install solution for their water storage needs but is best used in moderate climates or for smaller storage applications.

Metal

Metal tanks, typically made from steel or galvanized iron, are commonly used in both above-ground and underground applications. They offer strength and durability but come with certain limitations.

Pros of Metal:

- **Durable**: Metal tanks are sturdy and can hold large amounts of water without breaking or cracking.
- **Long Lifespan**: When properly maintained, metal tanks can last for many years.
- **Resistant to UV Damage**: Unlike plastic tanks, metal tanks are not susceptible to UV degradation, making them suitable for outdoor use.

Cons of Metal:

- **Corrosion**: Metal tanks are prone to **rusting** and **corrosion** if not properly maintained. Regular inspections and maintenance are necessary to prevent rust.
- **Cost**: Metal tanks tend to be more expensive than plastic alternatives, particularly when using **galvanized steel**.
- **Weight**: Metal tanks are heavier than plastic and may require more effort and tools to install.

Metal is a good choice for those who need strong, long-lasting storage solutions, particularly in areas where UV damage could be an issue. However, rust and maintenance can be a concern.

Natural Reservoirs

In some off-grid setups, especially those with large land areas, natural reservoirs such as ponds or streams can serve as water storage. These bodies of water may already be available on the property or can be created with minimal effort.

Pros of Natural Reservoirs:

- **Free**: If you have an existing pond or stream on your land, this option can be free or low-cost.
- **Large Storage**: Natural reservoirs can hold **vast amounts of water**, making them perfect for large-scale storage needs.
- **No Maintenance Costs**: Once the system is established, maintenance is generally minimal.

Cons of Natural Reservoirs:

- **Contamination Risk**: Natural bodies of water can be easily contaminated by **agricultural runoff**, **animals**, or **debris**.
- **Seasonal Fluctuations**: Water levels can fluctuate based on rainfall, evaporation, or drought conditions, making it an unreliable source during dry seasons.
- **Legal Restrictions**: Some regions have strict regulations on **water rights** and may restrict your ability to use or modify natural water bodies.

Natural reservoirs can work well in areas with plenty of rainfall and little human disturbance, but they are vulnerable to contamination and seasonal changes in water availability.

Choosing the right water storage system for your off-grid home involves balancing the pros and cons of each option with your unique needs, climate, and land conditions. Whether you choose above-ground tanks for convenience, underground cisterns for climate control, or natural reservoirs for large-scale storage, the key is to ensure sustainability, long-term durability, and water safety.

First-Flush Diverters and Sediment Filters Before Storage

First-Flush Diverters

The first-flush diverter is a critical component of a rainwater collection system. During the first few minutes of rain, the water flowing off your roof will carry dirt, dust, debris, and potentially harmful pollutants. The first flush of rainwater typically contains the most contamination, which is why it's important to divert it away from your storage tank.

A first-flush diverter works by capturing the first few gallons of runoff and redirecting them to a separate drain or holding area, while the rest of the water is allowed to flow into the storage tank. This ensures that the water entering your storage tank is cleaner and safer.

Installing a First-Flush Diverter

1. **Install the diverter on your downspout**: The first-flush diverter is installed between the downspout and the tank. It captures the initial flow of water and holds it back.
2. **Calculate the capacity**: The size of the diverter should be calculated based on the size of your roof and average rainfall. A general guideline is that the diverter should hold back about **1-2 gallons of water per 100 square feet of roof area**.
3. **Regular maintenance**: Check and clean the diverter regularly to ensure it's functioning correctly and not clogged with debris.

Sediment Filters

While a first-flush diverter can help prevent large debris from entering the system, a sediment filter is essential to capture finer particles like dust, sand, and dirt that may still make it into the tank. These filters are typically installed at the inlet of the tank, where the water enters.

1. **Install a mesh or cartridge filter**: A **mesh filter** or **cartridge filter** can trap smaller particles before they enter the tank.
2. **Check the filter regularly**: Depending on the amount of rain and debris, you may need to clean or replace the filter regularly to ensure efficient filtration.

Proper Tank Sealing, UV Protection, and Ventilation for Potable Water

Now that the water has passed through a first-flush diverter and sediment filter, it's time to focus on the storage tank. Proper sealing, UV protection, and ventilation are essential to keeping the water in your tank clean and safe for long-term use.

Tank Sealing

A well-sealed tank is essential for protecting your water from contamination. Without proper sealing, debris, insects, and bacteria can enter the tank and compromise water quality.

1. **Use a tight-fitting lid**: Make sure your tank has a **secure lid** that fits tightly to keep out debris, leaves, insects, and animals.
2. **Check seals regularly**: Inspect the lid and the tank for any signs of damage or wear that could allow contaminants to enter.

UV Protection

UV rays from the sun can damage both the water and the tank itself. UV radiation can break down certain contaminants and harmful chemicals, and it can also promote the growth of algae inside the tank.

1. **Use UV-resistant tanks**: Opt for tanks made from **UV-resistant materials**, or cover your tank with a **UV-protective tarp** or paint it with **UV-blocking paint** to prevent damage from sunlight.
2. **Shade your tank**: If you can, position the tank in a shaded area or build a simple structure to shield it from direct sunlight.

Ventilation

While sealing your tank is important for preventing contamination, ventilation is also essential to avoid the buildup of gases and to maintain airflow.

1. **Install a ventilation valve**: Add a small **ventilation pipe** or valve at the top of the tank. This allows air to circulate, preventing the formation of vacuum pressure as the tank empties.
2. **Ensure proper airflow**: Keep the venting system free from blockages, as this ensures that your tank remains balanced and functional.

9. Purification Techniques: Ensuring Safe Water Off the Grid

While some concepts have been introduced in an earlier section with regards to storage solutions here we will deep dive on the wider topic of purification techniques.

Understanding Water Contaminants

Water may seem like a simple necessity—clear, refreshing, and essential for life—but in reality, it can carry a range of contaminants that make it unsafe to drink or use. Whether you're sourcing water from a well, a river, or collecting rainwater, it's crucial to understand the types of contaminants that may be present. These contaminants can be biological, chemical, or physical, and each type requires different purification methods to make the water safe for drinking, cooking, or cleaning.

Biological Threats: Bacteria, Viruses, and Parasites

Biological contaminants, including bacteria, viruses, and parasites, are among the most common and dangerous threats to water quality. These microorganisms can make you sick, and they're often the reason why water from natural sources like rivers, lakes, or even wells needs thorough treatment before use. They can enter water through human or animal waste, agricultural runoff, or contamination from environmental sources. Understanding these threats is the first step in keeping your water clean and safe.

Bacteria

Bacteria are single-celled organisms that thrive in warm, moist environments. Some bacteria, like E. coli, Salmonella, and Cholera, can cause serious gastrointestinal diseases, leading to symptoms such as diarrhea, vomiting, and stomach cramps. These bacteria are often found in water that has been contaminated by human waste, agricultural runoff, or animal feces.

You can't see bacteria in water with the naked eye, so it's important to test the water for their presence if you suspect contamination. Even if the water looks clear, it might still harbor harmful bacteria that can make you sick.

To protect yourself from bacterial contamination:

1. Always filter water through a **ceramic filter**, **activated carbon filter**, or **UV sterilizer** to remove bacteria.
2. If you have access to clean water for filtering, use **boiling** as a simple and effective method to kill bacteria.
3. For larger-scale off-grid systems, consider a **reverse osmosis** filtration system to remove harmful bacteria from the water.

Viruses

Viruses are even smaller than bacteria, and they can be transmitted to water sources through human or animal waste, as well as from contaminated food or surfaces. Common viruses in water sources include the Norovirus, Hepatitis A, and Rotavirus. These viruses can cause symptoms ranging from mild nausea to more severe liver damage and gastrointestinal illnesses.

Viruses are harder to remove than bacteria, but certain filtration methods are effective in eliminating them.

To protect against viral contamination:

1. **UV light treatment** can deactivate viruses in water, making it a reliable method for disinfection.

2. **Boiling** water for at least **1–3 minutes** will also kill most viruses, making it safe to drink.
3. **Reverse osmosis systems** with the appropriate filtration capabilities are effective at removing viruses from drinking water.

Parasites

Parasites are organisms that rely on other living beings to survive. In water, parasites such as Giardia, Cryptosporidium, and Amoebas can cause illnesses like dysentery, giardiasis, and other intestinal infections. These parasites are commonly found in untreated surface water, such as streams, lakes, and rivers. Their cysts are tough and can survive in harsh environmental conditions, making them particularly difficult to remove.

Parasites are typically found in colder environments and can be difficult to eliminate without the right methods. Some filtration systems are designed to remove parasites, but proper disinfection is often necessary to ensure the water is completely purified.

To protect yourself from parasitic contamination:

1. **Ceramic filters** with **0.5 microns or smaller pore size** are effective at filtering out parasites.
2. **Boiling** water for at least **1–3 minutes** will kill most parasites and their cysts.
3. **UV light** treatment is also effective against many types of parasites, making it an excellent option for small-scale water purification.

Chemical Pollutants: Pesticides, Heavy Metals, and Toxins

While biological contaminants pose an immediate threat to your health, chemical pollutants can accumulate over time and lead to long-term health problems. These pollutants can come from industrial runoff, agriculture, household chemicals, and even natural sources. Pesticides, heavy metals, and other toxins are often present in water sources and can affect the taste, color, and safety of the water.

Pesticides

Pesticides are chemicals used to kill insects, rodents, and weeds, but they can also contaminate water supplies when they run off from agricultural land or leak from storage containers. Pesticides can cause a variety of health problems, including neurological damage, cancer, and reproductive issues.

To remove pesticides from water:

1. Use an **activated carbon filter**, which is effective at adsorbing a wide range of chemicals, including pesticides.
2. Consider a **reverse osmosis system**, which can remove many pesticides from water, though some chemicals may require specialized filters.
3. Boiling water will not remove pesticides but can kill bacteria and viruses, making it useful as a supplementary method.

Heavy Metals

Heavy metals such as lead, mercury, and arsenic are commonly found in contaminated water. These metals can enter water supplies through industrial discharges, mining activities, and even old lead pipes. Long-term exposure to heavy metals can lead to serious health issues, including kidney damage, brain damage, and developmental problems in children.

To protect against heavy metals:

1. Install a **reverse osmosis filtration system**, which is highly effective at removing heavy metals from water.
2. **Activated carbon filters** can remove some heavy metals, though they may not be as effective as reverse osmosis systems.
3. **Distillation** can be another method for removing heavy metals, though it is energy-intensive.

Toxins

Toxins in water may come from industrial waste, household chemicals, or environmental pollution. These toxins, such as pharmaceutical residues, chlorine byproducts, and industrial chemicals, can pose long-term health risks if consumed over time. Many toxins are tasteless and odorless, making them difficult to detect without proper filtration.

To remove toxins from water:

1. **Activated carbon filters** are excellent for removing a broad range of toxins, including **chlorine** and **volatile organic compounds (VOCs)**.
2. **Reverse osmosis** systems can remove many chemical toxins, though specialized filters may be necessary for certain pollutants.
3. **UV light treatment** does not remove toxins but can help with bacteria and viruses. For chemical pollutants, other filtration methods are necessary.

Physical Impurities: Sediments, Organic Debris

In addition to biological and chemical contaminants, physical impurities like sediment, sand, and organic debris can also affect the quality of your water. These impurities often make water look dirty or cloudy and can clog filters and pipes if not removed before use.

Sediment

Sediment consists of fine particles like sand, dirt, and silt that are often found in surface water or can enter well water through poor well casing. These particles can make the water look cloudy and may cause abrasion in pipes and pumps over time.

To remove sediment from water:

1. **Sediment filters** or **mesh filters** are effective at trapping large particles like sand and dirt.
2. **Sand filters** can be used for larger systems, providing more thorough removal of sediment over extended periods.
3. If you're using well water, consider installing a **sand separator** to prevent particles from entering the water system.

Organic Debris

Organic debris, such as leaves, twigs, and algae, can enter water sources during rainfall or through runoff. While organic debris does not usually pose a health risk by itself, it can clog filters, add odor to the water, and provide a breeding ground for bacteria.

To remove organic debris:

1. Install a **pre-filter** to catch larger particles before they enter your main filtration system.
2. **First-flush diverters** on rainwater collection systems can help reduce the amount of debris that enters the tank during initial rainfall.
3. Regularly clean your **storage tank** to prevent the buildup of organic material that could lead to contamination.

Filtration Methods

Water filtration is an essential part of living off the grid, ensuring that your water is safe for drinking and daily use. Regardless of whether your water comes from a well, a stream, or a rainwater collection system, it may contain harmful contaminants such as sediments, bacteria, viruses, and chemicals. Each filtration method serves a specific purpose depending on the type of contaminants you're dealing with.

Sand Filters

When it comes to basic filtration, sand filters are one of the oldest and most reliable methods. For centuries, people have used sand to filter water, relying on its natural properties to trap dirt, sediment, and other large particles. While sand alone cannot remove chemical contaminants or bacteria, it serves as an excellent first line of defense in the filtration process, especially when used as part of a multi-stage system.

How Sand Filters Work

Sand filters operate on a simple principle: gravity. Water is poured into the top of the filter, and as it moves down through layers of sand and gravel, larger particles and sediments are trapped, and the water becomes clearer. The effectiveness of a sand filter depends on the type and size of the sand used, as well as the design of the system.

To build a DIY sand filter for your off-grid system, you'll need a container such as a large bucket or barrel. The design is simple and involves layering gravel at the bottom, followed by coarse sand, and then fine sand at the top. As water flows through these layers, the larger particles are filtered out, and cleaner water exits from the bottom.

While sand filters do an excellent job at removing visible debris, they are not sufficient for removing bacteria, viruses, or chemical contaminants. For these, additional filtration methods, such as activated carbon or UV treatment, are needed. Sand filters are best used as a pre-filter or in combination with other filtration methods to improve their overall effectiveness.

Commercial Sand Filters

If you want something more advanced than a DIY setup, there are also commercial sand filters that are designed for off-grid or larger water systems. These filters often include backwash capabilities, which allow you to clean the sand and remove any accumulated particles without disassembling the system. Some commercial sand filters are built into multi-stage filtration systems, combining sand with other filtration technologies to ensure the water is as clean as possible.

Activated Charcoal Filters

Activated charcoal, also known as activated carbon, is another powerful filtration method. While sand filters remove physical debris, activated charcoal filters are particularly effective at removing chemicals, odors, and taste-altering substances like chlorine, pesticides, and volatile organic compounds (VOCs).

How Activated Charcoal Filters Work

Activated charcoal works through a process called adsorption, where contaminants bind to the surface of the charcoal. The carbon in the charcoal has a highly porous structure, which gives it an incredibly large surface area. This allows it to trap a wide range of chemicals, toxins, and pollutants as water passes through it.

Activated charcoal filters can be commercially purchased as cartridge filters, pitcher filters, or whole-house systems. For DIY setups, you can buy activated charcoal in bulk and pack it into a filter container. As the water passes through the charcoal, harmful chemicals are trapped, while clean water exits on the other side.

Types of Activated Charcoal Filters

There are two main types of activated carbon filters: granular activated carbon (GAC) and carbon block filters.

1. **Granular Activated Carbon (GAC)**: This type consists of loose granules of charcoal. GAC filters are great for general water filtration needs, such as removing **chlorine**, **pesticides**, and **bad taste**.
2. **Carbon Block Filters**: These filters are made of compressed charcoal and provide a finer filtration, effectively removing **bacteria**, **chlorine**, and **heavy metals**. Carbon block filters are generally more effective at purifying water than GAC filters, as they have a greater surface area and provide better contact time for the water with the charcoal.

Maintenance of Activated Charcoal Filters

While activated charcoal is effective, it does require regular maintenance. Over time, the charcoal will become saturated with contaminants, reducing its effectiveness. Depending on the system, you may need to replace the carbon filter every few months or backwash the filter to clear out any impurities. In off-grid settings, regular monitoring of your water quality is essential to ensure that the filtration system remains effective.

Ceramic Filters

Ceramic filters are another powerful tool in your off-grid water filtration arsenal. Ceramic filtration is a physical filtration process, where water is passed through a ceramic material with very fine pores. The microporous structure of ceramic filters is effective at removing bacteria, parasites, and larger particles like dirt and sand. These filters can also be combined with activated carbon for enhanced chemical filtration.

How Ceramic Filters Work

Ceramic filters work through filtration by size exclusion. The ceramic material has microscopic pores that are small enough to trap contaminants like bacteria and protozoa. As water passes through the ceramic material, larger particles are physically blocked, while clean water flows through. Some ceramic filters are also combined with activated carbon to adsorb chemicals, improving the water's taste and quality.

Many off-grid systems use ceramic filter candles (also called ceramic filter elements) to filter water.

These filters are shaped like hollow cylinders and are placed inside a filtration unit. Some DIY ceramic filters can be made by molding and firing clay and activated carbon together, making them both effective and affordable.

Benefits and Limitations of Ceramic Filters

Ceramic filters are excellent for removing bacteria and parasites from water, but they do not remove viruses or chemical pollutants as effectively. Therefore, ceramic filters are often used in combination with other filtration systems such as UV light or activated carbon for complete water purification.

Maintenance of ceramic filters is simple—cleaning is necessary every few weeks to remove any accumulated debris or bacteria. Depending on the design, you may need to scrub the filter with a brush or soak it in a disinfecting solution to remove built-up grime.

Reverse Osmosis—Feasibility for Off-Grid Settings

Reverse osmosis (RO) is one of the most effective water filtration systems available, and it's capable of removing nearly all types of contaminants from water, including salts, chemicals, bacteria, and viruses. This filtration method forces water through a semi-permeable membrane, which allows water molecules to pass but blocks contaminants. Although highly effective, reverse osmosis is more complex and can require significant energy to operate, especially in an off-grid setting.

How Reverse Osmosis Works

In reverse osmosis, water is pushed through a membrane that has tiny pores. These pores are small enough to prevent contaminants like heavy metals, pesticides, bacteria, and viruses from passing through. The clean water exits the membrane and is collected for use, while the contaminants are flushed away.

Feasibility for Off-Grid Settings

While reverse osmosis systems are highly effective, they can be challenging to use in an off-grid environment due to their energy requirements. RO systems typically require a high-pressure pump to force the water through the membrane. If you don't have access to electricity, you may need to use solar power or a battery bank to operate the system. Additionally, reverse osmosis systems are often wasteful, as they require several gallons of water to produce just one gallon of purified water.

That being said, reverse osmosis can still be a good option for off-grid systems where water quality is a significant concern and you have access to the necessary energy sources. If you're considering reverse osmosis for your system, you'll need to plan carefully for the energy consumption and water wastage.

Disinfection and Sterilization Methods

In an off-grid setting, ensuring the safety of your water is essential. Whether you're collecting rainwater, sourcing from a well, or drawing from a stream, the water you rely on might contain harmful microorganisms, such as bacteria, viruses, and parasites, that can make you sick. While filtration can remove many physical contaminants, disinfection and sterilization are critical to eliminate pathogens that might remain after filtration.

Boiling—When and How Long to Ensure Safety

Boiling water is one of the simplest and most effective ways to sterilize it. Heat kills bacteria, viruses, and parasites, making boiling a reliable method for making water safe to drink, especially in emergency situations. If you don't have access to complex filtration or disinfection systems, boiling is often the go-to method.

When to Boil Water

While boiling is a powerful method for disinfection, it's most useful when you suspect that water may be contaminated with pathogens or if it has been exposed to environmental contaminants like animal waste. For water that looks clear and has no obvious signs of contamination, you may still want to boil it as an added safety measure.

In off-grid settings, it's especially important to boil water if:

1. You collect rainwater, which could be contaminated by airborne particles or microorganisms.
2. You rely on surface water (e.g., from a river or stream) that could contain harmful bacteria, viruses, or parasites from surrounding areas.
3. You're uncertain about the cleanliness of well water or other sources that haven't been filtered.

How Long to Boil Water

The time required to safely boil water depends on the altitude and the clarity of the water:

1. **At sea level**: Boil water for **1-3 minutes** to ensure that most pathogens are killed. The higher you are above sea level, the lower the boiling point of water becomes, which means you'll need to boil it for longer.
2. **At higher altitudes** (above 6,500 feet or 2,000 meters): Increase the boiling time to **5 minutes** or more to ensure the water is properly sterilized.

Boiling is effective at killing harmful microorganisms, but it does not remove any chemicals or particulates from the water. For this reason, it's often used in combination with other filtration methods to provide the safest water.

UV Purification—Using Natural Sunlight or UV Lamps

UV (ultraviolet) purification is another highly effective method for disinfecting water, as it kills bacteria, viruses, and parasites by destroying their DNA. The beauty of UV purification is that it's a chemical-free process, leaving no taste or odor in the water. UV treatment can be done using natural sunlight or UV lamps.

UV Purification Using Natural Sunlight

One of the easiest and most accessible methods of UV purification is using natural sunlight to disinfect water. This method works by exposing clear water to the sun for an extended period, allowing the UV rays to destroy pathogens.

1. **Fill clear plastic bottles**: To use natural sunlight, fill **clear plastic bottles** (preferably PET bottles) with water. The clarity of the water is essential because UV light needs to pass through it to effectively treat the water. If the water is murky, the sunlight may not penetrate well enough to kill pathogens.
2. **Place the bottles in direct sunlight**: Lay the bottles flat on a surface where they will receive full sunlight. Aim for **6 hours of direct sunlight** in **full sun**. On cloudy days or in places with weak sunlight, you may need to leave the bottles out for **up to 2 days**.
3. **Water Quality**: This method is most effective in clear water. If the water has high turbidity (cloudiness), it will need to be filtered before UV exposure to ensure the sunlight can penetrate.

Natural sunlight is an effective method of purification, but it may not be sufficient in cloudy climates or in regions with limited sunlight. For more consistent results, you can invest in a solar-powered UV sterilizer.

UV Lamps

For more reliable and faster UV purification, UV lamps are a great investment, especially in off-grid settings where you may need to purify larger quantities of water. These lamps use electricity to produce UV light, which kills pathogens in a matter of seconds.

1. **Install a UV water purifier**: These systems are available as **small countertop units** or **larger systems** that can treat water for an entire household. The UV light is emitted inside a protective chamber where water flows through.
2. **Operating the system**: Adhere to the guidelines provided by the manufacturer to ensure correct and safe use. The water flows through the UV chamber, where the ultraviolet light destroys harmful microorganisms.
3. **Electricity requirements**: Depending on the size of the system, UV lamps may require **12V DC or 120V AC** power. In off-grid settings, solar power is often used to run the system.

Chlorination—Correct Ratios and Storage Safety

Chlorine is widely used in municipal water treatment systems, and it can also be used in off-grid settings to disinfect water. Chlorination works by adding chlorine to the water, which kills bacteria, viruses, and other pathogens. Chlorine is particularly effective at disinfecting large quantities of water and can be stored for extended periods.

Correct Ratios

The amount of chlorine needed depends on the quality of the water and its level of contamination. As a general guideline:

1. For **clear water**, use **2-4 drops of household bleach (5% sodium hypochlorite)** per gallon of water.
2. For **murky or turbid water**, use **4-8 drops per gallon**.
3. After adding the chlorine, **stir the water** well and allow it to stand for **30 minutes**. During this time, the chlorine will disinfect the water. You should be able to smell a slight chlorine odor when the water is properly treated. If you don't smell chlorine, add another dose and wait for 30 minutes more.

Safety and Storage

Chlorinated water should be stored in clean, covered containers to prevent re-contamination. Avoid using containers that have previously held non-food-grade substances, as residual chemicals could interfere with the purification process.

Be aware that chlorine-treated water may have an unpleasant taste and odor, especially when using household bleach. To reduce the taste, you can let the water sit uncovered for a few hours to allow the chlorine to dissipate or use activated carbon filters to improve the taste.

Iodine and Chemical Treatments—When and How to Use Them

Iodine is another chemical treatment used to disinfect water. Like chlorine, iodine can kill bacteria, viruses, and parasites. It's often used in emergency situations or when there are limited resources available. However, iodine has a distinctive taste and can cause reactions in some individuals if used long-term.

When to Use Iodine

1. **Emergencies**: Iodine is a great option when you need to purify water quickly and don't have access to boiling or UV treatments. It's also used in portable water purification tablets for hiking or camping.
2. **Contaminated water**: If you're drawing water from a questionable source, iodine can be used to disinfect it, especially if the water is not visibly polluted.

How to Use Iodine

1. **Use iodine tablets** or **liquid iodine**. A common dosage is **1 tablet per liter (1 quart)** of water.
2. Stir the water well and let it stand for at least **30 minutes**. Iodine takes a bit longer than chlorine to work, so give it the time it needs to disinfect effectively.
3. Iodine-treated water should be consumed within **24 hours**. After this time, the iodine may lose its effectiveness.

Testing Water Quality with DIY and Professional Kits

After purifying water, it's essential to test it to ensure that it's safe for consumption. Both DIY water testing kits and professional testing kits are available to check for the presence of harmful contaminants.

DIY Testing Kits

DIY water testing kits are affordable and easy to use. They typically come with test strips or colorimetric reagents that can detect a variety of common contaminants, including bacteria, chlorine, pH levels, nitrates, and hardness.

To use a DIY water testing kit:

1. **Fill a clean container with water**: Collect a sample from your water source.
2. **Use the test strips**: Dip the test strips into the water for the recommended amount of time.
3. **Compare the color**: After a few minutes, compare the color change on the strip to the color chart provided in the kit. This will show the levels of specific contaminants in your water.

While DIY kits are convenient and quick, they might not provide the level of detail required for detecting all contaminants, particularly in larger off-grid systems.

Professional Water Testing Kits

Professional water testing kits, typically used by environmental agencies and water treatment professionals, provide a more thorough analysis of water quality. These kits often include chemical reagents and more advanced tools, such as spectrometers or digital testers, for precise measurements.

To use a professional water testing kit:

1. **Collect a sample**: Collect water from the source you intend to test.
2. **Follow the instructions**: Professional kits will typically include a step-by-step guide for using the testing tools. Some kits may require you to mix water with chemicals that react with contaminants.
3. **Interpret the results**: The results from a professional kit will provide you with precise measurements of **pH**, **bacteria**, **heavy metals**, **nitrates**, and **other pollutants**.

Professional testing kits are more reliable than DIY kits but can be more expensive. They are useful for in-depth analysis, especially if you suspect water contamination from chemical sources.

Elements of Scaled Sustainable Purification Systems

When living off the grid, the long-term sustainability of your water purification system is just as important as its immediate effectiveness. A sustainable system ensures that you have continuous access to clean water, reducing the need for constant manual intervention or reliance on emergency methods.

Gravity-Fed Purification Setups

Gravity-fed purification systems rely on the simple concept of gravity to move water through filtration materials. These systems are low-maintenance, cost-effective, and ideal for off-grid living, as they don't require electricity or complicated machinery.

How Gravity-Fed Purification Works

In a gravity-fed system, water flows from a higher elevation to a lower one, typically through a series of filters that clean the water as it travels. These systems often use sand filters, activated carbon, and ceramic filters in stages to provide comprehensive purification. The key is ensuring that the water's flow is steady and that there's enough elevation difference to generate pressure.

1. **Source tank**: Collect water from a reliable source such as a well, rainwater catchment system, or stream.
2. **Filtration stages**: As water flows through the system, it passes through several filtration stages. The first stage might remove large particles, while later stages address finer contaminants like bacteria, parasites, and chemicals.
3. **Final storage**: After the water has been filtered, it enters a storage tank or is dispensed through a faucet or tap for use.

Gravity-fed systems are great for areas where high water pressure is not available and electricity is scarce.

Constructed Wetlands for Large-Scale Purification

Constructed wetlands and biofilters are sustainable, low-energy solutions for purifying large volumes of water. These systems mimic natural processes, using plants, microorganisms, and soil to remove contaminants from water.

How Constructed Wetlands Work

Constructed wetlands are artificial systems designed to filter and clean water in much the same way that natural wetlands do. They consist of a shallow pond or basin, planted with water-loving plants such as cattails and reeds. The roots of these plants provide surface area for microorganisms that break down contaminants, while the water is filtered through the soil and plants.

1. **Water enters the wetland**: Contaminated water flows into the constructed wetland, where it is filtered by the plants and soil.
2. **Natural filtration**: Microorganisms in the soil and plant roots break down **organic pollutants**, while plants absorb **nutrients** and **heavy metals**.
3. **Clean water exits**: After passing through the wetland, the purified water exits through a drain or filter for use.

Constructed wetlands can treat both greywater (wastewater from sinks and showers) and blackwater (wastewater from toilets) in a highly sustainable manner.

Developing a Multi-Layered Purification System for Redundancy

A multi-layered purification system combines various methods to ensure redundancy and increase the effectiveness of water purification. By using different types of filters and disinfection methods together, you can create a system that is robust enough to handle multiple types of contaminants.

1. **Layer 1 - Sediment filtration**: Use **sand filters** to remove larger particles and debris.
2. **Layer 2 - Chemical adsorption**: Add **activated carbon** to remove chemicals, odors, and improve the taste of the water.
3. **Layer 3 - Microbial filtration**: Use **ceramic filters** to remove bacteria and parasites.
4. **Layer 4 - Disinfection**: Finish with **UV purification** or **boiling** to kill any remaining pathogens.

By using a combination of these methods, you ensure that even if one part of your system fails, the other layers will continue to provide safe water.

10. Water Distribution

Water is the lifeblood of any homestead, and having a reliable distribution system is vital for maintaining both everyday household needs and agricultural requirements. Off-grid systems, by definition, require innovative approaches since you won't have access to the usual utility infrastructure. Fortunately, gravity-fed systems, pumps, and carefully designed piping layouts can provide an effective and reliable water distribution system. Here, we will explore the components of water distribution systems, focusing on **gravity-fed water systems**, **pumping solutions**, and the best **pipe layout and infrastructure** for off-grid living. Additionally, we'll cover how to integrate water systems for **household** and **agriculture** use, ensuring efficient distribution for all your needs.

Gravity-Fed Water Systems

Gravity-fed water systems have been used for centuries, harnessing the power of gravity to move water from one location to another. These systems are simple, energy-efficient, and highly reliable, making them ideal for off-grid settings where power may be scarce or unreliable.

Designing a System That Uses Elevation for Distribution

The fundamental principle behind a gravity-fed system is that **water flows downhill**. By placing your water source at a higher elevation than your home or agricultural area, gravity can do the work of pushing the water through pipes, reducing the need for pumps and external energy sources.

To design a gravity-fed system:

Determine the elevation difference: The first thing you need to do is determine how much elevation you have between your water source (e.g., well, rainwater collection system, or spring) and your home or agricultural area. The more height you have, the greater the pressure that gravity will provide to push the water through the pipes.

Calculate required pressure: The pressure generated by gravity is determined by the height difference between the water source and your point of use. A simple rule of thumb is that for every 1 foot (0.3 meters) of elevation, you get approximately 0.433 pounds per square inch (psi) of water pressure. A typical household system needs around 30 psi to function effectively, which would require approximately 70 feet (21 meters) of vertical height.

Design the pipe layout: With the right elevation, you can now design the layout for the water pipes. The system should be designed to minimize the number of turns and bends in the pipes to maximize the flow of water. Also, consider the flow rate of water needed by your household or farm, as larger systems will require larger pipes to deliver sufficient water.

Water Tower Placement and Pipeline Considerations

Water towers are key components of gravity-fed systems, providing the **elevation** needed for pressure. Proper placement of the water tower is crucial to ensure sufficient water pressure and flow.

Tower placement: The water tower should be located at the highest point of your property that is still easily accessible for maintenance and repair. Ideally, it should be at least 10-20 feet (3-6 meters) above the ground to provide adequate pressure.

Tank size: The size of the water tank at the top of the tower will depend on your water consumption needs. A larger tank provides greater water storage and reduces the frequency of refills. Consider the daily water usage for your home or farm when determining the size of the tank.

Pipe material: The pipes that connect your water tower to your home or farm should be strong and durable. Common materials for these pipes include PVC, PE (polyethylene), and copper. PVC and polyethylene are lightweight, affordable, and resistant to corrosion, while copper pipes are more durable but tend to be more expensive.

Slope and trenching: Ensure that the pipes are placed at a slight slope to allow water to flow freely. The trench should be deep enough to prevent freezing in cold climates and should be well insulated to avoid any water loss.

Pumping Solutions

While gravity-fed systems are excellent for distributing water using natural pressure, there are times when you'll need a pump to move water to higher elevations or across longer distances. Off-grid pumps come in several forms: **hand pumps**, **solar-powered pumps**, and **electric pumps**. Let's explore each of these options in detail.

Hand Pumps—Affordable, Durable, and Effective Backup

Hand pumps are an excellent, **low-tech option** for off-grid water systems, especially when you need a simple backup or a system that requires little maintenance.

How hand pumps work: Hand pumps are mechanical devices that use a **manual lever** to push water from a well or water source. When you pump the lever, it creates suction that pulls water up through the pipe and out of the spout. They're often used in emergency situations or as a supplementary source for small-scale water needs.

Advantages of hand pumps:

1. **Low cost**: Hand pumps are relatively inexpensive compared to solar or electric pumps.
2. **Durable**: These pumps are designed to last for many years and are often built to withstand harsh conditions.
3. **No power required**: Since they don't require electricity, they are perfect for off-grid systems where power may not be readily available.
4. **Simple maintenance**: Hand pumps have very few moving parts, so maintenance is simple and straightforward.

Installation: Installing a hand pump is relatively simple. It requires a **borehole** or **well**, and a pipe connected from the water source to the pump. The lever mechanism is mounted above ground for easy access.

Solar-Powered Pumps—Calculating Energy Needs and System Design

Solar-powered pumps are an ideal choice for off-grid systems, as they harness solar energy to pump water without relying on traditional electricity sources. These pumps are particularly well-suited for rural areas and homesteads with ample sunlight.

How solar-powered pumps work: Solar-powered pumps use **photovoltaic (PV) panels** to convert sunlight into electricity, which powers the pump to move water. The power generated is stored in **batteries** for use during cloudy days or at night. These pumps can be used to move water from wells, reservoirs, or surface water sources.

Calculating energy needs: The size of the solar array and pump will depend on your **water requirements**, the **depth of the water source**, and the **available sunlight**. To determine the **solar pump size**, calculate your daily water usage in gallons or liters, and consider how much sunlight your location gets per day. For example, a pump system may need a **500-watt solar panel** to deliver adequate water flow in a sunny area, but this might change based on local conditions.

System design: A typical solar pump system consists of the following components:

1. **Solar panels**: These capture sunlight and convert it to electrical energy.
2. **Charge controller**: This regulates the energy output and prevents overcharging of batteries.
3. **Batteries**: These store the energy for use during low-sunlight periods.
4. **Pump**: The pump moves water from the source to storage or the point of use.

Installation and maintenance: Solar-powered pumps are easy to install, but you'll need to consider the location for your panels (i.e., ensuring they are exposed to full sunlight). Routine maintenance includes **cleaning the panels** and checking the **battery condition** to ensure optimal performance.

Electric Pumps—Power Requirements and Efficiency Considerations

Electric pumps are another option for off-grid systems, particularly for larger-scale operations or where solar power is insufficient. These pumps can be used for **deep wells**, **high-flow needs**, or **long-distance water transport**.

How electric pumps work: Electric pumps are powered by AC or DC electricity and are used to push water through pipes. The pump moves water from a lower level to a higher one or over a distance, depending on the system design.

Power requirements: Electric pumps require a steady power source, which can be generated by solar panels, wind turbines, or a backup generator. The power requirement will vary depending on the pump's horsepower, flow rate, and the distance the water must be pumped.

Efficiency considerations: Electric pumps can be very efficient, but they require careful planning to ensure that your off-grid energy system can handle the power needs. In locations with limited sunlight or unreliable power sources, electric pumps may be best paired with a battery storage system to ensure continuous operation.

Installation and maintenance: Electric pumps require installation by professionals to ensure that wiring and power sources are correctly set up. Maintenance generally involves regular checks of the motor, cleaning the pump components, and ensuring the power system is functioning properly.

Pipe Layout and Infrastructure

Designing a water distribution system for your off-grid home involves much more than just laying pipes and connecting them to a water source. A well-thought-out pipe layout ensures that water is delivered effectively and efficiently to all points of use, from the kitchen to the garden. Whether you're using gravity-fed systems, pumps, or a combination of both, understanding pipe materials, installation techniques, and pressure considerations is essential for maintaining a reliable system. Additionally, in colder climates, special care must be taken to prevent freezing that could damage your pipes and disrupt water supply.

Choosing the Best Piping Materials—PVC, Copper, PEX, Polyethylene

The choice of pipe material significantly impacts the efficiency, cost, and durability of your water system. Each material has its own benefits and limitations, so it's important to select the one that suits your off-grid needs based on factors like budget, climate, and water quality.

PVC (Polyvinyl Chloride)

PVC is one of the most commonly used materials for water distribution due to its affordability, lightweight design, and easy installation. It's an excellent choice for systems that will be buried underground or used in non-pressure applications. PVC pipes are resistant to corrosion and UV rays, making them durable in outdoor settings.

- **Advantages**:

Cost-effective and widely available.

Easy to install and work with, especially in DIY setups.

Resistant to **corrosion** and **UV rays**.

- **Disadvantages**:

Not suitable for **extreme temperatures** or **high-pressure** applications.

Can **crack** under freezing conditions or with heavy impacts.

For off-grid homes in moderate climates, PVC is a great choice for most applications. However, in colder climates, you'll need to take extra precautions to prevent it from cracking due to freezing temperatures.

Copper

Copper has long been used for plumbing systems due to its durability and antimicrobial properties, making it an ideal material for drinking water systems. It is highly resistant to corrosion and provides excellent water pressure, making it suitable for both hot and cold water applications.

- **Advantages**:

Durable and resistant to corrosion.

Does not affect water taste or quality.

Resistant to UV radiation.

- **Disadvantages**:

More **expensive** than PVC and PEX.

Difficult to install, requiring special tools for cutting and joining pipes.

Prone to theft due to high scrap value.

Copper pipes are ideal for areas where durability is essential, especially for long-term use. However, the cost and difficulty of installation may make it less appealing for DIY off-grid projects, unless you need the durability and longevity it offers.

PEX (Cross-Linked Polyethylene)

PEX is a relatively new and flexible plastic piping material that has become popular for water distribution systems. It's ideal for off-grid homes due to its ability to expand without cracking when frozen, making it a great option for areas that experience freezing temperatures.

- **Advantages**:

Flexible and can be bent around obstacles without needing fittings.

Resistant to freezing, making it ideal for colder climates.

Easy to install with **fewer tools** required compared to copper.

Affordable and widely available.

- **Disadvantages**:

Less **durable** than copper or PVC in the long term.

UV sensitive and should not be exposed to direct sunlight for extended periods.

Can be **damaged** by **rodents**.

PEX is often the best choice for off-grid water systems in cold climates due to its resistance to freezing. It's also easy to install, making it a popular choice for DIYers.

Polyethylene (PE)

Polyethylene pipes are commonly used for large-scale water distribution in both residential and commercial applications. PE pipes are highly durable and resistant to abrasion and impact, making them ideal for underground installation.

- **Advantages**:

Durable and resistant to impacts.

Great for underground installations in areas with heavy foot or vehicle traffic.

Affordable and flexible.

- **Disadvantages**:

Not suitable for **hot water** applications.

Not as widely available in smaller sizes for household plumbing.

Polyethylene is most suitable for large-diameter water distribution, such as for delivering water from a well or storage tank to various parts of your homestead.

Preventing Freezing in Cold Climates—Burial Depth, Insulation, Heating Options

In cold climates, frozen pipes can be a major concern. Freezing can cause pipes to burst, leading to expensive repairs and water disruptions. Fortunately, there are several ways to prevent freezing and ensure that your pipes remain functional during the winter months.

Burial Depth

One of the most effective ways to prevent freezing is to bury the pipes deep enough underground so that they are insulated by the earth. The deeper the pipe, the more protected it is from freezing temperatures.

Depth recommendation: In areas with cold winters, pipes should be buried at least 36 inches (91 cm) deep to stay below the frost line. The frost line is the depth at which the ground freezes during the winter. In some areas, you may need to dig deeper if the frost line extends further into the earth.

Calculating the depth: Check with local guidelines or a local contractor to determine the appropriate depth for your pipes based on local weather patterns and frost depth.

Insulation

If burying pipes deep enough isn't feasible, insulating your pipes is another effective way to prevent freezing. Insulation slows the heat transfer between the air and the pipes, keeping them warmer.

Pipe insulation: Insulation materials such as foam pipe insulation or heat tape can be wrapped around the pipes to keep them from freezing. These materials are relatively inexpensive and can be easily installed over the pipes.

Insulating around foundations: If you have exposed pipes running along the exterior of your house, make sure to insulate both the pipes and the foundation around them to prevent cold air from reaching the water supply.

Heating Options

In particularly cold areas, additional heating may be necessary to keep your pipes from freezing. This can be achieved using electric heating cables, which are designed to wrap around pipes and provide heat.

Heat tape: Electric heat tape is a great solution for small sections of piping. It can be wrapped around the pipe and plugged into an electrical outlet to keep the pipes warm, even in freezing temperatures.

Heating cables: These cables are installed alongside pipes to provide continuous heat, and they automatically turn on when the temperature drops too low.

Calculating and Maintaining Optimal Water Pressure

Water pressure is a crucial factor in any plumbing system, and it's especially important when designing an off-grid water distribution system. Too much pressure can damage pipes and appliances, while too little pressure can result in weak water flow.

Water Pressure Calculation

To ensure that your water system operates efficiently, you'll need to calculate the pressure that your water source provides and the pressure needs for your household. Most residential water systems require between 30 psi (2 bar) and 50 psi (3.4 bar).

1. Pressure needs for household: You'll need to calculate the total flow rate for your household based on the number of faucets, showers, and appliances that will use water. The flow rate is typically measured in gallons per minute (GPM) or liters per minute (LPM).
2. Elevation and distance considerations: Keep in mind that water pressure decreases as the elevation increases, and pressure loss can also occur over long distances due to friction in the pipes.

Maintaining Optimal Pressure

If your system has insufficient pressure, you may need to install a pressure tank or boost pump. Conversely, if the pressure is too high, a pressure regulator can be installed to ensure a safe, consistent flow.

Integrating Water Systems for Household and Agriculture

An off-grid water system should not only provide water for drinking and sanitation but also support agricultural needs such as irrigation. To achieve this, you'll need to design a system that integrates household plumbing with agricultural irrigation in an efficient way that conserves water.

Connecting Storage Systems to Household Plumbing

The first step in integrating water systems is to connect your water storage systems (such as tanks or wells) to your household plumbing. You'll need a system that delivers water to all points of use, including sinks, showers, and appliances, and provides enough pressure for everyday activities.

1. Pump systems: Use a pump to deliver water from your storage tank to the house. Depending on the size of your system, you may need a submersible pump, surface pump, or pressure pump to ensure consistent flow and pressure.
2. Filtration: Ensure that the water delivered to your house is properly filtered and disinfected before it enters the plumbing system.

Efficient Drip Irrigation Systems for Water Conservation

1. For agricultural needs, **drip irrigation** is one of the most water-efficient systems available. Drip irrigation delivers water directly to the roots of plants, minimizing water waste by avoiding evaporation or runoff.
2. Setting up a drip system: Lay drip lines along rows of plants, ensuring that each plant gets a controlled amount of water.
3. System maintenance: Drip irrigation systems need regular checks for clogs in the lines, as debris can block the flow of water.

11. Backup and Emergency Storage Strategies

Having a primary water storage system is essential, but it's equally important to have backup and emergency storage strategies in place. These strategies will ensure that you have water available even if your primary system is compromised or running low.

Redundant Storage Solutions

A redundant storage system provides multiple layers of security in case one component fails.

1. **Multiple tanks**: Instead of relying on a single tank, set up several smaller tanks that are interconnected. If one tank runs out or becomes contaminated, the others can still provide water.
2. **Mobile storage**: Keep **mobile water storage containers**, such as **water barrels** or **portable tanks**, on hand in case you need to quickly relocate your water source or access additional supplies.
3. **Water rotation**: Rotate your stored water by using it regularly and replacing it with fresh water to ensure it remains safe to use. This practice prevents stagnation and contamination.

Ensuring Availability if Primary Storage is Compromised

In the event that your primary water storage system is damaged or compromised, you'll need a backup plan to ensure continued access to water.

1. **Install an emergency backup pump**: In case your main water pump fails, having a backup pump on hand will allow you to access water from your storage tanks without interruption.
2. **Create a water diversion system**: If your storage tank is compromised, have a system in place to divert water from an alternative source, such as a river or pond, into your storage tanks.
3. **Keep extra filters and treatment options**: Ensure that you have spare filters and water treatment options on hand in case your primary filtration system fails.

Freeze-Proofing in Cold Climates

If you live in a region where winter temperatures regularly drop below freezing, you face a unique challenge when it comes to water storage. Frozen pipes, tanks, and pumps can disrupt your water supply, causing significant problems. However, with the right strategies, you can protect your water storage and ensure it remains usable throughout the winter months.

Insulate Your Tank and Pipes

The first step in freeze-proofing your water system is ensuring that your tank and the pipes leading to it are insulated from the cold. This is especially important for above-ground tanks, which are most vulnerable to freezing temperatures.

1. **Wrap pipes with foam insulation**: Foam pipe insulation is widely available and can be used to wrap the exposed water pipes leading to and from your storage tank. This is an easy and cost-effective method for preventing freezing.
2. **Insulate the tank**: If you have an **above-ground tank**, consider using **insulation blankets** or wrapping the tank in a thick layer of insulation. Alternatively, burying the tank in the ground can also help insulate it naturally from the cold.
3. **Create an insulated shelter**: For above-ground tanks and pump systems, building a simple shelter around them can offer added protection from the elements. Use materials like **straw bales** or **insulated plywood** to keep the tank and pipes safe from freezing.

Keep Water Flowing

One of the most effective ways to prevent freezing in your water storage system is to ensure that water continues to flow. Water that is stagnant is more likely to freeze than water that is constantly moving.

1. **Use a circulating pump**: Installing a circulating pump helps to move the water around the system continuously, preventing it from freezing. This is especially important for **larger storage systems**.
2. **Run water periodically**: If you are using a **gravity-fed system**, you may need to run water through the system at regular intervals to keep the pipes from freezing. During extremely cold weather, running the water for a few minutes every few hours can prevent blockages from ice.
3. **Use a heat tape**: For areas with **extreme cold**, you can wrap **heat tape** (electrical heating cables) around pipes to provide constant warmth. This will keep the water flowing and prevent freezing even in the coldest conditions.

Expanding Storage Capacity for Drought Resilience

A key strategy in ensuring long-term water security is expanding your storage capacity to buffer against dry spells. Droughts can last for weeks, months, or even years, and having enough water stored can be a lifesaver when rainfall is scarce.

Calculate Your Water Needs

Before expanding your storage capacity, assess your daily water needs and seasonal consumption. This will help you determine how much additional water storage is necessary to withstand a drought.

1. Start by calculating your **daily water usage** for drinking, cooking, cleaning, sanitation, and irrigation. Include seasonal fluctuations such as higher irrigation needs during summer.
2. Multiply your daily water needs by the **length of the expected drought period** to determine the total water storage required.
3. Factor in the **efficiency of your water collection systems**, such as **rainwater harvesting**, to determine how much extra storage will be needed to complement your existing water supply.

Increase Storage Capacity

Once you have determined your water storage requirements, consider the following methods for expanding your capacity:

1. **Add more tanks**: Adding additional **above-ground tanks** or **underground cisterns** will increase your water storage capacity. The key is to create a system that is scalable and easy to maintain.

2. **Use larger storage containers**: If space allows, use larger tanks or cisterns to hold more water. Large **plastic tanks** or **concrete cisterns** can store several thousand gallons of water.
3. **Create auxiliary storage**: In addition to your main water storage system, consider using **mobile water containers** or **barrels** to supplement your water supply. These smaller storage units can be strategically placed around your property for easy access during a drought.

Monitor and Maintain

Once you have increased your storage capacity, it's important to monitor the water levels and the quality of the stored water to ensure it remains safe for use.

1. **Install water level indicators**: Use a **water level sensor** or **gauge** to monitor the amount of water stored in your tanks. This will help you track water usage and anticipate when additional resources will be needed.
2. **Check water quality**: Regularly test the water in your storage tanks to ensure it remains clean and free from contamination. Use a **water testing kit** to check for bacteria, chemical contaminants, and pH levels.

12. Water Self-Sufficiency Module Projects

Project: Assessing Water Needs for an Off-Grid Home

Securing a reliable water supply is one of the most critical aspects of off-grid living. Without access to a municipal water system, every drop must be accounted for, whether it comes from rain, a well, a spring, or surface water. A sustainable off-grid water system starts with knowing exactly how much water is needed daily, how seasonal changes affect availability, and what reserves must be in place for emergencies.

Failing to plan for water needs can lead to shortages, forcing reliance on expensive or unreliable backup solutions. On the other hand, overestimating requirements can lead to unnecessary infrastructure costs. Striking the right balance ensures that water is always available when needed while avoiding waste or excess costs.

Building an off-grid water system starts with a clear understanding of how much water is actually needed. Without accurate estimates, there is a risk of either running out of water or overinvesting in unnecessary infrastructure. This project walks through the process of calculating daily, seasonal, and emergency water needs based on household size, climate, and lifestyle. The goal is to develop a customized water budget that ensures a sustainable and efficient supply.

Materials Needed

- Notebook or spreadsheet for calculations
- Measuring container (gallon or liter-based)
- Timer (for tracking water use during activities like showering or dishwashing)
- Access to local climate data (for rainfall and drought trends)
- Calculator (optional)

Step 1: Calculate Daily Water Needs

Daily water use varies by household, depending on the number of people, habits, and efficiency measures in place. The first step is measuring how much water is used per day across different activities.

1. **Drinking Water:** Each person typically consumes **0.5 to 1 gallon (2 to 4 liters)** of drinking water per day. Multiply by the number of household members.
2. **Cooking and Dishwashing:** Measure how much water is used when preparing meals and washing dishes.
3. **Personal Hygiene:** Track shower or bath usage. A typical shower uses **2 to 5 gallons (8 to 20 liters) per minute**. Washing hands, brushing teeth, and other hygiene activities should also be considered.
4. **Laundry:** If using a washing machine, note the water consumption per load. Handwashing clothes will use less but requires additional effort.
5. **Toilet and Sanitation:** If using a flush toilet, calculate based on the number of flushes per day. Composting toilets require little to no water

but may require periodic rinsing of collection bins.
6. **Outdoor and Garden Use:** Irrigation for plants and water for livestock must be factored in. The amount varies significantly based on climate, plant species, and livestock size.

For one full day, record how much water is used for each category. If measuring every use is impractical, estimate based on averages and adjust based on lifestyle choices.

Category	Unit (Gallons per day)	Household Members / Usage
Drinking Water		Enter number of people
Cooking & Dishwashing		Estimate cooking & dishwashing water
Personal Hygiene (Showers, Handwashing, etc.)		Estimate personal hygiene water
Laundry		Number of laundry loads per week
Toilet & Sanitation		Number of toilet flushes per day
Outdoor & Garden Use		Outdoor water use estimation
Livestock Watering		Livestock type & quantity
Total Daily Water Use		Auto-calculated
Seasonal Adjustment (Summer/Winter)		Auto-calculated
Peak Season Water Demand		Auto-calculated
Emergency Water Reserve (14 days)		Auto-calculated
Emergency Water Reserve (30 days)		Auto-calculated

Step 2: Adjust for Seasonal Changes

Water needs change with the seasons. A water budget must reflect higher consumption in hotter months and reduced needs in colder seasons.

Identify Peak Demand Periods:

• In summer, irrigation demand increases, and personal consumption rises due to heat.
• In winter, indoor water use may increase, but irrigation may decrease.

Calculate Additional Summer Use:

• Garden and livestock watering increase significantly in hot months.
• If living in a dry climate, estimate double or triple water needs for peak summer use.

Plan for Winter Challenges:

• In freezing climates, stored water must be insulated or heated.
• If pipes freeze, alternative water access (such as indoor reserves) must be available.

These adjustments ensure that water availability is aligned with actual seasonal demand rather than relying on an unrealistic year-round average.

Step 3: Calculate Emergency Water Reserves

An off-grid system must be resilient enough to handle unexpected disruptions, such as droughts, well failures, or contamination events. Emergency reserves ensure that the household has enough water to survive short-term crises.

Determine Minimum Emergency Storage:

• The recommended minimum is **14 gallons (53 liters) per person** for two weeks.
• For long-term preparedness, aim for **one to three months' worth** of water storage.

Account for Critical Needs:

• Drinking and cooking should be prioritized over other uses in emergencies.
• Water-saving strategies (like reusing greywater for hygiene) can extend reserves.

Choose a Backup Source:

• If primary water collection fails, an alternative source such as a secondary well, rainwater storage, or a filtration system for surface water should be identified.

Step 4: Create the Final Water Budget

With all factors considered, the final water budget brings together daily usage, seasonal variations, and emergency reserves into a structured plan.

1. **Total Daily Water Requirement:** Sum up all household water uses.

2. **Peak Season Adjustment:** Increase storage capacity to match the highest seasonal demand.
3. **Emergency Reserve Calculation:** Plan backup supplies meet or exceed the recommended minimum.
4. **System Capacity Matching:** Use this info to design collection and storage systems that can provide sufficient supply.

With a clear budget in place, the household can design an efficient off-grid water system that meets real-world needs without excess cost or waste.

Project: Water Source Mapping and Feasibility Assessment

Access to water is the foundation of any off-grid home, but not all water sources are equally viable. Before committing to a specific water system, a thorough assessment must be conducted to determine what sources are available, how reliable they are, and what challenges they might present. This project involves mapping, analyzing, and ranking potential water sources to create a comprehensive plan for long-term water independence.

Materials Needed

- A notebook or digital spreadsheet for recording findings
- A printed or digital map of the property or planned location
- Access to online topographical maps or satellite imagery
- Soil testing kit (optional)
- A shovel or post-hole digger (for soil assessment)
- Tape measure or measuring wheel
- Water testing strips or a basic test kit (if sampling water sources)

Step 1: Identifying Potential Water Sources

Before assessing the viability of different water sources, the first step is identifying all possible options available in the area. This includes both primary and backup sources to ensure water security in different conditions.

Begin by walking the property or using an aerial map to mark potential water sources. The most common options to consider are:

- Groundwater sources such as wells, boreholes, and springs
- Surface water sources like lakes, rivers, creeks, or ponds
- Rainwater harvesting potential based on roof area and annual precipitation
- Alternative sources like condensation collection or atmospheric water generators

Each source should be recorded with its location and a brief description. If any existing water infrastructure is present, such as an old well or a natural spring, note its condition and accessibility.

Step 2: Evaluating Groundwater Availability

Groundwater is often the most stable and long-term water source, but it requires careful evaluation before use. If a well is already present, its depth, recharge rate, and water quality should be recorded. If no well exists, groundwater feasibility must be determined.

Check with local authorities or geological surveys to find information on average groundwater depths in the region. If other homes or farms nearby rely on wells, note their depth and whether they experience seasonal fluctuations.

If practical, perform a basic test for soil moisture and percolation. Using a shovel or post-hole digger, dig a hole at least two feet (0.6 meters) deep and observe the moisture content. If water pools at the bottom, the water table may be close to the surface. If the soil is extremely dry, deeper drilling may be required.

For a more detailed assessment, record:

- Approximate depth to groundwater based on local records
- Type of soil and rock formations (some soils, like sand, allow water to flow easily, while clay is more restrictive)
- Whether groundwater availability changes with the seasons

Step 3: Assessing Surface Water Viability

If a river, lake, or pond is nearby, it may serve as a potential water source. However, surface water requires purification, and its availability may fluctuate throughout the year.

Visit the water body and observe the following:

- Is the water flow consistent year-round, or does it dry up in the summer?

- Are there visible contaminants, such as algae blooms, industrial runoff, or animal waste?
- What is the distance from the home, and how difficult would it be to transport water?

If access to the water is possible, collect a small sample and use a basic water testing kit to check for pH, hardness, and contaminants like nitrates or bacteria. The results will help determine the level of filtration required.

If the water source is a small creek or pond, research whether it is fed by a natural spring or relies on seasonal rainfall. Spring-fed water bodies are generally more consistent, whereas rain-dependent sources may dry up unpredictably.

Step 4: Calculating Rainwater Harvesting Potential

For homes that receive moderate rainfall, rainwater harvesting can supplement or even replace other water sources. The potential amount of rainwater that can be collected depends on the surface area of the roof and the annual rainfall of the region.

To calculate the total rainwater collection capacity, measure or estimate the roof area in square feet. Use the following formula to determine how much rainwater can be collected per inch of rainfall:

Total Collection (gallons) = Roof Area (sq. ft) × Rainfall (inches) × 0.623

For example, if a home has a 1,000-square-foot roof and receives 20 inches of annual rainfall, the total collection potential would be:

1,000 × 20 × 0.623 = 12,460 gallons per year

If using metric units:

Total Collection (liters) = Roof Area (sq. meters) × Rainfall (mm) × 1

For a 93-square-meter roof receiving 500 mm of annual rainfall:

93 × 500 × 1 = 46,500 liters per year

This calculation provides an estimate of how much water can be captured and stored, helping determine the size of storage tanks and whether additional sources are needed.

Step 5: Identifying Seasonal and Emergency Backup Options

No water source is guaranteed to remain stable year-round. Droughts, pump failures, or contamination can occur unexpectedly, so it is essential to have backup water sources in place.

Compare each identified water source based on:

- Availability during dry seasons – Does the source dry up, or is it reliable all year?
- Ease of access – Is water easily collected, or does it require heavy equipment?
- Filtration and treatment needs – Does the water require significant purification?
- Long-term sustainability – Will the water source continue to be viable in 10, 20, or 50 years?

A primary water source should be dependable for at least 80% of water needs, while secondary and emergency sources should cover short-term disruptions or increased seasonal demand.

Step 6: Creating a Water Source Map and Feasibility Report

Once all water sources have been evaluated, the final step is to compile the findings into a clear, actionable reference.

Draw a simple water source map, marking each potential source on the property and noting key details such as:

- Distance from the home
- Estimated availability throughout the year
- Whether additional treatment is required
- Backup options if the source fails

If using digital tools, mapping applications or satellite images can be used to add location pins and measurements for more precise planning.

Alongside the map, summarize the key findings in a short feasibility report, outlining the strengths and weaknesses of each source and which will be prioritized for development.

By completing this project, a structured, data-driven approach is established for choosing the best water sources for an off-grid home.

Project: Water Sourcing and Collection - Building a Custom Above-Ground Water Tank System

Creating an above-ground water tank system is one of the most straightforward ways to store water for an off-grid home. It's a flexible solution that can be adapted to your specific needs, and it's relatively easy to set up. Whether you're collecting rainwater or drawing from a well, an above-ground tank allows for easy access and maintenance. This project will guide you through the process of designing and building a functional, custom water storage system for your off-grid water needs.

Materials Needed

- Large plastic, metal, or fiberglass tank (choose based on water storage needs)
- Gutters and downspouts (for collecting rainwater)
- PVC pipes and fittings (for connecting and transporting water)
- Mesh screens (for filtering debris)
- First-flush diverter (or DIY materials for one)
- Pump system (manual, electric, or solar-powered, depending on needs)
- Teflon tape or silicone sealant (for sealing connections)
- Drill and saw (for cutting holes in the tank and pipes)
- Pipe connectors and valves
- Gravel or sand (for base preparation)
- Concrete or concrete blocks (for tank stabilization and leveling)
- Water quality testing kit (optional)

Step 1: Choose the Location for Your Tank

The first step is to choose a location where the tank will be placed. You want to ensure the location is flat, stable, and easily accessible for maintenance. If you are collecting rainwater, make sure it is positioned near the downspouts from your roof to allow water to flow directly into the tank.

1. Ensure the area is **level** to prevent the tank from tipping over or becoming unstable over time.
2. The area should be well-**drained** to avoid standing water around the tank that could cause erosion or mold.
3. If you plan to use gravity to move water from the tank, **elevate the tank** slightly to facilitate water flow into the distribution system.

Step 2: Prepare the Tank Base

Before placing the tank, you need to create a stable and level foundation.

1. Start by clearing the area of any debris, rocks, or vegetation.
2. Lay a layer of **gravel or sand** to ensure proper drainage and prevent direct contact with the ground.
3. If necessary, build a **concrete slab** or place **concrete blocks** to further stabilize the base and provide an even surface for the tank.

Step 3: Install the Gutters and Downspouts

If you're using the system for rainwater harvesting, install gutters and downspouts to collect water from your roof. This system will direct the rainwater into the tank.

1. **Install gutters** along the edge of the roof, ensuring they are properly sloped toward the downspouts to prevent water from pooling.
2. Attach **downspouts** to the gutters to direct the rainwater to the tank.
3. Position the downspout so that the water flows easily into the tank, without splashing or causing debris to enter.

4. If you're using the system for surface water, ensure that the downspouts are **seamlessly connected** to the storage tank.

Step 4: Install the First-Flush Diverter

A first-flush diverter is used to remove the initial dirty water that comes off the roof during a rainstorm. This ensures that only clean water enters the tank.

1. **Install the diverter** where the downspout enters the tank system.
2. The diverter will **divert the first few gallons of rainwater** to a separate outlet, preventing debris from entering the tank.
3. After the initial flush, the clean water will flow into the tank, ready for storage.

Step 5: Install the Tank and Connect the Plumbing

Now it's time to position the tank and connect it to the collection system.

1. **Place the tank** in its final location on the prepared base.
2. **Connect the downspout** (with or without the first-flush diverter) to the tank's **inlet pipe**. This pipe will bring the rainwater or other collected water into the tank.
3. If your tank does not have pre-made inlet or outlet holes, **cut the holes** using a saw or drill and install **PVC pipe fittings**. Seal the connections with **Teflon tape** or **silicone sealant** to prevent leaks.
4. If using a **pump system**, install the **outlet pipe** at the bottom of the tank to allow water to be pumped out efficiently.

Step 6: Install the Pump and Water Delivery System

To move water from the tank to your home or garden, you will need to install a pump and water delivery system.

1. **Choose a pump system** based on the size of your tank and water needs. For small tanks, a **manual pump** may be sufficient, while larger systems may require an **electric or solar-powered pump**.
2. Connect the **outlet pipe** of the tank to the **inlet of the pump**.
3. From the pump, connect the **outlet pipe** to your distribution system. This could be **PVC pipe** or **flexible tubing** that carries the water to your faucet, irrigation system, or other points of use.
4. Test the pump to ensure that it is working properly and water is flowing from the tank as needed.

Step 7: Set Up Water Quality Measures

To ensure that your water remains clean and safe for use, install any necessary filtration systems or treatment options.

1. Consider adding a **mesh screen** to the top of the tank to prevent leaves, debris, and insects from entering the water.
2. If using the tank for **drinking water**, install a **filtration system** at the water outlet to ensure the water is free from bacteria and impurities. Options include **activated carbon filters**, **UV light systems**, or **reverse osmosis filters**.
3. Regularly **clean the tank** and any filtration components to prevent algae or mold growth, especially in warmer climates.

Step 8: Test and Monitor the System

After everything is set up, you'll want to ensure that the system is functioning properly and that water is flowing correctly.

1. **Test the system** by allowing the water to fill the tank, checking for any **leaks** at the connections or around the tank.
2. Turn on the pump and monitor the flow of water to make sure it is moving freely and with adequate pressure.
3. Test the **water quality** using a **water testing kit** to ensure that the water is safe for its intended

use. Check for any contaminants, especially if you're using the water for drinking.

Step 9: Regular Maintenance

For your system to continue functioning well, regular maintenance is necessary to ensure long-term effectiveness.

1. **Inspect the tank and connections** every few months to check for leaks or any signs of wear and tear.
2. Clean the **gutter system** and **first-flush diverter** periodically to ensure that rainwater is flowing into the tank without obstruction.
3. **Maintain the pump** by lubricating moving parts, checking electrical connections, and ensuring that the filter or pump components are in good condition.
4. **Monitor water quality** regularly and clean or replace filters as needed.

Project: Water Sourcing and Collection - DIY Manual Well Drilling and Pump Installation

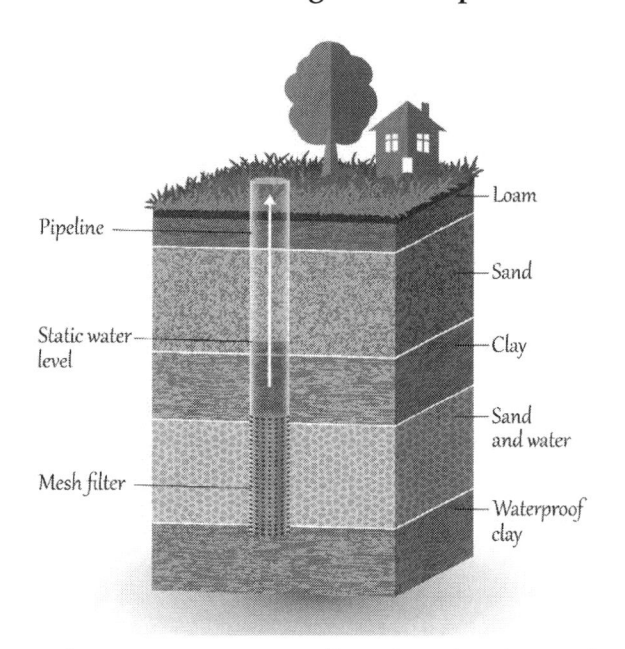

Creating your own well can be a deeply satisfying project that ensures you have reliable access to water, especially in areas where groundwater is available but other water sources are not sufficient. This project guides you through the steps of drilling a manual well and installing a basic hand pump system. The process may be labor-intensive, but with the right tools and careful planning, it is achievable.

Materials Needed

- Manual drilling tools (e.g., hand auger or post-hole digger)
- PVC pipe (for casing and pump)
- Shovel or post-hole digger
- Bucket and rope (for retrieving water)
- Gravel and sand (for filtration)
- Pump system (hand-powered pump or basic piston pump)
- Well casing (PVC or steel)
- Safety gear (gloves, goggles, etc.)
- A measuring tape or ruler
- A water quality testing kit (optional but recommended)

Step 1: Choose the Location for the Well

Before starting, choose an appropriate location for your well. Proximity to your home is important for convenience, but you must also consider safety and contamination risks.

- The well should be placed at least 100 feet (30 meters) away from potential contamination sources like septic tanks, animal waste areas, or industrial sites.
- The ground should be level, with no low spots that might trap runoff or lead to flooding.

You can also consult local water table data to determine the best depth for the well. If the water table is shallow, manual drilling is easier, but deeper wells may require a machine.

Step 2: Prepare for Drilling

Start by preparing the site for digging. You'll need to clear the area around the well to allow enough room for the drilling tools and any other equipment.

1. Mark the center of the well using spray paint or a stake.
2. Dig a small hole to make it easier to insert the auger or other manual drilling tool.
3. Have someone help with the drilling process or create a system to hold the auger in place while you turn it.

Step 3: Begin Drilling

Now comes the hard work. Drilling by hand will take some time and patience, but it's manageable with the right technique.

1. Insert the auger or post-hole digger into the ground and begin turning it.
2. Every few inches, pull out the auger to clear away the soil and debris.
3. Continue drilling until you reach the water table. You may hear or feel water entering the hole when you hit the right depth.
4. If drilling becomes difficult due to dense rock or clay, use a manual jetting system (using water to loosen soil and rock) or consider a borehole drilling system if the water table is too deep.

Step 4: Install the Well Casing

Once you reach the water table, it's time to protect the well from collapsing and prevent contamination. Installing a well casing ensures that the sides of the well remain stable and helps keep out dirt, debris, and pollutants.

1. Measure the depth of the well and cut your PVC casing pipe to size.
2. Lower the casing pipe into the hole. Ensure it is straight and reaches deep enough into the water table.
3. Seal the top of the casing with a cap to keep out debris and prevent contamination from surface runoff.

Step 5: Set Up the Pump System

Now that the well is properly cased, it's time to install a pump to draw water from the well. Since this is a manual system, you will be using a hand pump or piston pump.

1. Install the pump base over the well casing, making sure it's tightly sealed to prevent leaks.
2. Attach the pump rod to the pump base and lower it into the well casing.
3. Attach the pump handle and ensure that the piston moves up and down smoothly.
4. Test the pump by operating it a few times to ensure it draws water from the well properly. If the pump is installed correctly, you should be able to easily raise and lower the handle.

Step 6: Collecting Water

Once the pump is in place, you can begin using it to draw water. A bucket and rope system can be used as a backup if you don't need a dedicated pump. To operate the pump:

1. Lift the handle to pull water from the well.
2. Pour the water into a clean container for use.
3. Ensure the pump system is well-maintained to keep it operating smoothly over time. Check for any signs of wear or damage, especially to the seal and pump handle.

Step 7: Water Quality Testing

Before drinking water from your well, it's important to ensure it is safe. Use a water quality testing kit to check for contaminants such as bacteria, iron, or sulfur. Many well owners also choose to install a simple filtration system (such as a sand filter or activated carbon filter) to ensure water quality.

1. Test the well water for pH levels, bacteria, and turbidity (cloudiness).
2. Treat the water with appropriate filtration methods if any contaminants are found. Common filtration methods include boiling, UV light treatment, or using carbon filters.
3. Re-test the water periodically to ensure its quality remains high.

Step 8: Well Maintenance

A manual well system requires regular maintenance to keep the water clean and the pump working efficiently. Periodically check the following:

- Water quality: Test water for bacterial contamination and sediment buildup.
- Pump function: Check the pump mechanism to ensure it is not obstructed and operates smoothly.
- Casing and seals: Inspect the well casing and seals for cracks or damage.
- Debris: Clear any debris from around the well, especially if it's an open well.

Regular cleaning, maintenance, and water testing will keep your well running efficiently for many years, providing reliable water access to your off-grid home.

Project: Water Sourcing and Collection - Rainwater Harvesting System Design and Build

Rainwater harvesting is an essential off-grid water collection method that is both cost-effective and sustainable. By collecting and storing rainwater, you can ensure a reliable water supply for your home, garden, and other needs without relying on the grid. This project will guide you through the process of designing and building a **rainwater harvesting system** that is efficient, reliable, and tailored to your home's needs.

Materials Needed

- Gutter system (gutters, downspouts, brackets)
- First-flush diverter (or DIY materials for one)
- Rainwater storage tank (preferably food-grade or made from UV-resistant materials)
- Mesh screen (for filtering debris from rainwater)
- PVC pipes or flexible tubing for system connections
- Tools for cutting and fitting pipes (saw, drill, PVC cement)
- Teflon tape or silicone sealant (for sealing connections)
- Water quality testing kit (optional but recommended)
- Measuring tape or ruler
- Ladder (if working on a roof)

Step 1: Calculate Rainwater Collection Potential

Before you begin, it's important to estimate how much rainwater you can realistically collect. This calculation will determine the **size of your storage tank** and the **capacity of your gutters**.

Start by measuring your **roof catchment area** (the total surface area of the roof that will direct rainwater into the system). To do this:

1. Measure the length and width of each section of your roof.
2. Multiply the length by the width to get the area in square feet (or square meters).
3. Add up the areas of all roof sections that will collect rainwater.

Next, determine the **annual rainfall** in your area. You can find this information from local weather data or a simple online search. Once you have your roof area and annual rainfall, use the following calculation to estimate how much rainwater you can collect:

Total Collection (gallons) = Roof Area (sq. ft) × Rainfall (inches) × 0.623

For example, if your roof area is **1,000 square feet** and your area receives **20 inches of annual rainfall**, the calculation will be:

$1,000 \times 20 \times 0.623 =$ **12,460 gallons per year**

For metric units, use:

Total Collection (liters) = Roof Area (sq. meters) × Rainfall (mm) × 1

For a **93-square-meter roof** receiving **500 mm of rainfall**, the total collection would be:

$93 \times 500 \times 1 =$ **46,500 liters per year**

These calculations will give you an idea of how much water your system can collect and help you choose the appropriate **storage tank size**.

Step 2: Install the Gutter and Downspout System

The first step in setting up your rainwater collection system is to install gutters and downspouts on your roof to direct water into your storage tank. The system needs to be installed at the **lowest point of the roof**, where water naturally drains.

1. Install gutters: Attach gutters along the roof's edge to collect rainwater. Use brackets to secure the gutters to the roof's fascia board, ensuring they are sloped slightly towards the downspout for proper water flow.
2. Attach downspouts: Install downspouts at the ends of the gutters to direct water down to the ground and into your collection system. Make sure the downspouts are securely attached and that water flows freely without blockages.
3. Connect the downspouts to the tank: Use PVC pipe or flexible tubing to connect the downspouts to the storage tank. Depending on the setup, you may need a connector or elbow fitting to direct the flow into the tank.

Ensure that the gutters and downspouts are clear of debris and installed at an angle to prevent water pooling.

Step 3: Install the First-Flush Diverter

A first-flush diverter is an important component of a rainwater harvesting system, as it ensures that the first part of the rainfall, which may contain dust, leaves, and other debris, is directed away from the storage tank. Installing a first-flush diverter ensures that only clean water enters the tank.

You can purchase a pre-made first-flush diverter or create a **DIY version** using a length of pipe and a diverter valve. Here's how to install it:

1. Attach the diverter to the downspout, preferably near the base where water enters the storage tank.
2. When it rains, the first few gallons of water will fill the diverter pipe. The diverter will automatically redirect the dirty water into a designated drainage area, and only clean water will flow into the storage tank.
3. Set up the diverter's capacity to handle the amount of water you expect to flow from your roof during a typical rainfall. The diverter should be large enough to handle the initial volume of water without causing overflow.

Step 4: Prepare and Install the Storage Tank

Once the water is flowing through the gutters, downspouts, and diverter, it needs to be stored in a **sealed tank** to prevent contamination and algae growth. You can choose from a variety of tanks, including plastic, metal, or concrete. Make sure to choose a **food-grade tank** to ensure the water stays safe for consumption.

1. Choose a location for the tank that is both easily accessible and large enough to handle the volume of rainwater you expect to collect. Ideally, the tank should be installed on a flat, stable surface and elevated slightly to allow gravity to help with water flow.
2. Place the tank in a shaded area to reduce the chance of algae growth.
3. Install the tank's inlet and outlet connections. The inlet will connect to the downspout or diverter, while the outlet will be used to distribute water throughout the system.
4. If you're storing the tank above ground, consider building a raised platform to improve water flow and reduce contamination from debris and dirt.
5. Secure the tank with a tight-fitting lid to prevent insects, leaves, and other debris from contaminating the water.

Make sure the storage tank is **properly sealed** and equipped with an **overflow valve** to prevent spillage during heavy rain.

Step 5: Install the Filtration System

Once water is in the tank, it's essential to **filter it before use**, especially for drinking and cooking. Several filtration options are available depending on your needs and budget:

1. Mesh screen: Use a fine mesh screen to filter out large debris such as leaves and twigs before water enters the storage tank.
2. Activated carbon filter: These filters can remove chlorine, odors, and heavy metals from the water.
3. UV filtration or reverse osmosis: If you want to ensure your water is free of pathogens, you can install a UV purification system or reverse

osmosis system that removes bacteria, viruses, and other contaminants.

Set up the filtration system at the **outlet point** of the tank to ensure all water that leaves the tank is filtered.

Step 6: Testing and Monitoring

After your system is set up, it's time to test the water quality and check that the entire system is functioning as expected.

1. Test the water quality using a water testing kit. Test for common contaminants such as bacteria, pH levels, and sediment.
2. Test the first-flush diverter to make sure it is diverting the initial dirty water and not allowing it to enter the storage tank.
3. Monitor the system during the next rainfall to ensure the water flows smoothly through the system and into the tank. Check for any leaks or blockages.
4. Regularly check the filters and clean the gutters to keep the system functioning efficiently.

Step 7: Regular Maintenance and Seasonal Adjustments

Once the system is up and running, it's important to maintain it to ensure longevity and water quality.

• Clean the gutters regularly, especially before rainy seasons, to prevent debris buildup.
• Check the tank for leaks and ensure the lid remains sealed.
• Inspect the diverter and filters periodically to ensure they are functioning properly.

As rainfall patterns change with the seasons, be sure to **adjust storage expectations** and monitor how much water is being collected throughout the year. You might need to adjust the tank's overflow system or consider adding a second tank if your needs increase.

Project: Water Sourcing and Collection - Surface Water Collection and Filtration Setup

Materials Needed

Collection System:

• Rainwater harvesting system (gutters, downspouts, first-flush diverter)
• Surface water intake system (pond, lake, or river intake pipe)
• Water catchment basin (natural or constructed)
• Gravel, sand, and rocks for a sedimentation area

Filtration System:

• Large storage tanks or barrels (food-grade plastic or metal)
• Screen filters (mesh, fine stainless steel)
• Charcoal or activated carbon
• Fine sand and gravel
• Ceramic or bio-sand filter
• UV purification unit (optional)
• Chlorine or iodine tablets (as a backup method)

Tools & Miscellaneous:
- Shovel, pickaxe, and rake
- PVC pipes or food-grade hoses
- Hose clamps and fittings
- Bucket or scoop
- Sealing tape and waterproof sealant
- Solar pump (if needed for water transport)
- Protective gloves and safety glasses

Step 1: Select and Prepare the Water Source

1. Identify a **reliable surface water source** (pond, stream, river, or rainwater collection).
2. Ensure the **location is uphill or level for easy gravity-fed transport** to the filtration area.
3. If using rainwater, install **gutters and downspouts** on a roof to channel water into collection barrels.
4. If using a natural water source, create a **protected intake area**:
   4.1. Dig a **shallow basin** near the water's edge to allow sediment to settle.
   4.2. Line the basin with **gravel and sand** to pre-filter debris.
   4.3. Use a **screen filter** at the intake point to block large contaminants.

Step 2: Build the Water Collection and Pre-Filtration System

1. Install a **first-flush diverter** for rainwater harvesting to discard the initial runoff.
2. For natural water sources, extend a **PVC intake pipe** into the water, securing it with rocks.
3. Place a **gravel and sand pre-filter** around the intake to reduce sediment.
4. Connect the intake pipe to a **settling tank** or **sediment bucket** to allow heavier particles to settle.
5. Ensure all connections are **sealed tightly** to prevent contamination.

Step 3: Construct the Multi-Layer Filtration System

1. Use a **large drum or barrel** as the primary filtration unit.
2. Layer filtration media inside the barrel in the following order:
   2.1. **Bottom layer:** Large gravel for structural support.
   2.2. **Middle layer:** Fine gravel and coarse sand to trap debris.
   2.3. **Top layer:** Activated carbon or charcoal for chemical filtration.
3. Insert a **ceramic filter or bio-sand filter** at the outlet for final purification.
4. Attach a **sealed spout or pipe** to direct clean water into a storage container.

Step 4: Set Up Water Storage and Final Treatment

1. Position a **large, covered water storage tank** downhill or near the filtration system.
2. Use a **UV purification lamp or chlorine tablets** for extra treatment.
3. Ensure tanks are **sealed** to prevent contamination from insects, debris, or algae.
4. If necessary, install a **solar-powered pump** to transport water to elevated areas.

Step 5: Maintain and Monitor the System

1. Clean the **intake filter** weekly to remove debris.
2. Replace **activated carbon** every 3-6 months to maintain filtration efficiency.
3. Inspect and clean the **storage tanks** periodically to prevent algae growth.
4. Regularly test water for **contaminants**, especially after heavy rains or seasonal changes.

Project: Water Sourcing and Collection - Solar-Powered Water Pump System Installation

A solar-powered water pump system is an excellent choice for off-grid living, providing a reliable and sustainable way to pump water without the need for electricity or fuel. Whether you are using a well, a pond, or a river as your water source, a solar pump can help you efficiently move water into your home or storage tanks. This project will guide you through the process of installing a solar-powered water pump system, from choosing the right pump to setting up the solar panels that will power it.

Materials Needed

- Solar pump (appropriate for your water source depth and flow rate)
- Solar panel(s) (appropriate wattage for your pump)
- Charge controller (to protect the battery from overcharging)
- Battery (for energy storage)
- Wiring (for connecting components)
- PVC pipes or flexible tubing (for water transport)
- Pipe fittings (elbows, T-joints, and connectors)
- Water storage tank or cistern
- Mounting brackets for solar panels
- Basic tools (drill, saw, wrenches, tape measure)
- Batteries (deep cycle, as required)
- Water filter (optional for filtration)

Step 1: Choose the Right Solar Pump

The first step in setting up a solar-powered pump system is to choose the right pump for your needs. There are several factors to consider when selecting a solar pump:

1. Water source type: Whether you are drawing water from a well, river, or pond, make sure the pump is designed for that specific application.
2. Flow rate: Calculate the amount of water you need per day. A pump with a higher flow rate will move more water in a shorter time, while a smaller pump will be sufficient for a low-volume application.
3. Pump depth: If you are using a well, check how deep the water table is. Some pumps are designed for shallow wells, while others can reach deeper aquifers.
4. Solar panel compatibility: Ensure the pump you select is compatible with the size and wattage of your solar panel system.

Once you have these details, select a solar pump that matches your requirements. There are many reliable pumps on the market, from submersible pumps for deep wells to surface pumps for shallower applications.

Step 2: Install the Solar Panels

The solar panels are the heart of the system, converting sunlight into energy to power the pump. Installing the solar panels correctly ensures maximum energy generation.

1. Choose an optimal location: Install the solar panels in a location that receives uninterrupted, full sunlight for the majority of the day, typically facing south (in the northern hemisphere) and at an angle that matches your local latitude for best performance.
2. Mount the panels: Securely attach the solar panels to a mounting bracket or frame that is weather-resistant. You may need to tilt the panels slightly to ensure optimal sunlight absorption. Ensure they are mounted in a way that prevents them from being shaded during the day.
3. Wiring: Connect the solar panels to the charge controller using weatherproof wiring. The charge controller regulates the power going into the battery to prevent overcharging.

Step 3: Install the Charge Controller and Battery

The charge controller regulates the energy coming from the solar panels to the battery, ensuring that it does not overcharge or over-discharge.

1. Choose a charge controller: Make sure the charge controller is rated for the total wattage of your solar panels and pump system.
2. Mount the charge controller in a location that is easily accessible but protected from the elements (e.g., a weatherproof box).
3. Connect the charge controller to the battery: The charge controller should be connected to the battery so that it can manage the flow of energy. Deep-cycle batteries are best for off-grid water systems, as they are designed for long-term energy storage and can handle frequent charge and discharge cycles.
4. Install the battery: Place the battery in a dry, cool, and ventilated location. It should be in a position where it will not overheat, as this can damage its lifespan.

Step 4: Connect the Pump to the System

Now that you have your solar panels, charge controller, and battery set up, the next step is to connect the pump to the system.

1. Position the pump: If you are using a submersible pump for a well, carefully lower it into the well according to the manufacturer's instructions. Ensure the pump is securely placed at the appropriate depth, typically just above the water level to prevent dry running.
2. Install piping: For surface pumps or for transferring water from a surface source, install the necessary PVC pipes or flexible tubing to move water from the pump to the storage tank. Secure all pipe connections with pipe fittings to prevent leaks.
3. Connect the pump to the charge controller: Wire the pump to the charge controller, ensuring all connections are weatherproof and secure. The pump should be connected to the battery through the charge controller so that it can draw energy when needed.

Test the system to ensure that the pump activates properly when sunlight is available and that water flows smoothly into the storage tank.

Step 5: Install the Water Storage Tank

Water storage is an important consideration in off-grid systems. You need to ensure that the tank is large enough to store the water you collect, especially during dry periods.

1. Choose the right tank: The size of the storage tank depends on your household's water consumption and the pump's flow rate. A larger tank can store more water for use during periods when the pump isn't running (such as cloudy days or at night).
2. Place the tank: Set the tank in a stable, well-drained location to prevent contamination and allow for easy access. If you plan to use gravity-fed water systems, place the tank on an elevated platform or structure so that water can flow naturally into your home.
3. Connect the tank to the system: Install a pipe or flexible tubing from the pump to the storage tank, ensuring the water flows freely. If the pump's output is higher than the tank's capacity, install an overflow valve to allow excess water to escape safely.

Step 6: Test the System

Once all components are connected, it's time to test the solar-powered pump system to ensure that it's functioning properly.

1. Power on the system: Make sure that the solar panels are receiving adequate sunlight.
2. Activate the pump: The pump should begin drawing water as soon as sunlight hits the panels. Observe the flow of water from the pump to the storage tank.
3. Check for leaks: Inspect all pipe connections for leaks or loose fittings and tighten them as necessary.
4. Monitor pump performance: Ensure that the pump is operating smoothly and efficiently. If the water level in the tank isn't rising, verify the wiring and pump connections for any issues.

Step 7: Maintenance and Troubleshooting

Once the solar-powered water pump system is up and running, regular maintenance is necessary to keep it operating efficiently:

1. Clean the solar panels: Dust and dirt can accumulate on the solar panels, reducing their

efficiency. Clean them regularly using a non-abrasive cloth and mild soap.
2. Inspect the pump: Check the pump for any debris or blockages that could impair performance. Clean the pump regularly to ensure efficient water flow.
3. Monitor battery health: Ensure the battery is properly charged and functioning. Replace the battery if it shows signs of deterioration.

By completing these steps, you will have effectively installed a solar-powered water pump system, ensuring a dependable and eco-friendly water source for your off-grid living. The system will run independently, drawing energy from the sun to pump water without the need for external power sources.

Project: Complete Filtration System for a Water Tank

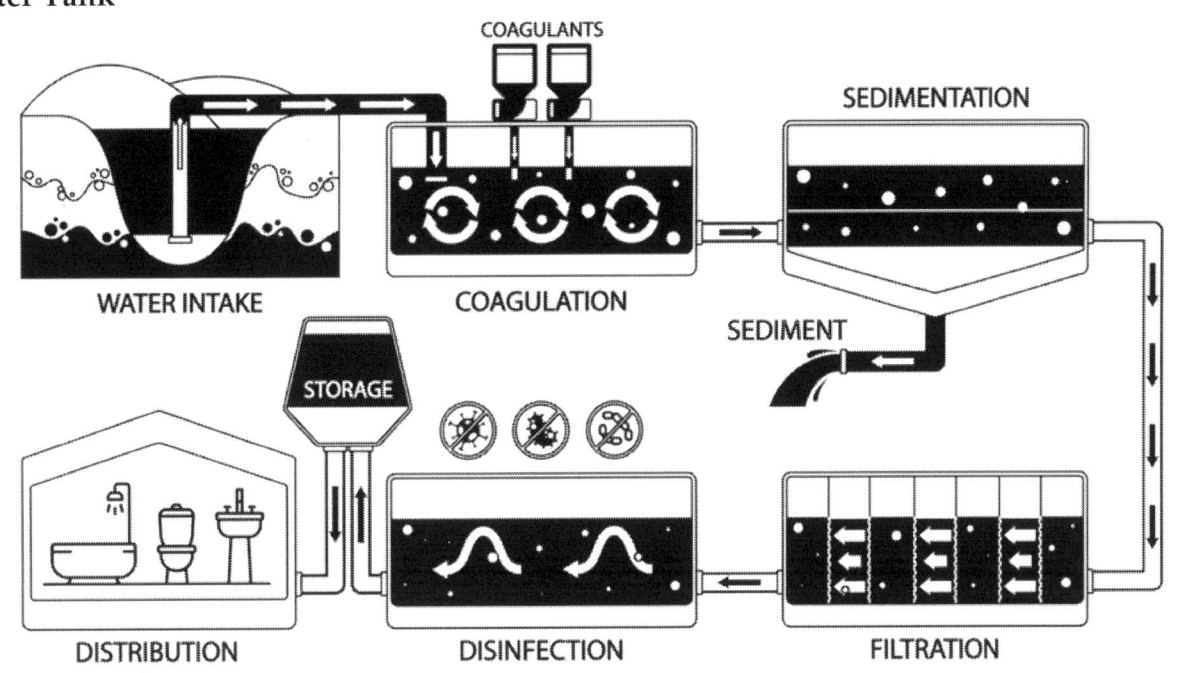

Water filtration is a crucial part of ensuring your stored water remains clean and safe for use, whether it's for drinking, cooking, or irrigation. A properly designed filtration system removes debris, chemicals, bacteria, and other contaminants from the water before it enters your home. This project will guide you through the steps to design and install a multi-stage filtration system for your water tank. By combining sediment filters, activated carbon filters, and UV light treatment, you'll be able to provide safe water for your off-grid living.

Materials Needed
- Sediment filter (mesh or cartridge type)
- Activated carbon filter (for removing chlorine, odors, and some chemicals)
- UV sterilizer (portable or installed system)
- PVC pipes and fittings (for connecting filters and tank)
- A clean container (for filtered water storage)
- Teflon tape or silicone sealant (for sealing pipe connections)
- Water quality testing kit (optional but recommended)
- Drill and saw (for cutting and fitting pipes)
- Protective gloves and eyewear (for safety)

Step 1: Assess Your Water Quality

Before starting, it's essential to assess the quality of the water you're filtering. Whether it's from a rainwater collection system, surface water, or a well, testing for common contaminants will help you design an effective filtration system.

1. **Test the water**: Use a water quality testing kit to check for **bacteria, pH levels, turbidity (cloudiness), and chemical contaminants**.
2. **Identify the contaminants**: Based on the results, determine the specific filtration needs. For example, if the water contains a high level of sediment or large particles, you'll need a more robust sediment filter. If bacteria are present, UV sterilization will be necessary.

Step 2: Plan Your Filtration Stages

A multi-stage filtration system works best for ensuring clean water. It involves filtering out larger particles first, followed by the removal of chemicals and bacteria.

1. **Stage-0 Coagulation (optional):** Collect water in a large bucket or tank before filtration. Add a

coagulant to the water to bind suspended particles: Use aluminum sulfate (alum): Add 1 teaspoon per gallon (4 liters) of water and stir vigorously for 2-3 minutes, then let it sit for 30 minutes. For natural alternatives, grind Moringa seeds into powder, mix with water, and stir for the same duration. As particles bind together, they form larger clumps (floc) that will settle at the bottom. Carefully pour the clearer water from the top, avoiding the sediment.

2. **Stage-1– Sediment filtration**: The first filter should remove **large particles**, such as dirt, sand, and leaves. Use a **mesh filter** or **cartridge filter**.
3. **Stage-2 – Activated carbon filtration**: This filter removes **chemicals**, such as chlorine, pesticides, and some heavy metals, as well as odors.
4. **Stage-3 – UV sterilization**: After filtering out solids and chemicals, UV light will **disinfect the water** by killing bacteria and other microorganisms.

Each stage works together to ensure the water is clean and safe for use.

Step 3: Install the Sediment Filter

The first step in the filtration process is to remove larger particles from the water. This is accomplished with a sediment filter, which can be either a mesh filter or a cartridge-type filter.

1. **Attach the sediment filter** to the **inlet pipe** where water enters the filtration system from the storage tank.
2. Use **PVC pipe fittings** to secure the filter and ensure that the flow of water is directed through it.
3. Seal all connections with **Teflon tape** or **silicone sealant** to prevent leaks.
4. Ensure that the filter is easily accessible for cleaning or replacement as needed.

Step 4: Install the Activated Carbon Filter

After the water passes through the sediment filter, it will be much clearer, but it may still contain chemicals and odors. The activated carbon filter will address this issue by removing contaminants like chlorine, heavy metals, and pesticides.

1. **Attach the activated carbon filter** after the sediment filter in the system's flow path.
2. If using a **cartridge filter**, install it in a housing that is easy to access for replacement.
3. **Connect the filter to the outlet pipe** using PVC pipes or flexible tubing.
4. Check all connections to ensure they are secure and leak-free.

Step 5: Install the UV Sterilizer

The UV sterilizer will disinfect the water by killing bacteria and viruses. This step ensures that any remaining pathogens are neutralized before the water is used.

1. **Position the UV sterilizer** at the end of the filtration system, just before the water is dispensed for use.
2. Connect the outlet from the activated carbon filter to the **input of the UV sterilizer**.
3. Connect the **outlet of the UV sterilizer** to the water delivery system, whether that's to a tap, faucet, or irrigation system.
4. Follow the manufacturer's instructions for the installation of the UV light unit. Typically, this involves **mounting it** securely and ensuring that the light chamber is sealed.
5. Ensure the **UV sterilizer's lamp** is exposed to sufficient power, and make sure the unit is positioned to avoid direct sunlight, which could affect its efficiency.

Step 6: Connect the System to the Water Storage Tank

Now that you have the filtration system set up, you need to connect it to your water storage tank so that filtered water can be used.

1. **Connect the input of the system** to the outlet of the water storage tank using **PVC piping** or **flexible tubing**.

2. Ensure that all connections between the tank, filters, and pump system are tight and leak-free.
3. Check that the **output** of the filtration system is directed to a clean **storage container** or directly to your household water system.

Step 7: Test the System

Once everything is connected, it's important to test the system to make sure it's working properly.

1. **Run water through the system** to ensure it flows freely and is filtered correctly.
2. **Check the water quality** by using your water testing kit. This will confirm that the filtration system has removed contaminants like **bacteria**, **sediment**, and **chemicals**.
3. Monitor the flow of water through each filter and the UV sterilizer. Ensure that each stage is effectively removing contaminants and that the system is functioning as expected.

Step 8: Regular Maintenance and Monitoring

To keep the filtration system functioning effectively over time, it's essential to perform regular maintenance and monitoring.

1. **Clean or replace the sediment filter** regularly to remove accumulated dirt and debris. This will prevent clogging and ensure efficient filtration.
2. **Replace the activated carbon filter** every few months or according to the manufacturer's recommendation to ensure it is effectively removing chemicals.
3. **Test the UV sterilizer's functionality** by checking the lamp's performance. Most UV systems have an indicator light to show if the bulb is working.
4. **Check the water quality periodically** to ensure the filtration system is still working effectively.

Project: Building a Hand Pump Backup System

A hand pump is a simple and effective way to provide water in off-grid environments, especially as a backup to your main water source. Setting up a hand pump system doesn't require complex equipment and is a reliable solution for emergency situations or when power is unavailable. In this project, you will install a hand pump on your well or water source to ensure that you always have access to water, regardless of external conditions.

Materials and Tools Needed
- Hand pump (manual pump)
- PVC or metal pipe (depending on the type of hand pump)
- Pipe fittings (couplings, elbows, etc.)
- Pipe wrench or adjustable spanner
- Pipe cement or Teflon tape for sealing
- Water source (well, cistern, or any groundwater source)
- Gravel (optional, for stabilizing the pump base)
- Concrete or stone (optional, for reinforcing the well casing)

Instructions

1. Choose the Right Location for Your Hand Pump

First, identify the most practical location for your hand pump. Ideally, it should be close to your water source (like a well or cistern) and positioned so you can comfortably operate the pump. Consider these factors:

- **Accessibility**: The pump should be easily accessible year-round, whether it's near the house or close to a garden.
- **Safety**: Ensure the pump area is free from hazards like sharp objects or overgrown vegetation.
- **Elevation**: If you are using a well, the hand pump should be placed above the water table to function effectively.

2. Install the Pump Base (if required)

If your water source is a well or deep cistern, you'll need to stabilize the base for your pump. For deeper wells, the pump will need to be securely anchored to withstand the force of pumping. Here's how you can do it:

- Dig a shallow hole around the well or water source (if necessary).
- Place gravel or small stones at the bottom of the hole to prevent shifting or settling.
- For additional stability, you can pour concrete around the well casing, ensuring it's level and the pump installation is secure.

3. Connect the Pipe to the Hand Pump

Once you've prepared the installation area, begin by connecting the hand pump to the pipe that leads to the water source.

- Measure the distance from the water source to the pump location and cut the pipe accordingly.
- Attach the pipe to the pump's intake port. Ensure the connection is tight and secure. If necessary, use Teflon tape or pipe cement to ensure there are no leaks.
- For **deep well** pumps, you may need an additional long **pump rod** that connects the handle to the pump mechanism.

4. Place the Pump in Position

- Carefully lift the hand pump and position it on the base you've prepared.
- Use the pipe fittings to connect the pump to the main pipe that leads to your water source.
- Make sure that the handle is within a comfortable height range for easy operation.

5. Test the Pump

Once everything is connected, it's time to test the pump to ensure it's working correctly. Follow these steps:

- Check all connections for leaks before you begin.
- Begin pumping the handle slowly to prime the system. If the system is correctly installed, water should start flowing through the pump within a few pumps.
- Adjust the pipe fittings as necessary to ensure smooth flow and proper pressure.

6. Maintain and Protect the Pump

After successfully installing the hand pump, maintain it regularly to ensure long-term performance. Some simple maintenance steps include:

- **Cleaning the pump** regularly to prevent clogging, especially in areas with hard water or high mineral content.
- **Inspecting the seals** and fittings every few months to ensure there are no leaks or deteriorations.
- **Winterizing**: In cold climates, it's essential to winterize the hand pump to avoid freezing. Remove any water left in the pump after each use, and consider covering it during the winter months to protect it from the cold.

7. Optional: Install a Water Reservoir

To make your hand pump even more practical, you might want to connect it to a small **water reservoir** or **storage tank**. This is especially useful in off-grid systems where you may want to store water for later use.

- Use gravity to move the water from the hand pump into a **storage container** or **rainwater barrel**. If you plan to use this water for non-drinking purposes, it can also serve for irrigation, washing, or general cleaning tasks.

Module C. Energy & Power Systems for Total Independence

In an off-grid lifestyle, achieving energy and power independence is essential for ensuring self-sufficiency and resilience. This module will guide you through the process of generating, storing, distributing, and optimizing energy in a way that minimizes your reliance on external sources. Whether you're harnessing the power of the sun, wind, or water, this section covers practical and sustainable methods for powering your home and farm. By understanding different energy production systems, creating efficient storage solutions, and mastering the distribution of power, you'll have the tools to create a reliable and robust off-grid energy system that meets your needs year-round.

13. Generating Power: Effective Off-Grid Energy Production

Generating your own power is one of the most important steps in creating an off-grid lifestyle. In this section, we'll dive into the various renewable energy sources available, hybrid systems for increased efficiency, and the emerging technologies in energy harvesting that can further optimize your off-grid power production. Whether you choose solar, wind, hydropower, biomass, or backup generators, this guide will help you select and set up the best systems for your needs.

Renewable Energy Sources

Solar Power: Photovoltaic (PV) Systems

Solar energy is one of the most widely used forms of off-grid energy production. With a photovoltaic (PV) system, you harness the power of sunlight to generate electricity through solar panels. The basic components of a solar power system include:

- **Solar Panels**: These capture sunlight and convert it into electricity. The output is in the form of **direct current (DC)** electricity.
- **Inverters**: These convert the DC electricity from the solar panels into **alternating current (AC)**, which is what most home appliances require.
- **Charge Controllers**: These regulate the amount of electricity flowing into the battery storage system to prevent overcharging, ensuring the longevity of the batteries.

When designing your solar power system, it's important to calculate the total energy demand of your household and size your panels accordingly. You'll also need to factor in the amount of **sunlight** your location receives, as that directly impacts the system's efficiency. For example, if you live in a sunny area, your panels will produce more power than if you're in a region with frequent cloud cover or shorter days.

Wind Power: Wind Turbines

Wind turbines offer another reliable and renewable energy source, especially in areas with consistent wind speeds. Installing a wind turbine requires careful consideration of several factors to ensure you're optimizing your power production.

- **Siting**: A wind turbine needs to be installed in an area with consistent winds. Typically, wind speeds of around **10-15 mph (16-24 km/h)** are required for optimal performance. This means you'll need to assess the wind conditions of your location.
- **Height**: The higher the turbine is placed, the better. Wind speeds tend to increase at greater heights, so positioning your turbine on a tall pole or structure can increase its efficiency.
- **Turbine Size**: Wind turbines come in a range of sizes, from small, residential turbines to larger, industrial-scale models. Your energy requirements will dictate the size of the turbine you need, which will also determine the installation complexity and costs.

Wind turbines are often used in combination with solar systems, as they can provide power at different times—solar during the day and wind at night or during stormy weather. Wind is also highly effective in areas where the sun is less reliable.

Hydropower: Micro-Hydro Systems

If you have access to a stream, river, or other flowing water sources on your property, hydropower can be an incredibly efficient and reliable source of energy. Micro-hydro systems are small-scale hydropower systems that can provide continuous electricity for off-grid homes.

• **Site Selection**: The most important factor in a micro-hydro system is the **flow** of water and **elevation** drop. The greater the difference in height between the water source and your turbine, the greater the pressure available to drive the turbine. Look for a site where there's both **consistent flow** and enough **elevation** to produce reliable energy.
• **Flow and Elevation Requirements**: Ideally, you need at least **1-2 feet (0.3-0.6 meters)** of water fall to generate sufficient power. You will also need to ensure that the stream or river has consistent water flow throughout the year to provide continuous energy.

Hydropower systems are highly efficient but often require professional installation and careful site selection to ensure they meet your energy needs.

Biomass Energy: Wood Stoves, Biomass Heaters, and Biogas Production

Biomass energy, derived from organic materials like wood, plant matter, or waste, is another option for off-grid power. This type of energy is particularly useful for heating and cooking.

• **Wood Stoves**: Wood stoves can be used for space heating or cooking. They are **fuel-efficient** and easy to operate, making them an excellent choice for off-grid homes. You'll need to plan for a sustainable source of firewood.
• **Biomass Heaters**: Biomass heaters use organic material, such as wood chips or pellets, to generate heat. These systems can be used for **hot water** and **space heating**, and they are relatively easy to set up in off-grid homes.
• **Biogas Production**: If you generate enough organic waste (such as food scraps, animal manure, or plant matter), you can create a **biogas system** that produces **methane** for cooking or heating. This system uses anaerobic digestion, where bacteria break down organic matter in the absence of oxygen to produce methane gas.

Biomass is renewable and has the added benefit of being able to use local, sustainable resources, but it also requires regular maintenance and fuel sourcing.

Backup Generators: Diesel, Propane, and Natural Gas Generators

While renewable energy sources are fantastic for off-grid living, backup generators are essential for times when the sun isn't shining, the wind isn't blowing, or water sources are low. Off-grid generators are usually powered by **diesel**, **propane**, or **natural gas**. These generators can ensure that you always have access to power, even in the most challenging conditions.

• **Diesel Generators**: Diesel-powered generators are efficient and reliable, especially in rural areas. They are typically used as backup systems and require a steady supply of diesel fuel.
• **Propane Generators**: Propane is cleaner than diesel and often used for small or medium-sized off-grid systems. Propane is stored in **liquid form**, making it easy to keep fuel reserves.
• **Natural Gas Generators**: If you have access to a natural gas line, this is a convenient and **cleaner fuel option**. However, it's less commonly available in remote off-grid settings.

A generator is a necessary **backup** option but should be used sparingly to avoid relying too much on fossil fuels.

Hybrid Systems

Hybrid systems combine renewable energy sources to create a more reliable and efficient energy system. Combining **solar power**, **wind turbines**, or **hydropower** with a **backup generator** provides continuous power, even when one system isn't producing energy.

• **Solar and Wind**: Combining **solar** and **wind** energy allows you to generate power during the day and night, ensuring reliability. This system is particularly useful in areas with fluctuating sunlight or wind.
• **Solar and Backup Generator**: A solar system can provide daily energy needs, while a backup generator kicks in during times of low sunlight or high demand.
• **Wind and Hydro**: If you live in an area with both consistent wind and a reliable water source, a wind-hydro hybrid system can provide a steady stream of power throughout the year.

Hybrid systems require **careful planning** to ensure that all components work together seamlessly. They may also need **battery storage** to ensure that power is available during times when generation is low.

Energy Harvesting Technologies

Energy harvesting is a growing field that focuses on collecting energy from everyday activities or natural phenomena. These technologies can complement traditional power systems by capturing small amounts of energy from the environment.

- **Thermoelectric Generators (TEGs)**: TEGs convert **heat energy** into electricity. They can be used in off-grid systems to capture waste heat from stoves or engines and convert it into usable power.
- **Piezoelectric Energy Harvesting**: This technology captures energy from mechanical stress, such as vibrations or movement. Piezoelectric devices can be integrated into small systems, like self-powered sensors or emergency lighting.
- **Energy Storage Systems**: As energy harvesting systems often produce small amounts of power, they can be paired with **battery storage** to ensure that the collected energy can be stored and used when needed.

These low-power devices aren't suitable for running large appliances but can be an excellent addition to a fully integrated off-grid energy system, providing supplemental power for lights, sensors, or small devices.

14. Storing Energy: Power Reserves for Low-Output Times

One of the most important aspects of an off-grid system is the ability to store energy for times when renewable sources like solar, wind, or water are not producing enough power. Whether it's for cloudy days, calm nights, or unexpected power surges, energy storage ensures you have a reliable supply of electricity when needed most. This section will guide you through the various energy storage options, how to size a battery bank, and the benefits of energy management systems (EMS) to optimize your energy usage.

Battery Storage Systems

Energy storage primarily relies on **batteries**, which store electricity generated by your renewable energy sources. There are different types of batteries to consider, each with its own benefits, lifespan, and costs. When selecting the right battery, you'll need to think about your budget, how much space you have for storage, and how long you want your batteries to last.

Lead-Acid Batteries: Flooded, Sealed, AGM

Lead-acid batteries are one of the most common and cost-effective types of batteries used in off-grid systems. They come in several varieties, each suited to different needs and environments.

- **Flooded Lead-Acid Batteries**: These are the traditional type of lead-acid battery. They are affordable and can handle deep discharges, making them suitable for large off-grid systems. However, they demand consistent upkeep, such as monitoring water levels and maintaining adequate ventilation to prevent the accumulation of harmful gases.
- **Sealed Lead-Acid Batteries**: These batteries are maintenance-free because the acid is sealed inside the battery. They are less prone to leakage, which makes them ideal for areas with high humidity or where maintenance might be difficult. They typically come at a higher cost compared to flooded batteries.
- **Absorbed Glass Mat (AGM) Batteries**: AGM batteries are a type of sealed lead-acid battery with a higher energy density and faster recharge time. They offer greater efficiency compared to flooded lead-acid batteries and require no maintenance. They are a good option if you're looking for something more efficient, but they come at a higher price.

While lead-acid batteries are relatively cheap and widely available, they are heavier, have a shorter lifespan compared to newer battery technologies, and are less efficient in deep discharge situations.

Lithium-Ion Batteries: Benefits, Lifespan, and Cost Considerations

Lithium-ion batteries are gaining widespread popularity in off-grid systems because of their superior energy density, extended lifespan, and enhanced efficiency.

- **Efficiency**: Lithium-ion batteries are **more efficient** than lead-acid batteries. They can discharge deeper without causing damage and can hold a charge longer. This makes them more suitable for off-grid living, where energy needs fluctuate.
- **Lifespan**: Lithium-ion batteries last significantly longer than lead-acid batteries, typically between 10 to 15 years compared to 3 to 5 years for lead-acid batteries. This makes them a more long-term investment for off-grid systems.
- **Cost Considerations**: Although lithium-ion batteries have a higher initial cost, their extended lifespan and superior efficiency often result in greater cost-effectiveness in the long run. However, for some off-grid homeowners, the initial cost may be prohibitive.

Lithium-ion batteries are great for users who are looking for higher performance and don't mind the higher initial cost. They are ideal for systems that need to store energy for longer periods and provide reliable performance with fewer maintenance needs.

Flow Batteries: Advantages in Scalability and Off-Grid Applications

Flow batteries represent a relatively recent advancement in energy storage technology that store energy in liquid electrolytes rather than solid materials. They have some unique benefits for off-grid systems.

- **Scalability**: Flow batteries are highly scalable. You can increase your storage capacity simply by adding more electrolyte liquid or increasing the size of the system. This makes them perfect for large off-grid systems where your energy storage needs might increase over time.
- **Long Lifespan**: Flow batteries last a long time, with the potential for over **20 years** of use. They are also more resilient to deep discharges than traditional batteries, meaning they can be drained and recharged frequently without a significant loss in capacity.
- **Efficiency**: They are efficient at storing large amounts of energy and can provide steady power over long periods of time, making them a good option for users with higher power demands.

However, flow batteries are still relatively new and may come with a higher upfront cost, making them more suitable for large-scale systems rather than smaller residential ones.

Sizing Battery Banks

Sizing your battery bank correctly is crucial to ensure you have enough storage for your energy needs without overloading or underloading your system. The goal is to have enough capacity to handle your **average daily energy usage** as well as provide a buffer for days with low energy generation.

Calculating the Required Storage Capacity Based on Energy Consumption

To size your battery bank, start by calculating your total daily energy consumption in watt-hours. You need to take into account all the appliances, lights, and systems you will run off the batteries. Here's how to do it:

- **List the appliances** you want to run from the battery bank, along with their **wattage**.
- **Calculate the daily consumption** for each appliance by multiplying its wattage by the number of hours it runs per day. For example, a 100-watt light bulb running for 5 hours per day uses **500 watt-hours (Wh)** of energy.
- **Add up all the daily consumption** from all your appliances to get your total energy usage for the day in watt-hours.

Once you know your total daily energy consumption, you can calculate the size of the battery bank required by dividing the total watt-hours by the nominal battery voltage (e.g., 12V, 24V, 48V systems) to get the total amp-hour (Ah) rating of the battery bank.

Determining the Right Number of Batteries and Arrangement (Series vs. Parallel)

To reach the required capacity, you may need multiple batteries. How you arrange the batteries depends on the system voltage and desired energy storage.

- **Series Configuration**: If you need to increase the voltage (e.g., from 12V to 24V or 48V), you connect the batteries in series. For example, to create a 24V system, you would connect two 12V batteries in

series. The total voltage is the sum of the individual voltages, but the amp-hour rating remains the same.
• **Parallel Configuration**: If you need to increase the total **amp-hour** capacity (the amount of energy stored), you connect the batteries in parallel. This configuration keeps the voltage the same, but increases the total storage capacity by adding the amp-hours of each battery.

The combination of series and parallel connections allows you to tailor the battery bank to the voltage and storage needs of your system.

Depth of Discharge (DoD) and Its Effect on Battery Longevity

One important consideration when sizing your battery bank is the **depth of discharge (DoD)**. The DoD refers to how much of the battery's capacity you can safely use before needing to recharge it.

• **For Lead-Acid Batteries**: The recommended DoD is typically **50%**, meaning you should only use half of the battery's capacity before recharging. Exceeding this can drastically reduce the lifespan of the battery.
• **For Lithium-Ion Batteries**: These batteries can generally handle a deeper DoD, typically **80-90%**, without significant degradation, making them more efficient for off-grid applications.

Keep in mind that using a higher percentage of the battery's capacity means you'll need a larger battery bank to avoid running out of power on days with low generation.

Energy Management Systems (EMS)

Energy Management Systems (EMS) allow you to monitor, control, and optimize your battery bank and the distribution of energy throughout your off-grid system. EMS can help you track the health of your batteries, automate recharging processes, and ensure that your energy use is as efficient as possible.

Monitoring and Controlling Energy Storage and Distribution

An Energy Management System (EMS) delivers real-time insights into the state of charge, power generation, and energy consumption. This information enables you to make informed decisions to maximize the efficiency of your energy use.

• **State of Charge (SoC)**: The EMS will track how full or empty your battery bank is, giving you insight into when to conserve power or switch to backup systems.
• **Energy Distribution**: An EMS can automatically switch between energy sources, such as between solar and stored energy, to ensure the most efficient use of your available resources.

Battery Monitoring and Health Systems

Battery monitoring systems (BMS) are critical for maintaining the health of your battery bank. These systems track the voltage, temperature, and overall condition of each battery in the bank.

• **Voltage Monitoring**: A BMS ensures that the voltage levels remain within safe operating ranges, preventing overcharging or deep discharge, both of which can reduce battery life.
• **Temperature Control**: Excessive heat or cold can damage your batteries. The BMS can monitor temperature and alert you to take action if the system is at risk.

Automation for Recharging and Managing Reserve Power

An advanced EMS system can **automate** the process of recharging and managing power reserves. When the battery bank discharges beyond a specified threshold, the EMS can automatically switch to a backup generator or additional energy source to recharge the batteries without manual intervention.

Alternative Storage Options

While battery banks are the most common energy storage solution, several alternative methods can be used in off-grid systems, depending on your energy needs and environment.

Thermal Energy Storage: Using Water or Rocks for Heat Storage

Thermal energy storage systems capture excess energy and store it in the form of heat. Water or rocks can be used as storage mediums.

- **Water storage**: Water has a high thermal mass and can store significant amounts of heat. You can use a large tank of water to store thermal energy, which can then be used for space heating or hot water needs.
- **Rocks or concrete**: These materials also have high thermal mass and can be used to store heat. They are typically used in systems like **masonry heaters**, where heat is captured in stones or concrete during the day and released during colder periods.

Flywheel Energy Storage: Applications in Off-Grid Power Systems

Flywheels store energy mechanically, using a rotating disk to store kinetic energy. Flywheel systems are highly durable and can handle large power fluctuations, making them suitable for off-grid applications.

Compressed Air Energy Storage (CAES): Feasibility and Considerations

CAES stores energy by compressing air into underground caverns or large tanks. When electricity is needed, the compressed air is released and used to drive turbines.

- **Feasibility**: While CAES is an efficient storage method for larger off-grid systems, it requires a **large infrastructure** and access to underground storage, making it less suitable for small-scale residential use. However, it could be considered for larger off-grid communities or remote industrial applications.

By integrating various storage options and monitoring systems, you can optimize your off-grid energy setup and ensure ensuring you maintain a consistent and dependable power supply, even during low-output times or emergencies.

15. Distributing Power Safely & Efficiently

In an off-grid energy system, ensuring that power is safely and efficiently distributed across your home is essential. Proper wiring, choosing the right inverters and converters, and adhering to electrical safety standards all play a crucial role in maintaining a reliable and secure power setup. This section covers key considerations for distributing power within your off-grid system, including wiring systems, inverters, electrical safety, and backup strategies to ensure you have a fail-safe energy distribution plan.

Wiring Systems for Off-Grid Homes

The wiring system in an off-grid home is responsible for distributing electricity from your power generation sources (solar, wind, hydropower, or backup generators) to the appliances and devices that need it. Proper wiring and electrical connections ensure that your system operates safely and efficiently, minimizing energy loss and preventing dangerous electrical issues.

AC vs. DC Power: Choosing the Right System Based on Application

In off-grid systems, you will encounter both **AC (alternating current)** and **DC (direct current)** power. The choice between AC and DC depends on the appliances you plan to run and the type of power generation you use.

- **DC Power**: Solar panels, wind turbines, and batteries generally produce DC power. DC is more efficient for charging batteries and storing energy. Many off-grid systems operate on DC power because it's direct and more easily stored. DC-powered appliances are also available, but they are typically more expensive and less common than AC-powered appliances.
- **AC Power**: AC power is commonly used in household appliances like refrigerators, washing machines, and lights. It's the standard form of electricity delivered by the grid and is easily converted from DC power using inverters. While AC power is more versatile and widely available, converting DC to AC introduces some energy loss, but this can be mitigated with efficient inverters.

The decision to use DC or AC depends largely on your energy consumption and available appliances. Many off-grid homes choose a **hybrid system**, where solar panels charge a battery bank using DC power, and the energy is then converted to AC for use with common household appliances.

Wiring Considerations: Wire Gauge, Fuse Protection, and Grounding

When wiring your off-grid home, you need to consider the correct **wire gauge**, fuse protection, and grounding to ensure the system is safe and operates efficiently.

- **Wire Gauge**: The size of the wire you use depends on the amount of current (amperage) flowing through it. Larger currents require thicker wires. For example, a wire gauge of **12 AWG** is suitable for low-power circuits, while **8 AWG** or **6 AWG** might be required for higher current flows, like those from inverters or large battery banks. Always refer to electrical codes and standards when choosing wire sizes for your system.
- **Fuse Protection**: Fuses are crucial to protect your system from overcurrent. If too much current flows through the wire, the fuse will blow, preventing overheating and potential fires. Install fuses at **key points** in your system, such as at the battery bank, the inverter, and at the distribution panel.
- **Grounding**: Grounding ensures that in the event of a fault, the electrical system can safely discharge any excess power. It also prevents electric shocks. Proper grounding involves connecting the metal parts of the system (inverters, generators, batteries) to a **grounding rod**. The grounding system should be **well-documented** to ensure easy maintenance and troubleshooting.

Battery-to-Load Connections: Safe and Reliable Methods

When connecting your batteries to the load (appliances), it's important to use reliable, safe, and efficient methods to distribute power. The connections from the **battery bank** to the **inverter** and ultimately to your appliances need to be designed for both performance and safety.

- Direct connection: In off-grid systems, you'll typically use DC wiring to connect your batteries directly to the inverter. The inverter subsequently transforms the DC power into AC power, making it suitable for operating household appliances. Ensure all connections are tight, corrosion-free, and use appropriate gauge wire for the amount of current.
- Connection points: Use battery terminals or bus bars to connect the battery bank to the rest of the system. Bus bars make it easy to connect multiple batteries or systems in parallel or series.

Proper connection techniques and maintenance are critical for ensuring the longevity and efficiency of your power distribution system.

Inverters & Converters

Inverters and converters are critical components in any off-grid system. They are responsible for converting the electricity produced by your renewable energy sources into a usable form, whether that's AC or DC power. These devices ensure that power flows efficiently between your batteries and appliances.

DC to AC Inverters: Choosing the Correct Inverter Size

If your off-grid system produces DC power, you will need an **inverter** to convert it to AC power for most home appliances. Selecting the correct inverter size is essential to ensure that the inverter can handle your energy demands.

- Determine peak power needs: The first step in selecting an inverter is determining your peak power demand. This is the highest amount of electricity you expect to use at any one time. Look at the total wattage of all the appliances you plan to use simultaneously (e.g., refrigerator, lights, TV, etc.). Your inverter must be able to handle this peak power demand without shutting down or causing issues.
- Consider continuous wattage: Along with peak power, inverters also have a continuous wattage rating, which tells you how much power they can supply over an extended period. Choose an inverter with a continuous rating that matches your average power consumption.

When sizing your inverter, it's always better to slightly **oversize** to avoid overloading the inverter during high-demand situations.

Grid-Tied vs. Off-Grid Inverters

Inverters come in two main types: **grid-tied** and **off-grid** inverters.

- **Grid-Tied Inverters**: These are used in systems that are connected to the grid. The inverter synchronizes the output with the grid, sending excess energy back to the utility company. While grid-tied inverters are useful for hybrid systems or when you have net metering, they are not ideal for off-grid setups.
- **Off-Grid Inverters**: These inverters are designed specifically for off-grid systems, where energy is generated, stored, and consumed entirely within the home. Off-grid inverters do not require synchronization with the grid, and they often include additional features like **battery charging** and **power distribution**.

Select an **off-grid inverter** if you're not connected to the grid, and ensure it's capable of handling your system's energy load.

Power Converters: Ensuring Compatibility Between Energy Sources and Appliances

Power converters are used to adjust the voltage or type of current to ensure it's compatible with your energy sources and appliances.

- **DC-DC converters**: These are used when you need to change DC voltage from one level to another. For example, you may need to reduce the voltage from your battery bank to power small appliances that require a lower voltage.
- **AC-DC converters**: In some off-grid systems, you might need to convert AC power back to DC for charging batteries or running specific devices.

Converters are important for ensuring that each part of your off-grid system is compatible and running efficiently.

Electrical Safety

Electrical safety is crucial in any off-grid power system, where improper wiring or faulty components can lead to dangerous situations, including fires or electric shocks. You must take appropriate precautions to protect yourself and your system.

Proper Grounding and Surge Protection for Off-Grid Systems

- **Grounding**: Proper grounding of your system ensures that, in the event of a fault, electricity is safely diverted to the earth. This prevents electrical shocks and damage to your equipment. Install a **grounding rod** and connect it to your electrical system to protect the entire off-grid setup.
- **Surge protection**: Off-grid systems are vulnerable to **power surges** from lightning, wind, or other disturbances. Installing surge protectors will help safeguard your appliances, batteries, and other equipment from sudden spikes in voltage.

Overcurrent Protection and Circuit Breakers

- **Circuit breakers** are designed to cut off power when there is too much current flowing through the wires. They help prevent fires caused by electrical overloads. Install circuit breakers at key points in your system, such as between the battery bank and the inverter, or before connecting to your household electrical panel.
- **Fuses and disconnects**: Use fuses or **disconnect switches** to isolate parts of the system for maintenance or when there's an overload.

Isolation of High-Voltage Systems (If Applicable)

If you are running **high-voltage systems**, such as **solar systems with high voltage inverters**, proper isolation is necessary. High-voltage electricity can be dangerous, so isolating these systems from lower voltage ones using **transformers** or **isolating switches** will protect you and your equipment.

Backup Systems and Redundancy

To ensure a continuous power supply and to avoid system failure, backup systems and redundancy are essential.

Creating a Fail-Safe Energy Distribution Plan

In the event of a failure in your primary power system (solar, wind, etc.), you should have a fail-safe mechanism in place to maintain power to your home.

- **Multiple power sources**: Combining different energy sources, like solar and wind, ensures that if one system fails, the other can continue providing power. Having a backup generator as a final fail-safe will ensure that you have power even during long periods of low generation.

- **Automatic transfer switches**: Set up an automatic transfer switch (ATS) that will automatically switch from your primary power source (e.g., solar) to the backup generator when needed. This ensures that the transition is seamless and without interruption.

Importance of Backup Circuits and Power Feeds

Backup circuits are essential for ensuring that critical systems such as lighting, water pumps, and communication equipment continue to function in the event of a failure in your main power system.

- **Critical load circuits**: Set up separate circuits for essential appliances or equipment, such as refrigeration, water filtration, and lights. These circuits should be powered by backup sources when the main system goes down.

By following these steps and ensuring that each aspect of your electrical system is well planned and executed, you can safely and efficiently distribute power throughout your off-grid

home. Properly designed wiring, inverters, electrical safety protocols, and backup systems will ensure that you have reliable power for all your needs.

16. Optimizing Energy Usage for Maximum Efficiency

Off-grid living offers a unique opportunity to take control of your energy consumption, but with this power comes the responsibility to use energy wisely. By optimizing your energy usage, you can make the most of your resources, reduce waste, and maintain a sustainable lifestyle. In this section, we'll explore practical strategies for enhancing the efficiency of your off-grid system, including choosing energy-efficient appliances, managing energy use, improving insulation, and optimizing renewable energy systems.

Energy-Efficient Appliances

Choosing the right appliances is one of the most important decisions in building an energy-efficient off-grid home. While the initial investment may be higher for energy-efficient models, the long-term savings and environmental benefits make them a wise choice.

Identifying the Most Energy-Efficient Household Appliances

The first step is to select appliances that consume as little energy as possible while meeting your needs.

- **Lighting**: Traditional incandescent bulbs use much more energy than their alternatives. **LED bulbs** are an excellent choice for off-grid homes because they use up to 75% less energy and last significantly longer than incandescent or even compact fluorescent (CFL) bulbs. Installing LED lights in every room will reduce your home's overall energy consumption.
- **Refrigerators**: A refrigerator is one of the most energy-hungry appliances in any home, and this is especially true in an off-grid setting. Look for **Energy Star**-rated refrigerators that consume less power, and avoid models with energy-wasting features like automatic ice makers or water dispensers. **Chest freezers** are often more efficient than upright freezers due to the way they store cold air.
- **Washing Machines**: Modern washing machines offer significant improvements in energy efficiency compared to older models. **Front-loading washers** typically use less water and electricity than top-loading models. Additionally, consider washing clothes in cold water to further reduce energy use.

Energy-Efficient Cooking Solutions

Cooking can consume a significant portion of your off-grid energy, but with the right equipment and techniques, you can minimize energy use in the kitchen.

- **Solar Ovens**: Solar cooking is one of the most efficient methods of preparing food off the grid. **Solar ovens** use the sun's energy to cook food, meaning no need for gas or electricity. All you need is a sunny day, and you can cook everything from stews to baked goods without using any power from your battery bank. Solar ovens are particularly beneficial for slow cooking, roasting, and baking, allowing you to free up energy for other uses.

- **Induction Cooktops**: **Induction cooktops** heat up faster and are more energy-efficient than traditional electric or gas stoves. They use electromagnetic energy to directly heat the cooking pot, which means less wasted heat and faster cooking times. Since the heat is concentrated in the pot, induction cooktops also avoid heating up the kitchen, which can reduce the need for additional cooling in warmer months.

High-Efficiency Heating and Cooling Systems

Maintaining comfort in your off-grid home involves efficient heating and cooling systems, particularly in regions with extreme temperatures.

- **Wood Stoves**: If you live in an area with access to firewood, **wood stoves** can provide efficient heating with minimal energy consumption. Modern wood stoves are highly efficient, with some models featuring secondary combustion chambers that burn smoke for extra heat. This allows you to heat your home with a renewable resource while reducing the amount of wood you need.
- **Passive Solar Heating**: One of the most cost-effective and sustainable ways to heat your off-grid home is through **passive solar design**. By positioning your home to capture maximum sunlight during the winter months, you can reduce the need for additional heating systems. Large south-facing windows, thermal mass walls, and overhangs that block the summer sun will help regulate your home's temperature year-round.
- **Cooling**: In areas with hot summers, **natural cooling techniques** can help you maintain comfort without relying on energy-intensive air conditioners. Use **strategic shading** through trees or awnings, **cross-ventilation** by opening windows on opposite sides of your home, and **insulated roofs** to keep the interior cool. Incorporating a **cool roof** or using reflective materials on your roof can also reduce heat absorption.

Energy Management Strategies

Once you've chosen efficient appliances, the next step is to optimize how and when you use energy. Smart energy management can help you stretch your available energy resources further, particularly when your renewable energy generation is limited.

Using Timers and Smart Technology to Minimize Energy Waste

By automating your energy use, you can ensure that appliances only operate when needed, and that they are powered off during times of low energy availability.

- **Timers**: Use timers to control when appliances turn on and off. For example, set your water heater to only heat water during the hours when solar power generation is at its peak. Similarly, you can schedule the operation of a washing machine or dishwasher to coincide with high solar output.
- **Smart Plugs and Switches**: Install smart plugs or smart switches to control the power supply to your appliances. You can program them to turn on or off remotely, ensuring that devices aren't left running unnecessarily. This is particularly useful for appliances that draw a standby power load, like TVs or chargers.

Time-of-Use Management: Prioritizing High-Energy Tasks During Peak Generation Hours

When relying on renewable energy, particularly solar, the energy you generate throughout the day fluctuates based on weather and sunlight hours. To make the most of your system:

- **Identify peak energy production times**: Track when your renewable energy systems produce the most power. Typically, solar panels will generate the most energy between late morning and mid-afternoon, so try to schedule energy-intensive tasks (like laundry or using a water heater) during those hours.
- **Shift consumption**: Avoid running energy-hungry appliances during the early morning or late evening when solar or wind power generation is low. If your system includes a battery bank, it can help to store energy during peak hours and use it later during low-output times.

Load Management: Balancing Power Consumption Based on Available Supply

Effective load management is about understanding how much power your system can provide and using it wisely.

- **Monitor consumption**: Keep track of the energy consumption of various appliances in your home. This will allow you to prioritize essential devices and avoid running multiple high-demand appliances at the same time.
- **Energy-efficient alternatives**: If you have large energy-consuming appliances, consider whether there are energy-efficient alternatives. For example, replacing an electric oven with a solar oven or switching to a high-efficiency LED lighting system can significantly reduce overall energy use.

Thermal Insulation & Heat Retention

Good insulation and heat retention are key to reducing the energy required to heat and cool your off-grid home. Proper insulation ensures that the heat you generate stays inside during the winter, while the cooler air remains during the summer.

Insulating Off-Grid Homes: Materials, Techniques, and Heat Retention Strategies

Proper insulation keeps your home comfortable without requiring excessive energy use. Different materials and techniques can be used based on climate and building style.

- **Insulation materials**: Use materials with high R-values, such as **fiberglass, foam, or cellulose**, for walls, ceilings, and floors. These materials help keep the temperature inside your home stable by reducing heat transfer. In colder climates, thicker insulation and double-glazed windows are particularly important.
- **Insulation techniques**: Ensure that **air sealing** is also a part of your insulation strategy. Seal gaps around doors, windows, and vents to prevent drafts. Insulation can only be effective if the home is sealed tightly.

Improving Air Sealing to Reduce Heating and Cooling Needs

Air leaks can cause significant heat loss in winter and cooling loss in summer, making your heating and cooling systems work harder. Air sealing techniques include:

- **Weatherstripping** doors and windows to eliminate gaps.
- **Caulking** cracks in the walls, foundation, and around windows and doors.
- **Installing door sweeps** at the bottom of exterior doors to prevent air leaks.

Passive Solar Design to Capture and Store Natural Heat

Passive solar design uses the sun's energy to heat your home in winter and reduce the need for heating.

- **Positioning**: Orient your home with large windows facing south (in the northern hemisphere) to maximize solar heat gain during the winter months.
- **Thermal mass**: Incorporate materials like concrete, brick, or stone that absorb heat during the day and release it at night, helping to maintain a stable indoor temperature.

Renewable Energy Optimization

To maximize the performance of your renewable energy systems, you need to fine-tune their setup and ensure they're working at peak efficiency.

Positioning and Tilting Solar Panels for Maximum Sun Exposure

The angle and direction of your solar panels will greatly affect their efficiency.

- **Tilt angle**: Solar panels should be angled to capture the most sunlight throughout the day. In most locations, the ideal tilt angle is equal to the **latitude** of your location. However, adjusting the tilt seasonally can optimize performance.
- **Orientation**: In the northern hemisphere, solar panels should face **true south** for optimal sunlight exposure, while in the southern hemisphere, they should face **true north**.

Sizing Wind Turbines Based on Local Wind Conditions

Wind turbines need to be sized according to the local wind conditions.

- **Wind speed**: Determine the **average wind speed** in your area. Wind turbines require a minimum wind speed of 5-7 miles per hour (8-11 km/h) to generate power. For regions with lower average

wind speeds, consider smaller turbines with higher efficiency.
- **Sizing**: Based on the wind data, choose a turbine size that can generate the required power for your off-grid needs.

Water Heating Optimization Using Solar and Wood-Burning Systems

- **Solar water heaters** can be an efficient way to reduce the need for electric or propane water heating. Install **flat-plate collectors** or **evacuated tube collectors** on your roof to harness solar energy for heating water.
- **Wood-burning stoves** with integrated water heaters provide a reliable source of hot water in areas with access to wood, and can be used in combination with solar water heaters to create a hybrid system.

Energy Auditing

Regular energy audits help identify inefficiencies in your off-grid system and can help you make adjustments to further reduce consumption.

Assessing Energy Consumption Patterns in the Home or on the Property

- **Monitor energy usage**: Use an **energy meter** to track how much electricity is consumed by various appliances throughout the day. This will give you a clear picture of where you are using the most energy.
- **Identify peak usage times**: Take note of when energy consumption spikes and identify any patterns that could be modified to reduce power demands.

Identifying Energy Inefficiencies and Implementing Corrective Actions

Once you've identified the most energy-hungry appliances, look for ways to reduce their consumption:
- **Upgrade inefficient appliances**: If you find that certain appliances are draining more energy than others, consider replacing them with energy-efficient models.

- **Consider alternative energy solutions**: For instance, using **solar ovens** instead of an electric stove can save significant amounts of energy.

Regular Review and Updates of Energy-Saving Practices

Efficient off-grid living is a continuous process. Make it a habit to periodically check your energy consumption, especially after installing new equipment or adjusting your energy production. Revisiting your energy-saving practices regularly will help you stay efficient and keep your off-grid system running smoothly.

This ongoing effort will help ensure that you are making the most of the resources available and creating a sustainable, low-impact lifestyle.

17. Energy and Power Independence Projects

Project: Solar Panel System Design and Installation

SOLAR PANEL SYSTEM FOR HOME

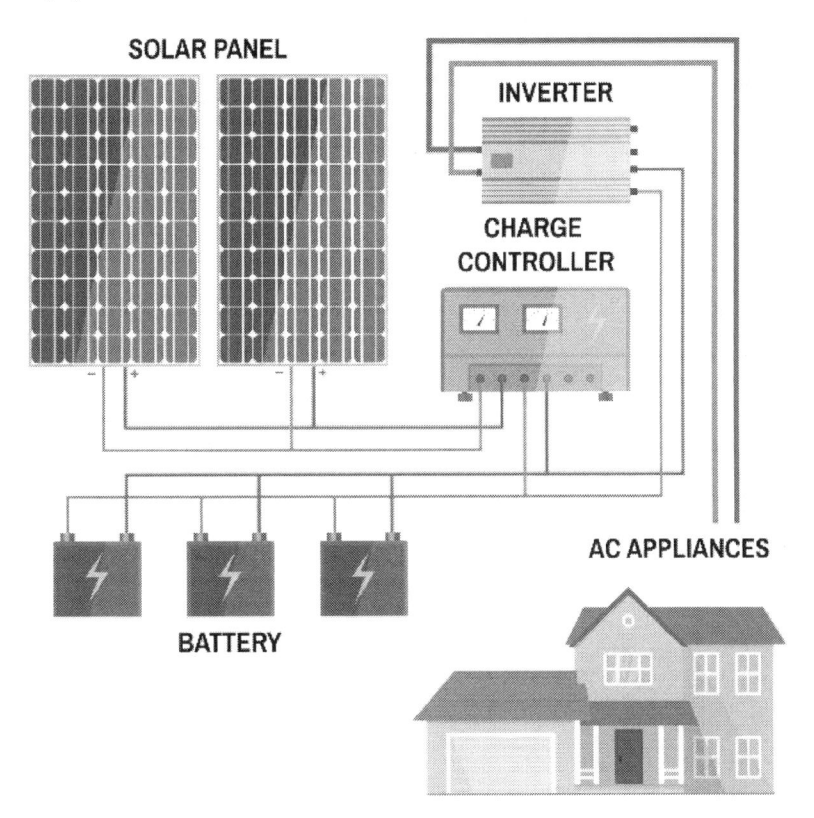

Setting up a solar power system is an essential part of achieving off-grid energy independence. By harnessing the sun's energy, you can generate clean, renewable electricity for your home or homestead. In this project, you'll design and install a photovoltaic (PV) solar panel system, which includes selecting the right components, calculating your energy needs, and setting everything up to power your off-grid living.

Materials and Tools Needed:
- Solar panels (based on your energy needs)
- Inverter (DC to AC converter)
- Charge controller
- Battery storage system (optional, for energy storage)
- Mounting brackets for panels
- Wiring and connectors
- Grounding equipment
- Electrical tape
- Battery or electrical panel (for connecting the system)
- Voltage meter for testing
- Drill and screws for mounting
- Roof or ground space for panel installation

Instructions

1. Determine Your Energy Needs

Before purchasing any solar equipment, you need to know how much energy your home requires to operate. This will allow you to calculate how many solar panels you need.

- Make a list of all the appliances and systems you plan to power with solar (e.g., lights, refrigerator, pump).
- For each appliance, check the **wattage** (usually listed on the appliance label).
- Multiply the wattage by the number of hours each appliance runs per day. For example, if a 100-watt light runs for 5 hours, it uses 500 watt-hours of energy per day (100W x 5 hours).
- Add up the total watt-hours for all appliances to determine your total daily energy requirement in watt-hours (Wh).

2. Select the Right Solar Panels

Now that you know how much energy you need, you can determine how many solar panels to buy. Each solar panel has a **wattage** rating (e.g., 300 watts). To calculate how many panels you need:

- Take your total daily energy requirement in watt-hours and divide it by the expected daily watt-hours of a single panel. For example, if your total daily need is 18000 Wh i.e. 18KWh and every panel produces 300watt-hours and the average peak sun hours are 6, your panel would generate 1800Wh per day. Therefore 18000/1800 → 10 solar panels needed.

3. Choose the Right Inverter

The inverter converts the **DC** electricity generated by the solar panels into **AC** electricity, which your home appliances use. The size of the inverter should match the wattage of your solar panels.

• Calculate your **total wattage** by adding the wattages of your appliances.
• Choose an inverter with a slightly higher wattage than your total load to ensure it can handle peak usage times. For example, if your total load is 2000W, consider an inverter rated for 2500W.

4. Install the Solar Panels

Now that you have your solar panels, inverter, and charge controller, it's time to install the system.

• **Select a location**: Panels should be installed where they will receive maximum sunlight exposure. Typically, this is on the roof of your house or a ground-mounted system. Make sure the area is free from obstructions like trees or buildings that could cast shadows on the panels.
• **Mount the panels**: Use mounting brackets to secure the panels to the roof or ground structure. The panels should be tilted at an angle that optimizes sun exposure, which varies by location. In general, the angle should match the latitude of your location (e.g., for a location at 30° latitude, tilt the panels at 30°).
• **Ensure proper alignment**: Make sure the panels are oriented to face **south** in the northern hemisphere (or **north** in the southern hemisphere) to capture the most sunlight throughout the day.
• **Connect the wiring**: Connect the **positive** and **negative** wires from each solar panel to form a series or parallel circuit (depending on your system design).

5. Set Up the Charge Controller and Battery Storage (Optional)

The charge controller regulates the amount of energy flowing from the panels to the battery to prevent overcharging.

• **Connect the charge controller**: Wire the output from the solar panels to the charge controller, and then connect the charge controller to the battery bank (if using).
• **Battery storage**: If you plan to store energy for utilization at night or on overcast days, connect **deep-cycle batteries** to store the energy. Ensure to follow the manufacturer's instructions for connecting batteries in parallel or series, depending on the voltage of your system.
• **Monitor the system**: Some charge controllers come with monitoring systems to check battery levels and overall system performance.

6. Connect the Inverter to the Battery or Electrical Panel

Now that the solar panels are generating power and the charge controller is managing the flow to the battery bank (if installed), you need to connect the inverter.

• **Inverter connection**: Connect the output of the battery or charge controller to the inverter's DC input. Then connect the AC output from the inverter to your household electrical panel.
• **Test the system**: Turn on the inverter and check the voltage and current with a **voltage meter**. Make sure the output matches the system requirements.

7. Final Testing and Optimization

Once everything is connected, it's time to test your system.

• **Turn on the system**: Power up your solar panels and the inverter. Test the voltage output to ensure the system is generating the correct amount of electricity.
• **Check the performance**: Use a **voltmeter** to monitor the voltage at various points in the system (solar panels, charge controller, battery, and inverter). Ensure the system is performing within the expected parameters.
• **Optimize efficiency**: Ensure your panels are angled correctly, and regularly clean them to remove dust and debris that can block sunlight.

By following these steps, you've successfully designed and installed a solar power system for your off-grid home. Regular maintenance and monitoring of the system will help you maintain efficient power generation, storage, and use over time.

Project: Wind Turbine Installation and Siting

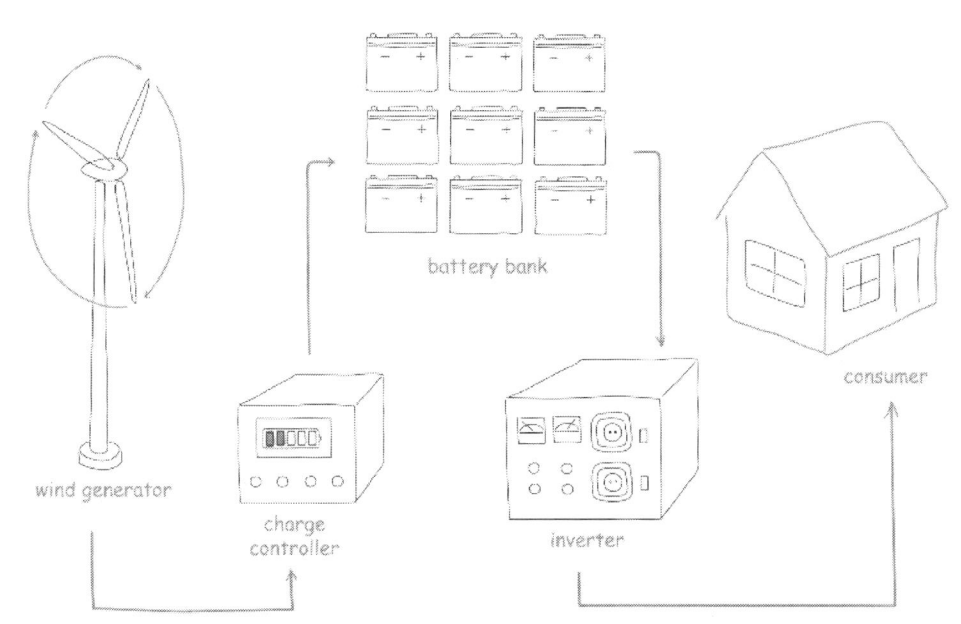

Installing a wind turbine is a great way to supplement your off-grid power generation system. Wind energy can provide a reliable and sustainable source of electricity, especially if you live in an area with consistent wind speeds. This project will guide you through selecting the best location for your wind turbine, installing it, and connecting it to your power system.

Materials and Tools Needed
- Wind turbine (size based on energy needs)
- Mounting pole or tower
- Guy wires for stabilization (if using a tower)
- Inverter (to convert DC to AC if needed)
- Electrical wiring (for connections to the system)
- Battery bank (for energy storage)
- Charge controller (optional, if storing energy)
- Tools for assembly (wrenches, screwdrivers, pliers)
- Concrete (for securing the base, if necessary)
- Grounding equipment (for electrical safety)
- Voltmeter for testing the system

Instructions

1. Choose the Right Location for Your Wind Turbine

The performance of your wind turbine will largely depend on the **location** where you install it. Follow these steps to find the optimal placement:

- **Assess wind conditions**: The most critical factor in choosing a location for your wind turbine is the wind speed. Wind turbines typically need **consistent wind speeds** of at least **10-15 mph (16-24 km/h)** for efficient operation. Ideally, the location should experience wind speeds of **12-20 mph (19-32 km/h)** for optimal performance.
- **Avoid obstructions**: Wind turbines work best when they are placed in open areas with minimal obstructions. Trees, buildings, and hills can block the wind and reduce the efficiency of the turbine. Aim to install your turbine at least **300 feet (90 meters)** away from obstacles, such as trees or other structures, to ensure smooth airflow.
- **Height considerations**: The height of your turbine is key to accessing better wind speeds. Most turbines should be installed at least **30 feet (9 meters)** above ground level, but higher placements can further increase performance. The higher the turbine, the less interference from surface turbulence.

2. Select the Right Size and Type of Wind Turbine

Wind turbines come in various sizes and types, each with different power output capabilities. To determine the right turbine for your needs:

- **Calculate your energy needs**: Before purchasing a turbine, calculate your daily energy consumption (similar to how you would for a solar panel system). Determine how many watts of electricity you need, then select a wind turbine with a capacity that meets or exceeds that demand.

- **Match turbine size to wind conditions**: Make sure the turbine is rated for the average wind speed in your area. A smaller turbine might be suitable for regions with lower wind speeds, while larger turbines are better for areas with higher winds.

3. Install the Mounting Pole or Tower

The mounting structure is essential for holding the turbine in place and ensuring it's elevated to catch the wind.

- **Prepare the site**: Clear the area where you will be installing the tower. Ensure that there are no underground utilities, and remove any obstacles that may interfere with the turbine's rotation.
- **Install the base**: For a **ground-mounted pole** or **tower**, begin by digging a hole about **3-4 feet (0.9-1.2 meters)** deep and **2 feet (0.6 meters)** in diameter. Pour **concrete** into the hole to create a solid base for the pole.
- **Assemble the tower**: Follow the manufacturer's instructions to assemble the tower. Typically, the turbine is mounted at the top of a steel tower or a tall pole. The height of the pole should be based on the wind conditions and local regulations.
- **Secure the tower**: Use **guy wires** to stabilize the tower. These wires should be attached to the top of the pole and anchored to the ground at a wide angle, ensuring that the tower remains steady in high winds.

4. Install the Wind Turbine on the Pole or Tower

Once the tower is in place, you can mount the wind turbine.

- **Lift the turbine**: With assistance, carefully lift the wind turbine and attach it to the top of the pole or tower. Make sure the turbine is securely fastened using bolts or brackets.
- **Check alignment**: Ensure that the turbine is facing the correct direction—typically towards the **prevailing wind direction**. You may need to install a **wind vane** to help the turbine orient itself automatically to the wind.

5. Wire the Wind Turbine to Your Power System

Now that your turbine is installed, it's time to connect it to your power system.

- **Connect the turbine to the charge controller**: The wind turbine will generate DC electricity, so it should be connected to a **charge controller** (if you are using battery storage) to prevent overcharging the batteries.
- **Inverter connection**: If you are using an **inverter** to convert DC power to AC for household appliances, connect the turbine to the inverter.
- **Battery storage (optional)**: For off-grid systems, the electricity generated by the wind turbine should be stored in a **battery bank** for use when wind speeds are low. Wire the charge controller to the battery bank, ensuring proper connections.

6. Test the System

Once everything is connected, it's time to test the wind turbine and the power system.

- **Check the voltage**: Use a **voltmeter** to check the output from the turbine. Ensure that it is producing the expected voltage and current.
- **Monitor the system**: Turn on any appliances or devices you intend to run with the wind turbine, and monitor how they perform. The turbine should generate enough power to run your off-grid systems.

7. Perform Ongoing Maintenance

Wind turbines require regular maintenance to ensure they operate efficiently.

- **Lubricate moving parts**: Periodically lubricate the moving components of the turbine, such as the bearings, to prevent wear and tear.
- **Check for debris**: Make sure there is no debris such as leaves or twigs around the turbine. These can obstruct the movement of the blades and reduce efficiency.
- **Inspect the tower**: Check the tower and guy wires for any signs of wear or rust, especially after storms.

Project: Micro-Hydro Power System Setup

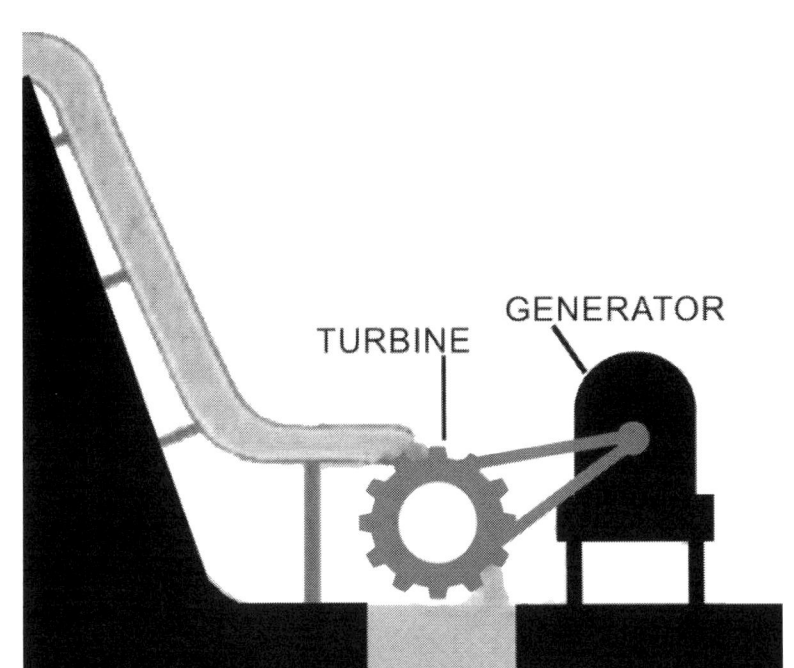

If you have access to a flowing water source such as a stream or river on your property, setting up a micro-hydro power system is a reliable and efficient way to generate off-grid electricity. Micro-hydro systems can provide a constant, renewable energy supply, especially in areas with a consistent water flow. In this project, you will design and install a small-scale hydroelectric system that uses the natural flow of water to generate power for your off-grid homestead.

Materials and Tools Needed
- Micro-hydro turbine
- PVC or steel pipes (for water intake and delivery)
- Powerhouse or enclosure for the turbine and generator
- Electrical components (wires, inverter, charge controller)
- Concrete or mounting base for the turbine (if needed)
- Voltage meter for testing
- Flow meter (for measuring water flow)
- Shovel and digging tools (for trenching the intake and output pipes)
- Safety equipment (gloves, goggles, etc.)
- Tools for assembly (wrenches, screwdrivers, etc.)

Instructions

1. Assess Your Water Source and Site

Before purchasing any equipment, you need to determine whether your water source is suitable for a micro-hydro system. Here's how to evaluate the location:

- **Water Flow**: Measure the **flow rate** of your stream or river. The flow rate is crucial because it dictates how much energy your system can generate. Typically, a micro-hydro system requires a **minimum of 1-2 cubic feet per second (cfs)** of water flow to operate efficiently.
- **Elevation Drop (Head)**: The **head** refers to the vertical distance the water falls between the intake (where water is collected) and the turbine. The greater the elevation drop, the more power you can generate. A minimum head of **2-3 feet (0.6-0.9 meters)** is ideal for small-scale systems, but higher elevations will result in more efficient energy production.
- **Access and Location**: Ensure that the water source is accessible, and there is enough **elevation drop** for the system. The location should be free from debris, and the stream should have a consistent flow year-round.

2. Select the Right Micro-Hydro Turbine

Choosing the correct turbine for your site is crucial for generating the required amount of power. Turbines come in different sizes and types, including Pelton wheels, Turgo wheels, and Francis turbines. The turbine should match your flow rate and head to ensure optimal efficiency.

- **Flow rate**: The higher the flow rate of your water, the larger the turbine required. For small residential applications, turbines range from **300W to 5kW**.
- **Head**: The more elevation you have, the more efficient your turbine will be. Typically, **Pelton wheels** are best for high-head, low-flow sites, while **Turgo turbines** work well for moderate head and flow combinations.

Once you've selected the appropriate turbine for your needs, consider its durability and maintenance requirements. Opt for a model that is specifically designed for off-grid, small-scale applications.

3. Set Up the Water Intake System

The water intake is responsible for diverting water from the source to your turbine. The system should allow a consistent flow of water while filtering out debris that could damage the turbine.

- **Dig a trench**: Start by digging a trench to lay down the intake pipe. The pipe should run from the water source to the turbine, ensuring that the water flows downhill to utilize gravity. Ensure the pipe is **slightly sloped** to allow for efficient flow.
- **Install the intake**: The intake pipe should be submerged in the water to ensure a steady supply. You can create a simple intake using **PVC pipe** with a **filter** at the end to block debris from entering the system. Ensure the filter is fine enough to catch larger particles but not so fine that it restricts water flow.
- **Install the pipe**: Lay the intake pipe down the trench and ensure it's **securely fastened**. If you're using a steel pipe, make sure it's tightly sealed at each connection. If you're using PVC, be careful to glue the joints properly to avoid leaks.

4. Install the Turbine and Powerhouse

Now that you've set up the intake system, you can install the turbine and other electrical components in the powerhouse.

- **Build the powerhouse**: Choose a location close to your water intake where you can safely install the turbine. The powerhouse needs to be **weatherproof** to protect the generator from the elements. You can either build a simple structure or use an existing shed.
- **Mount the turbine**: Secure the turbine onto a solid base or platform inside the powerhouse. Follow the manufacturer's instructions for mounting the turbine securely. The turbine must be placed at the correct height to ensure it receives adequate water flow.
- **Connect the turbine to the generator**: Depending on the model, your turbine may come with an integrated generator or require a separate generator. Connect the turbine shaft to the generator shaft using the appropriate coupling. Make sure the generator is positioned correctly to generate electricity efficiently.

5. Connect the Turbine to the Electrical System

After setting up the turbine, you will need to connect the electrical components to complete the system.

- **Install the inverter**: The inverter converts the DC power generated by the turbine into AC power that can be used in your home. Install the inverter inside your powerhouse and connect the output from the turbine's generator to the inverter's input.
- **Connect to the charge controller**: If you are using a battery bank for storage, connect the inverter to a charge controller. The charge controller will regulate the voltage going into the batteries to prevent overcharging.
- **Set up the battery bank**: Connect a **deep-cycle battery** or several batteries to store the power generated by the turbine. These batteries will provide electricity when there is little or no water flow (e.g., during dry spells).
- **Connect to the electrical panel**: Once the energy is converted to AC, it can be routed to your household electrical panel. From there, you can use the power to run appliances, lights, or charge devices.

6. Test the System

Once everything is connected, you can test the system to ensure it's working properly.

- **Turn on the turbine**: Allow water to flow into the intake pipe and watch as the turbine starts to spin. Ensure the turbine is generating the expected voltage and current.
- **Monitor the output**: Use a **voltmeter** to test the electrical output from the turbine. Verify that the system is producing the desired amount of energy. If you're storing power, check the battery bank to ensure it is charging correctly.
- **Test the load**: Turn on household appliances or devices that you plan to run with the micro-hydro system and verify they receive power. Ensure that the turbine generates enough power to meet your needs.

7. Maintain the Micro-Hydro System

Your micro-hydro system will require regular maintenance to ensure it continues to operate efficiently.

- **Check for debris**: Periodically clean the intake filter to remove debris, leaves, or algae that could

block water flow. Also, check the turbine and generator for any signs of wear or damage.
• **Monitor water flow**: If you experience changes in water flow due to seasonal changes, be sure to adjust the intake system or redirect the water flow to maintain power generation.
• **Inspect the electrical components**: Regularly inspect wiring and connections for wear, corrosion, or damage. Tighten connections and replace any damaged components as necessary.

Project: Backup Generator Integration

A backup generator is a vital component of an off-grid power system, ensuring that you have a reliable source of energy when renewable sources like solar or wind are insufficient. In this project, you will integrate a backup generator into your off-grid energy system. This generator will act as an emergency power source, automatically kicking in when your primary renewable systems cannot meet your energy needs.

Materials and Tools Needed

- Backup generator (diesel, propane, or natural gas)
- Transfer switch or automatic transfer switch (ATS)
- Electrical wiring (appropriate gauge for your system)
- Circuit breaker panel (for connecting generator power)
- Fuel source (diesel, propane, or natural gas)
- Electrical connectors and terminals
- Tools for assembling the system (wrenches, screwdrivers, pliers)
- Voltmeter for testing
- Extension cords (if necessary for initial testing)
- Exhaust venting system (if applicable, especially for gasoline or diesel generators)

Instructions

1. Choose the Right Backup Generator

The first step is selecting the correct generator for your off-grid power needs. There are three common types of generators used for off-grid systems: diesel, propane, and natural gas. Choose the one that best suits your available fuel supply and energy requirements.

• **Diesel Generators**: Diesel is a common choice due to the high energy density of diesel fuel and the availability of generators that can provide high power output. Diesel generators are often more fuel-efficient than propane or natural gas.
• **Propane Generators**: Propane generators are clean-burning and typically used for smaller systems. They are ideal if you already use propane for other off-grid needs like heating or cooking.
• **Natural Gas Generators**: If your property is connected to a natural gas line, this is a convenient and relatively clean option. However, it may not be suitable in remote areas without gas infrastructure.

The size of the generator depends on the total power consumption of your home. Calculate your **total power load** (in watts) and choose a generator that can handle your peak demand.

2. Select a Location for the Generator

The generator should be installed in a location that is easily accessible for maintenance and operation, but also safely positioned away from windows, vents, and other openings in your home.

• **Ventilation**: Make sure the location is well-ventilated, especially if you are using a diesel or gasoline generator, as exhaust fumes can be harmful. A **well-ventilated outdoor shed** or **enclosure** works well for this purpose.
• **Distance from the House**: Place the generator at least **10 feet** away from your house to avoid harmful fumes entering the living space, ensuring that exhaust is directed away from doors and windows.
• **Level Ground**: Ensure the ground is level to avoid instability during operation.

3. Install the Transfer Switch

A transfer switch is an essential component that allows you to safely connect the generator to your home's electrical system. It automatically transfers power between your off-grid system and the generator in case of power shortages.

- **Choose the type of transfer switch**: The two main types of transfer switches are the **manual transfer switch** (MTS) and the **automatic transfer switch** (ATS). An ATS will automatically detect when the power goes out and switch to the generator, while an MTS requires manual operation.
- **Install the transfer switch**: Follow the manufacturer's installation instructions carefully. The transfer switch should be installed **near the main electrical panel**. Make sure the wiring is properly rated for the generator's output, and connect the input wires from the transfer switch to the generator.
- **Wiring**: The transfer switch will need to be connected to the **generator's output terminals** and to the electrical panel of your home. Use the correct gauge electrical wire for the current that will pass through the system. Tighten connections to ensure a secure installation.

4. Connect the Generator to the Transfer Switch

Once the transfer switch is installed, you will connect the backup generator to the system.

- **Wiring the generator**: Connect the output terminals of the generator to the input terminals of the transfer switch. This connection will carry the generator's power to the electrical panel when needed. Ensure the generator is properly grounded and that the connections are tight.
- **Check the voltage**: Before finalizing the connection, use a **voltmeter** to ensure the generator is producing the correct voltage (usually 120V or 240V, depending on your system's requirements). If needed, adjust the generator settings to match the power requirements.

5. Test the Generator and Transfer Switch

Once everything is connected, it's time to test the system to ensure it works correctly.

- **Power off your off-grid system**: To simulate a power outage, turn off the primary off-grid power system (solar, wind, or hydro).
- **Start the generator**: Turn on the generator and verify that it is running smoothly. The transfer switch should automatically engage, and the generator should begin supplying power to your home.
- **Check the electrical panel**: Go to your main electrical panel and confirm that the generator is supplying power to your home. If using an automatic transfer switch, it should have switched automatically; if using a manual transfer switch, switch it manually to the generator power.
- **Test load devices**: Run a few appliances to verify that the generator can handle your home's power load. If everything runs smoothly, you can proceed to the next step.

6. Maintain the Generator

Proper maintenance is essential to ensure the longevity and efficiency of your backup generator.

- **Fuel and oil changes**: Follow the manufacturer's guidelines for fuel and oil changes. Diesel and gasoline generators require regular maintenance to avoid breakdowns.
- **Check the air filter**: Regularly inspect and clean or replace the **air filter** to ensure proper airflow and prevent damage to the engine.
- **Run the generator periodically**: Even if you don't need to use the generator often, run it once a month for 30 minutes to keep it in good working condition.
- **Inspect exhaust systems**: Ensure that exhaust pipes are clear of any debris and not clogged. Regular inspection will help prevent dangerous fumes from leaking.

Project: Designing a Complete Wiring System for an Off-Grid Home

Setting up an off-grid electrical system requires careful planning and precise wiring to ensure that your power is distributed efficiently and safely. This project will walk you through the process of designing and installing a complete wiring system for your off-grid home. From choosing between AC and DC power to making safe battery-to-load connections, you will gain hands-on experience with the essential wiring components of an off-grid energy system.

Materials and Tools Needed
- DC and AC wiring
- Wire gauge (12 AWG, 10 AWG, 8 AWG, etc.)
- Fuse protection (fuses or circuit breakers)
- Grounding rod and wire
- Battery bank
- Inverter or charge controller (depending on the system)
- Wire connectors and terminals
- Electrical tape
- Voltage meter
- Screwdrivers, wrenches, pliers
- Cable ties for organizing wires
- Conduit or cable sheath for protection
- Safety gloves and goggles

Instructions

1. Determine the Power Requirements for Your Home

The first step in designing a wiring system for your off-grid home is to understand how much power you'll need. Calculate your total energy consumption by listing all the appliances and devices you want to run from the system. For each item:

- **Identify the wattage** of the appliance or device.
- **Estimate usage time**: Determine how many hours per day each appliance will be used.
- **Calculate the daily energy consumption** for each item (Wattage x hours per day = watt-hours).
- **Add up the total daily watt-hour usage** to get the overall energy consumption for your home.

For example, if you have a refrigerator (200 watts) running for 8 hours a day, it uses 1600 watt-hours (200 x 8 = 1600Wh) per day.

2. Choose Between AC or DC Power

Decide whether to use AC or DC power throughout your home. The decision depends on the appliances you plan to use:

- **DC Power**: Typically, solar panels and battery systems generate DC power. DC systems are more efficient for charging batteries and powering DC appliances like LED lights, water pumps, or refrigerators designed to run on DC.
- **AC Power**: Most household appliances (like TVs, microwaves, and large kitchen appliances) are designed to run on AC power. In an off-grid system, you'll need an **inverter** to convert DC power from your battery bank into usable AC power for these appliances.

You can also design a hybrid system, where DC is used for some devices and AC is used for others. In this case, you'll need both a DC system for charging and running some appliances, and an inverter to supply AC power to larger appliances.

3. Select the Proper Wiring Gauge

Once you know the power requirements and have decided between DC and AC, the next step is to choose the correct wire gauge for your system.

- **DC Wiring**: For DC circuits, you will need wires that can safely handle the current your system will produce. For small circuits (e.g., lighting), 12 AWG wire might be sufficient, but for larger systems (e.g., battery banks), you may need 6 AWG or 4 AWG wire.
- **AC Wiring**: For AC wiring, use thicker wire for larger appliances. For example, 12 AWG wire is sufficient for most small AC appliances, while larger

circuits (e.g., for dryers or stoves) may require 10 AWG or 8 AWG wire.

Be sure to use wire that is rated for outdoor use or for the specific conditions in your off-grid setup, as weatherproof or underground cable might be needed depending on the installation.

4. Install Fuses or Circuit Breakers

Safety is paramount when working with electricity. Fuses and circuit breakers protect your wiring from overloads or short circuits, which can cause fires.

- **Install fuses or circuit breakers** at key points in your system, such as at the battery bank, the inverter, and at the main distribution panel.
- **Choose the correct size fuse or breaker** based on the wire gauge and the equipment connected to the circuit. For example, a 20A fuse may be appropriate for a 12 AWG wire.

5. Grounding Your System

Proper grounding is essential to prevent electric shock and equipment damage. All electrical systems should be grounded to ensure that in the event of a fault, the current is safely diverted to the ground.

- **Install a grounding rod** near the inverter or battery bank and connect it to your electrical system using an appropriate wire gauge (typically 6 AWG or 8 AWG).
- **Connect the grounding wire** to the metal parts of your system, such as the inverter frame and battery bank. This ensures that all components are properly grounded and can safely discharge electrical energy in the event of a fault.

6. Wiring the Battery Bank to the Inverter

Next, you'll need to wire the battery bank to the inverter (if using AC) or charge controller (if working with DC-only circuits).

- **Use heavy-duty DC wiring to connect the positive terminal** of the battery bank to the positive input terminal on the inverter or charge controller.
- **Using appropriate gauge DC wiring, connect the negative terminal** of the battery bank to the negative terminal on the inverter or charge controller, ensuring all connections are firmly tightened and secure.

7. Install the Inverter and AC Wiring

If you're using AC power, you'll need an inverter to convert the Direct current (DC) power from the batteries is converted into alternating current (AC) power for your household appliances.

- **Install the inverter** in a cool, dry location, preferably near the battery bank to minimize power loss in the cables.
- **Wire the inverter to the AC distribution panel**: Use proper gauge wire to connect the inverter's AC output to your off-grid home's electrical system.
- **Secure the inverter and AC wiring**: Use mounting brackets or platforms to securely fasten the inverter in place. Avoid any stress on the wires, as this can cause damage and safety hazards.

8. Install and Test the Circuitry

Now that your wiring is in place, it's time to test everything to ensure it's functioning correctly.

- **Check the wiring connections**: Ensure that all wire connections are properly tightened and secured. Double-check that each connection is made according to the system design (i.e., positive to positive, negative to negative).
- **Test the DC system**: If using DC appliances, check that the power flows from the battery bank to the load, and the voltage is correct.
- **Test the AC system**: If using an inverter, plug in a small AC appliance and check if it's running properly. Monitor the inverter and battery bank to ensure that the system isn't overloaded.

9. Organize and Protect the Wiring

Once the system is functioning, you can organize the wires and protect them from potential damage.

Bundle the wires using cable ties or wire loom, ensuring that there's no tension on the wires. This will make the system neat and help avoid accidental damage.

Install conduits if necessary, particularly if the wiring runs outdoors or through areas where it might be exposed to the elements or physical wear.

10. Ongoing Monitoring and Maintenance

Regularly check the system to ensure that it's operating efficiently. Monitor the battery bank, inverter, and wiring for any signs of wear or overheating.

- **Check voltage levels** and ensure the system is properly charging and discharging. Regularly monitor the battery bank's state of charge to ensure optimal performance and prevent over-discharge or overcharging.
- **Inspect wires and connections** at least once a year for signs of corrosion, wear, or loose connections, and replace any damaged components as needed.

Project: Selecting and Installing the Right Inverter for Your Off-Grid Home

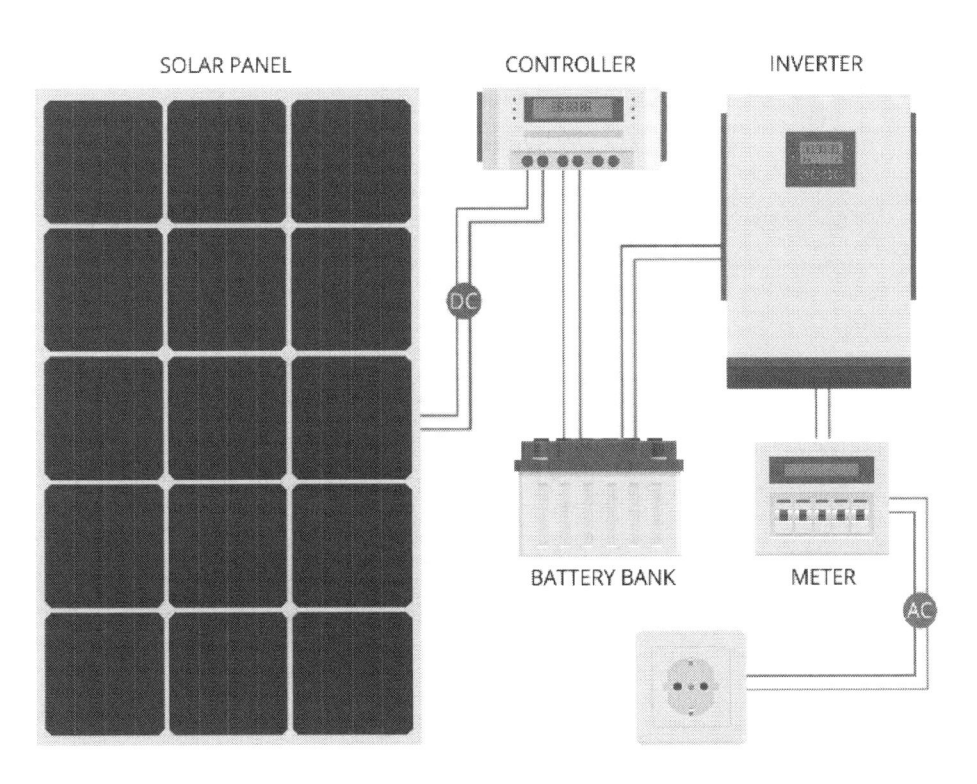

- Off-grid inverter (choose between a **pure sine wave** or **modified sine wave** inverter based on your needs)
- Battery bank (DC power source)
- Appropriate wiring (DC and AC wiring)
- Circuit breakers for safety
- Battery monitor
- Voltage meter
- Screwdrivers, pliers, wrenches
- Mounting hardware or platform for the inverter
- User manual for the inverter

An inverter is a crucial part of your off-grid power system, responsible for converting the direct current (DC) electricity stored in your battery bank into usable alternating current (AC) power, which is used by most household appliances. This project will guide you through the process of selecting and installing the correct inverter for your off-grid system. The right inverter will ensure that your energy is used efficiently and safely.

Materials and Tools Needed

Instructions

1. Determine Your Power Requirements

The first step in selecting the right inverter is understanding how much power you need to supply your home or off-grid setup. This involves calculating your total energy consumption:

- **List all appliances** that you intend to run on the inverter. Include the wattage of each appliance and how many hours per day each one will be used. For example, if you have a refrigerator that uses 200 watts and it runs for 8 hours per day, the daily energy consumption for that appliance is **200W x 8h = 1600Wh**.
- **Calculate the total energy consumption** by adding up the watt-hours (Wh) for each appliance. This will give you the total energy required per day.
- **Consider peak power**: Some appliances, like refrigerators and pumps, require more power when they start up. This is called "surge power." Choose an inverter that can handle the **peak surge power** of all your devices at once.

2. Choose the Right Type of Inverter

There are two primary types of inverters to choose from:

- **Pure Sine Wave Inverter**: This type of inverter produces a smooth, high-quality AC signal that is virtually identical to the electricity provided by the grid. Pure sine wave inverters are ideal for sensitive electronics like computers, medical devices, or audio equipment. They come at a higher cost compared to modified sine wave inverters, but offer greater versatility and improved efficiency.
- **Modified Sine Wave Inverter**: This inverter produces a stepped AC signal that is less smooth. While it's sufficient for many off-grid applications like powering lights and basic appliances, it may cause some electronic devices to operate inefficiently or generate noise.

For most off-grid systems, a pure sine wave inverter is recommended, especially if you plan to use modern electronics.

3. Select the Correct Size for Your Inverter

Choosing the right size for your inverter is critical to ensure it can handle the power needs of your home.

- **Determine continuous wattage**: This is the amount of power your inverter can supply continuously. The inverter should be sized to handle your **daily load**, which is the total wattage of all the appliances you want to run at once.
- **Account for surge wattage**: Inverters also need to handle **surge power**, which is the extra power required when appliances like refrigerators, pumps, or air conditioners start up. Check the manufacturer's specifications for both **continuous wattage** and **surge wattage**.

For example, if your total daily energy consumption is 3000Wh, and your peak surge power is 4000W (for an appliance like a pump), choose an inverter with a continuous wattage capacity of at least 3000W and a surge capacity of 4000W.

4. Prepare the Inverter Installation Site

The inverter needs to be installed in a well-ventilated area that's cool and dry to avoid overheating. Here are some considerations for setting up:

- **Space**: Ensure the area has enough room for the inverter and any associated components like fuses and circuit breakers.
- **Ventilation**: Inverters generate heat during operation, so proper ventilation is crucial to prevent overheating. Avoid installing it in tight spaces or near heat sources.
- **Mounting**: Most inverters come with mounting brackets or hardware. Secure the inverter on a wall or platform to keep it stable and easy to access for maintenance.

5. Install the Inverter

Now that you have selected the right inverter and prepared the site, it's time to install it:

- **Connect the inverter to the battery bank**: Using appropriate DC wiring, connect the positive and negative terminals of your inverter to the corresponding terminals of your battery bank. The positive (usually marked red) terminal of the inverter should connect to the positive terminal of the battery, and the negative (usually marked black) should connect to the negative terminal.
- **Install the circuit breaker**: For safety, always install a **circuit breaker** between the inverter and the battery. This will protect the system from overloads and short circuits. Make sure the circuit breaker matches the inverter's capacity.
- **Connect to AC system**: If your inverter is capable of AC power output, it will have AC outlets or terminals to connect to your household electrical system. Using proper gauge wiring, connect the AC output of the inverter to your home's electrical panel or to the specific circuits you want powered.
- **Ensure proper grounding**: Ground the inverter properly to prevent electrical shocks and to protect the system from electrical surges. Adhere to the manufacturer's instructions on how to ground the inverter. This is a crucial safety step.

6. Test the Inverter System

After installation, it's important to test the inverter to ensure that it is functioning properly:

- **Check the connections**: Double-check all connections to ensure they are secure and correct.
- **Power on the system**: Turn on your inverter and allow it to start. You should see the display panel light up, and the inverter should begin supplying power to your connected appliances.
- **Monitor the voltage**: Use a voltmeter to check the voltage output of the inverter. Make sure it matches the rated output (e.g., 120V or 240V AC).
- **Test with appliances**: Plug in a few small appliances and check if they're working correctly.

Start with appliances that don't require a high surge of power and gradually increase the load.

7. Monitor the System

After installation and testing, it's important to monitor your system regularly to ensure it is operating efficiently.

- **Battery monitoring**: If your inverter has a built-in battery monitor, use it to track the charge level and health of the battery bank.
- **Energy monitoring**: Many modern inverters have energy monitoring features that allow you to track how much energy is being used and stored. Regularly check these readings to ensure your system is operating within safe limits.
- **Maintain the system**: Periodically inspect the inverter, wiring, and connections. Clean the inverter's filters (if applicable) and check for signs of wear or corrosion. Regular maintenance will extend the life of your system and improve its efficiency.

By following these steps, you will have installed an inverter that enables you to convert stored DC power into usable AC power, making your off-grid system functional and efficient. With proper care and attention, this inverter will serve as a critical component of your off-grid home, ensuring you have power when you need it most.

Project: Designing Your Own Battery Bank for Off-Grid Use

A battery bank serves as an essential part of any off-grid energy system, enabling you to store electricity produced by renewable sources such as solar panels or wind turbines. This stored energy ensures you have power available during periods when the sun isn't shining or the wind isn't blowing. In this project, you'll determine your household's energy requirements, choose the right type of batteries, and design a battery bank capable of delivering consistent, reliable power to support your off-grid lifestyle.

Materials and Tools Needed:
- Battery (Lead-Acid, Lithium-Ion, or Flow Battery based on your selection)
- Battery cables and connectors
- Charge controller
- Inverter (if needed for AC power)
- Battery storage box or battery rack
- Battery monitor
- Voltmeter
- Screwdrivers, wrenches, and other basic tools
- Electrical tape
- Wire cutters and strippers
- Safety gloves and goggles

Instructions:

1. Determine Your Daily Energy Consumption

The first step in designing a battery bank is to calculate how much electricity your household uses on a daily basis. To do this:

- **List all appliances** that will be powered by the battery bank, including lights, refrigerator, water pump, and any other devices.
- **Find the wattage** of each appliance (this is typically labeled on the device or in the manual).
- **Calculate daily usage**: Multiply the wattage of each appliance by the number of hours per day it will be used. For example, if a 100-watt light runs for 5 hours, it uses 500 watt-hours (Wh) per day (100W x 5 hours).
- **Add up all the watt-hours** to get your total daily energy requirement. This will give you an idea of how much energy you need to store in your battery bank to meet your daily needs.

2. Choose the Right Type of Battery

The next step is selecting the appropriate type of battery for your system. The three most commonly used options are lead-acid, lithium-ion, and flow batteries. Each comes with its own set of benefits and limitations, which should be carefully considered based on your specific energy needs and budget.

- **Lead-Acid Batteries**: These are the most cost-effective options and come in flooded, sealed, or AGM types. They are a good choice for smaller systems but require maintenance and have a shorter lifespan.
- **Lithium-Ion Batteries**: These batteries are more efficient and offer a longer lifespan compared to lead-acid batteries. They can handle deeper discharges without damage and typically require less maintenance, but they are more expensive.
- **Flow Batteries**: These are ideal for large off-grid systems due to their ability to scale up easily. Flow batteries are more expensive and less common for residential systems but can be a good option if you need a large capacity.

3. Calculate the Required Battery Capacity

To calculate the size of your battery bank, you need to know your total energy usage and how much of that energy you want to store.

- **Energy Storage Calculation**: The battery bank needs to store enough energy to power your home for a set period (usually for a couple of days during bad weather or low output from renewable sources). A general rule is to have at least **2-3 days of storage** to ensure you have enough power during cloudy days or low-wind periods.
- **Consider Depth of Discharge (DoD)**: The depth of discharge is the amount of the battery's capacity that can be safely used before recharging is needed. For lead-acid batteries, the DoD is typically **50%**, while lithium-ion batteries can safely discharge up to **80-90%**. This means if you need **3000Wh** of usable energy, you should choose a battery that is **twice** that size for lead-acid batteries (6000Wh) or **1.2 times** for lithium-ion (3600Wh).

4. Determine the Number and Arrangement of Batteries

After calculating the total required capacity, you need to determine how many batteries you'll need to achieve that capacity. You'll also need to decide whether to arrange the batteries in series or parallel.

- **Series Arrangement**: When you connect batteries in series, you increase the voltage. For example, connecting two **12V batteries** in series will give you a **24V system**. This is useful for high-power applications.
- **Parallel Arrangement**: When you connect batteries in parallel, you increase the total **amp-hour** (Ah) capacity, but the voltage remains the same. For example, connecting two **12V 100Ah batteries** in parallel will give you **12V 200Ah**, which increases your total energy storage without changing the system voltage.

5. Set Up the Battery Bank

Now it's time to set up the battery bank. Follow these steps carefully:

- **Choose a location**: The batteries should be placed in a cool, dry, and well-ventilated area. Ensure that the space is secure and that the batteries are protected from extreme temperatures.
- **Install a battery rack**: If necessary, install a rack or storage box to securely house and stabilize the batteries in position. This will prevent movement, which could damage the batteries or the electrical connections.
- **Connect the batteries**: Using appropriate battery cables and connectors. Connect the batteries in series or parallel configurations, according to the specific requirements of your system design. Use a **voltmeter** to check the voltage as you go to ensure proper connections.
- **Wire to charge controller and inverter**: Connect the positive and negative terminals of the battery, ensuring secure and correct polarity based on your system's configuration bank to the **charge controller** (if you're using one) and **inverter** to distribute power throughout your home. Make sure all connections are tight to prevent power loss or potential hazards.

6. Install the Battery Monitoring System

To ensure your batteries are working efficiently and to extend their lifespan, it's important to have a battery monitoring system (BMS).

- **Install a battery monitor**: This system will track the state of charge (SoC), depth of discharge (DoD), and health of each battery. It will give you a real-time overview of your battery bank's performance.
- **Set up alarms**: Many BMS systems allow you to set alarms for when the voltage is too high or low, helping prevent overcharging or over-discharging of the batteries.

7. Test the Battery Bank

Once everything is connected, it's time to test your system.

- **Check voltage**: Use a voltmeter to check the voltage of the battery bank. Make sure it matches the intended system voltage.
- **Power test**: Turn on some small appliances to see if the system is providing the necessary power. Regularly monitor the battery voltage and state of charge to confirm the batteries are operating efficiently and within safe parameters.
- **Monitor over time**: Over the next few days, observe how well the battery bank powers your off-grid system. Make adjustments to the setup if necessary, such as recalculating your daily energy usage or tweaking the battery connections.

Module D. Waste & Recycling Solutions

Managing waste efficiently and sustainably is a key component of off-grid living, ensuring that you reduce your environmental footprint while maintaining a healthy and functional home. In this module, you will explore eco-friendly solutions for dealing with waste, including composting toilets, greywater recycling, biogas production, and safe disposal of hazardous materials. These methods will not only allow you to recycle and repurpose resources, but they also contribute to your self-sufficiency by reducing reliance on external services. By the end of this module, you will have a comprehensive understanding of how to handle waste in an off-grid setting, ensuring a sustainable, safe, and environmentally responsible lifestyle.

18. Composting Toilets: Eco-Friendly Off-Grid Sanitation

As an off-grid homeowner, managing waste efficiently is one of the most important steps toward creating a self-sufficient, sustainable living space. Composting toilets offer a highly effective and environmentally friendly way to manage human waste without relying on municipal systems or the need for water-intensive septic tanks. This system not only reduces water consumption, but it also transforms waste into valuable compost that can be used for fertilizing plants and gardens, closing the loop of resource consumption. Understanding how composting toilets work, the various types available, and how to design and maintain your own system will help you create a system that is both practical and eco-friendly for your off-grid lifestyle.

Introduction to Composting Toilets: Understanding the Basics and Benefits

Composting toilets are systems that treat human waste by breaking it down into compost over time. Unlike traditional toilets that flush waste into a sewage system or septic tank, composting toilets use natural processes—primarily aerobic bacteria and other microorganisms—to decompose waste. This process transforms the waste into compost, a nutrient-rich material that can be safely returned to the earth, enriching soil and promoting plant growth.

One of the key benefits of composting toilets in an off-grid setting is their low environmental impact. In many off-grid homes, water is a precious resource, and traditional toilets use a significant amount of water for each flush. A composting toilet, on the other hand, uses little to no water, making it an ideal solution for regions where water is scarce. Additionally, since composting toilets do not require a septic system or connection to municipal waste infrastructure, they reduce the need for costly and complex waste management solutions.

In terms of sustainability, composting toilets support the natural recycling of nutrients. When waste is decomposed in a controlled environment, it is turned into rich, organic material that can be safely returned to the soil, reducing the need for chemical fertilizers and promoting healthy, fertile ground.

Types of Composting Toilets: Dry vs. Waterless, Self-Contained vs. Central Composting Systems

There are several types of composting toilets to choose from, each with its own set of advantages and considerations. Understanding the differences between them will help you select the system that best suits your off-grid home.

Dry vs. Waterless Toilets

The primary distinction between composting toilets lies in how they manage waste and whether they use water.

- **Dry Toilets**: These composting toilets do not use water at all. Waste is directly deposited into a composting chamber, and it is decomposed through aerobic processes that rely on oxygen. Dry toilets are ideal for off-grid situations where water is in short supply. In these systems, sawdust, wood chips, or peat moss is typically added after each use to help absorb moisture and encourage the composting process.
- **Waterless Toilets**: While waterless toilets still manage waste in a dry environment, some systems include a small amount of water to help facilitate decomposition. However, they are still considered

water-efficient compared to traditional toilets. These toilets often include a urine diversion mechanism, which separates liquid waste from solid waste to prevent excess moisture and odors in the composting chamber.

Self-Contained vs. Central Composting Systems

Another distinction in composting toilet designs is whether they are self-contained or part of a larger, central system.

- **Self-Contained Systems**: In a self-contained system, the entire composting process occurs within a single unit, typically located directly over the toilet. These systems are compact, easy to install, and ideal for smaller off-grid homes or cabins. They can be moved or relocated with relative ease, making them suitable for seasonal or mobile off-grid living.
- **Central Composting Systems**: These systems involve a central composting unit, often located in a basement, shed, or other separate location from the toilet. In this design, waste is transported from the toilet to the composting chamber via pipes or chutes. Central systems are generally more suitable for larger off-grid homes or communities with multiple toilets, as they can handle greater volumes of waste.

Design Considerations: Sizing, Location, Ventilation, and Maintenance for Composting Toilets

When designing your composting toilet system, several factors must be taken into account to ensure it operates effectively and efficiently.

Sizing and Location

The size of your composting toilet will depend on the number of people using it and the amount of waste it will process. In smaller households or for individual use, a compact self-contained system may be sufficient. However, larger off-grid homes or communities may require a more robust central composting system with a larger composting chamber.

Location is another key factor. Ideally, you'll want to place the toilet in a well-ventilated space, away from areas where waste or odors could create issues. For self-contained systems, this typically means positioning the toilet near a window or vent. Central systems should be placed in an area where waste can be transported easily to the composting chamber and where ventilation can be maximized.

Ventilation

Proper ventilation is crucial for a composting toilet to function effectively. Without adequate airflow, the composting process can become sluggish, and you may encounter unpleasant odors. Ventilation systems can include simple exhaust fans, vent pipes, or even solar-powered fans to keep air circulating. Ensuring proper airflow also helps keep the temperature inside the composting chamber at the right level for microbial activity, which is essential for the breakdown of waste.

Maintenance

Regular maintenance is essential for keeping your composting toilet system functioning properly. For self-contained systems, this includes adding the right amount of carbon material (like sawdust) to absorb excess moisture and maintain proper composting conditions. The composting chamber should be emptied periodically, depending on usage, and the compost should be processed into usable soil after the appropriate amount of time.

Central systems may require additional maintenance, such as monitoring the flow of waste through pipes or chutes, checking for clogs, and ensuring that the central composting chamber is kept at the correct temperature and moisture levels.

Composting Process: Breakdown of Waste, Composting Times, and Required Conditions

Understanding the composting process will help you ensure that your system operates efficiently. The breakdown of waste is a biological process driven by microorganisms, including bacteria, fungi, and worms. These microorganisms break down organic matter into compost by consuming the waste and converting it into humus.

Waste Breakdown

The composting process begins when solid waste is deposited into the composting chamber. Waste material is typically broken down into carbon (from organic materials like sawdust or straw) and nitrogen (from the waste itself). The ratio of carbon to nitrogen (C:N ratio) is important for encouraging the right microbial activity. Generally, a 30:1 ratio of carbon to nitrogen is ideal.

Microbial activity produces heat, which accelerates the composting process. Over time, the waste will break down into a fine, dark, soil-like substance. This process usually takes anywhere from several months to a year, depending on temperature, moisture, and the frequency with which the composting chamber is turned or stirred.

Required Conditions for Composting

To ensure successful composting, certain conditions must be maintained in the composting chamber:

- **Temperature**: Ideal composting temperatures range between 130°F to 160°F (54°C to 71°C), which helps accelerate microbial activity. Ensuring that the composting chamber is insulated or placed in a location with consistent temperatures can help maintain these conditions.
- **Moisture**: The composting chamber should be kept moist, but not wet. Too much moisture can lead to anaerobic conditions and unpleasant odors, while too little moisture can slow down the decomposition process. A good rule of thumb is to keep the moisture content between 40-60%.
- **Aeration**: Turning or stirring the compost regularly helps introduce oxygen into the chamber, which is essential for aerobic bacteria to thrive. Aeration also helps prevent the buildup of odors and ensures even composting.

Maintenance and Troubleshooting: Handling Odors, Regular Cleaning, and Potential System Failures

Like any system, composting toilets require regular care and maintenance to function properly. Below are some common issues and their solutions.

Handling Odors

If your composting toilet starts to produce unpleasant odors, it's usually a sign of improper conditions. Odors are often caused by excess moisture, lack of aeration, or an imbalance in the C:N ratio.

- **Excess Moisture**: If the chamber is too wet, add more carbon material like sawdust or wood chips to absorb the moisture and restore balance.
- **Poor Aeration**: Make sure the composting material is being stirred or turned regularly to allow oxygen to flow through the chamber.
- **Imbalance in C:N Ratio**: Add more carbon material to balance the nitrogen content. A good mix of carbon-rich and nitrogen-rich materials is key.

Regular Cleaning

Although composting toilets are relatively low-maintenance, you will need to clean certain components regularly, especially the collection chamber or the urine diverter (if you're using a dual-chamber system). The frequency of cleaning depends on the amount of use, but generally, the collection chamber should be emptied and cleaned every 6-12 months, or when full.

System Failures

Occasionally, you may encounter system failures, such as clogs, leaks, or breakdowns in the composting process. Regular inspections can help identify these issues early. If you're using a self-contained system, check for signs of excess moisture, odors, or slow decomposition. For central systems, ensure that waste is being transported properly and that the composting chamber is functioning as it should.

Environmental Benefits: Reducing Water Use, Preventing Contamination, and Creating Nutrient-Rich Compost

Composting toilets are one of the most sustainable solutions for managing human waste in off-grid living. By using little to no water, composting toilets significantly reduce the amount of water required for waste disposal, helping to conserve a precious resource, especially in water-scarce areas. In addition to reducing water consumption, composting toilets also prevent contamination of local water supplies, which can occur with traditional septic systems.

Perhaps one of the greatest benefits is the creation of nutrient-rich compost. The breakdown of organic waste results in a humus-like material that can be safely used as fertilizer for gardens, reducing the need for chemical fertilizers and helping to close the loop in a self-sustaining off-grid system. By utilizing human waste as a resource

rather than a pollutant, composting toilets contribute to a more circular, eco-friendly system of living.

19. Greywater Recycling & Wastewater Treatment

When living off-grid, water conservation becomes a critical aspect of your daily life. With limited access to external water supplies, every drop counts. Greywater recycling offers a sustainable solution that allows you to reuse water from household activities—such as showers, sinks, and laundry—for non-potable purposes like irrigation. By implementing a greywater system, you can reduce your water usage, decrease your reliance on external sources, and help preserve precious resources.

Understanding Greywater: Definition, Sources, and Potential Uses for Irrigation

Greywater is the relatively clean wastewater that comes from household activities such as bathing, washing dishes, and doing laundry. Unlike blackwater, which comes from toilets and contains human waste, greywater is generally considered less contaminated and can be safely treated and reused with the right filtration and treatment methods. It is an excellent resource for irrigation, landscaping, and even for flushing toilets, provided it is properly filtered and treated.

The sources of greywater in your home include:

- **Showers and bathtubs**: Water from washing, shampooing, and bathing.
- **Sinks**: Wastewater from washing hands, dishes, or food preparation.
- **Laundry**: Water used for washing clothes, especially if you avoid using harsh chemicals or bleach.

These sources provide an abundant supply of water that can be reused for a variety of non-potable purposes. However, it's important to understand the limitations and potential contaminants in greywater. For example, water from laundry may contain detergents or fabric softeners, while water from showers could contain soap residues or oils. Proper filtration and treatment are crucial to make the water suitable for reuse, especially in food gardens.

Greywater System Design: Components of a Greywater System

A greywater system is made up of several components designed to collect, filter, and redistribute wastewater. Each component works together to ensure that the water is properly treated and safe for use. Here's a breakdown of the key elements of a greywater system:

1. Piping and Collection

The first step in a greywater system is the collection of water from the various sources within the home. Typically, a diverter valve or a dedicated plumbing line is used to separate greywater from blackwater. This allows the greywater to be routed to a storage tank or filtration system.

2. Filters

Filters are used to remove large particles and debris from the greywater. These can range from simple mesh filters to more advanced multi-stage systems that include sand, charcoal, and biological filters. Depending on the level of filtration required, you may need a coarse filter for removing solids, as well as a fine filter for smaller particles.

3. Storage Tanks

Once greywater is filtered, it's often stored in a holding tank before being redistributed for reuse. These tanks should be made of durable, non-corrosive materials such as plastic, fiberglass, or stainless steel. The size of the storage tank will depend on your household's water usage and the space available for installation.

4. Pumps

In some systems, pumps may be necessary to move water from the storage tank to the intended application, such as a garden or irrigation system. If you are installing a gravity-fed system, a pump may not be required, but for larger systems or situations where water needs to be lifted or moved over long distances, a pump will be necessary.

System Types: Simple Gravity-Based vs. Pumped Systems

There are different types of greywater systems that you can choose from based on your off-grid needs. Each type has its own advantages and challenges, and the choice you make will depend on factors like your location, water usage, and available space.

Gravity-Based Systems

A gravity-fed system is one of the simplest and most cost-effective types of greywater recycling systems. In this system, greywater is directed from the household plumbing directly to a filtration system, and the water is then distributed to plants via gravity. This type of system is ideal if you have a property with a natural slope, as gravity can move the water without the need for electrical pumps.

Pumped Systems

In areas without a natural slope or where you need to move water long distances, a pumped system may be necessary. These systems use a pump to move water from the collection point through the filters and into the irrigation or storage system. While more complex and requiring more maintenance than gravity-based systems, pumped systems are ideal for larger properties or areas where gravity alone cannot facilitate water movement.

Filtering and Treatment Methods: Sand, Charcoal, and Biological Filtration

The filtration process is essential in making greywater safe for reuse. Without it, the water may contain contaminants that could harm plants or contaminate soil. There are several filtering methods you can use to clean greywater.

Sand Filters

Sand filters are one of the most common and affordable methods for filtering greywater. They work by allowing water to flow through layers of sand that trap particles, debris, and organic matter. You can build a simple sand filter using a large container, filling it with multiple layers of coarse and fine sand. The water passes through the sand, and the particles are trapped in the layers.

Charcoal Filters

Activated charcoal is another excellent filtering material for greywater systems. Charcoal is highly porous, which allows it to adsorb impurities and chemicals, such as oils, detergents, and solvents. Charcoal filters are often used in combination with sand filters to provide a multi-stage filtration system that removes both solid particles and chemicals.

Biological Filters

Biological filtration uses microorganisms to break down organic material in the greywater, further purifying it. Biological filters can be built using natural materials like rocks, gravel, or plant roots. These systems often involve a tank or bed where the greywater is filtered and exposed to microorganisms that consume organic pollutants, turning them into harmless by-products. Constructed wetlands are an example of biological filtration systems.

Legal and Environmental Considerations: Local Regulations for Greywater Reuse

Before setting up your greywater system, it's essential to check local regulations regarding greywater reuse. Laws governing greywater use vary by region, and they are often in place to ensure public health and safety. In some places, greywater systems must be registered or inspected, and there may be restrictions on how the water can be used.

For example, in many areas, greywater cannot be used for direct human consumption or irrigation of edible crops unless it has been treated to a certain standard. Some regions may have specific rules about how greywater is stored, such as requiring the water to be used within 24 hours to prevent bacterial growth.

In addition to regulatory concerns, you must consider the environmental impact of greywater reuse. If you plan to use greywater for irrigation, ensure that it does not contain chemicals or detergents that could harm plants or the environment. Using biodegradable, non-toxic soaps and detergents can minimize the environmental risks associated with greywater recycling.

Maintenance and Monitoring: Regular System Checks, Cleaning Filters, Preventing Clogging, and Addressing Contamination

Greywater systems require regular maintenance to ensure that they continue to function effectively and efficiently. Neglecting to maintain your system can lead to clogs, leaks, odors, and reduced water quality. Below are the key tasks you'll need to perform regularly:

Cleaning Filters

Filters will need to be cleaned periodically to prevent clogging and ensure proper water flow. Depending on the type of filter system you have, you may need to clean sand filters by washing them or replacing the charcoal in charcoal filters. Biological filters may require periodic maintenance to remove excess organic material and ensure that the microbial activity remains active.

System Inspections

Perform regular system inspections to check for leaks, cracks in the piping, or any blockages that could cause water to back up or flow inefficiently. If you're using a pumped system, ensure that the pump is functioning correctly, and check the power source (whether solar, wind, or another off-grid source) to make sure everything is running smoothly.

Preventing Clogs

Clogging is a common issue in greywater systems, especially in the filters and pipes. To prevent this, ensure that only appropriate greywater (free from oils, fats, and food scraps) enters the system. Install fine mesh screens in showers or sinks to catch larger particles before they enter the system.

Monitoring Water Quality

Regularly test the quality of the greywater being reused to ensure it meets the necessary standards. This may involve simple testing kits to check for pH levels, bacterial content, or the presence of harmful chemicals. By staying on top of these checks, you can maintain a safe and effective greywater recycling system.

With a properly designed and maintained greywater recycling system, you can efficiently reuse water from your daily activities, reducing your reliance on external water sources and minimizing waste. This system can be a valuable component of your off-grid lifestyle, contributing to water conservation, environmental sustainability, and resource efficiency.

20. Biogas Production: Converting Organic Waste into Usable Energy

When living off-grid, self-sufficiency in energy production becomes a central goal. While solar panels and wind turbines often come to mind, one other powerful and sustainable energy source that many overlook is biogas. By converting organic waste into usable energy, biogas can power stoves, generators, and even heat water. Through the anaerobic digestion process, biogas systems offer an efficient, eco-friendly solution to off-grid energy needs. In this section, you will learn how to harness organic waste to produce biogas, and how to set up a biogas production system that can fuel your off-grid lifestyle.

What is Biogas?

Biogas is a mixture of gases, primarily methane, produced by the anaerobic (without oxygen) decomposition of organic material. It is typically produced in sealed containers known as digesters, where bacteria break down organic waste, releasing methane gas. The process is natural and occurs in landfills, animal manure piles, or natural wetlands, but it can be harnessed in a controlled setting to provide a renewable energy source.

The main benefits of biogas production are its sustainability and the reduction of waste. Instead of letting organic waste decay and produce harmful greenhouse gases, you can capture the methane for use as an energy source. Biogas can be used for cooking, heating, and even running small electricity generators. It's an ideal solution for off-grid living, where traditional energy sources may be limited or unavailable.

Biogas is a clean-burning fuel that, when used properly, has little to no negative environmental impact. Additionally, the byproduct of the digestion process, known as digestate, can be used as a rich fertilizer for gardens or agricultural use, further reducing waste.

Biogas System Components

A biogas system consists of several components, each playing a crucial role in the production, storage, and utilization of biogas. These components include the digester, gas storage tank, and gas utilization system.

1. Digester

The digester is the heart of a biogas system. It's where the anaerobic digestion process takes place. The organic waste (such as food scraps, animal manure, or plant material) is placed into the digester, where bacteria break it down and release methane gas.

The size and design of the digester depend on how much waste you plan to process and how much gas you want to generate. Digesters can range from small, simple setups to larger, more complex systems.

2. Gas Storage

Once methane is produced, it needs to be stored for later use. This is done in a gas storage tank, typically a flexible, airtight container that can expand as the gas is generated. The storage tank should be positioned above the digester to allow gas to flow into it by gravity.

The storage tank needs to be well-sealed to prevent leaks, and it must be sized based on your expected gas production. For small-scale systems, the storage tank may be as simple as a plastic or rubber bag, while larger systems might use steel or other durable materials.

3. Gas Utilization

After the biogas is stored, it can be used in a variety of ways. The most common uses are for cooking or running a generator.

- **Stoves**: You can use biogas directly to power a stove for cooking, much like you would with natural gas. The system is connected to the stove using a gas pipeline and a pressure regulator.
- **Generators**: Biogas can also be used to generate electricity through a biogas-powered generator. The generator works by burning the methane to create energy, which can be used to power lights, appliances, or other off-grid equipment.

Building a Biogas Digester

The biogas digester is the central component of the system, and building one requires careful planning and consideration of the materials, size, and location. Here's a general step-by-step guide on how to build a simple biogas digester:

1. Sizing the Digester

The size of your digester will depend on how much organic waste you plan to process daily and how much biogas you want to generate. For example, if you are planning to use biogas for cooking, you will need a smaller digester than if you want to generate electricity.

A common calculation for sizing a digester is that you need around 1 cubic meter of digester space for each person in the household. However, larger systems can be built for farms or larger properties, where more organic waste is available.

2. Material Selection

The materials you choose for the digester should be durable and capable of withstanding the pressure

and conditions inside the system. Common materials for digesters include:
- **Concrete**: Durable and long-lasting, ideal for permanent installations.
- **Steel or Metal**: Strong and corrosion-resistant, but can be more expensive.
- **Plastic or Rubber**: Flexible and affordable, but may require more maintenance over time.

3. Installation Process

To install the digester, start by digging a hole in the ground or setting up a concrete pad if you're building a stationary system. The digester should be placed in a location that allows for easy waste input and gas storage, with good ventilation and drainage.

Once the hole or base is prepared, assemble the digester structure according to your design and material choices. You'll need to ensure that the digester is airtight to prevent the escape of gas and that it can be sealed to maintain the anaerobic conditions needed for digestion.

Organic Waste Sources

The organic waste you use in the digester is crucial to the quality and quantity of the biogas produced. Common sources of organic waste include:

1. Kitchen Scraps

Food scraps like vegetable peels, fruit rinds, and leftover food can be added to the digester. Avoid adding oils, dairy, or meat, as these can affect the digestion process and lead to unwanted odors.

2. Animal Manure

Manure from livestock such as cows, pigs, chickens, and goats is an excellent source of organic material for biogas production. It's rich in the microorganisms that help break down waste in the digester, making it a valuable addition to the system.

3. Other Organic Materials

You can also add plant waste, yard clippings, and grass to the digester. Some systems also accept small amounts of paper or cardboard, as long as it is shredded or broken down.

Gas Collection and Storage

Once the anaerobic digestion process is underway, methane gas will begin to accumulate. This gas can be used for cooking, heating, or electricity generation. Proper storage is essential to ensure the gas is available when needed.

1. Techniques for Storing Biogas

Biogas is typically stored in a flexible storage bag or tank, which allows the gas to expand and contract as it is generated and used. It is important to ensure the storage container is airtight to avoid gas leaks, which can be dangerous. The tank should be positioned above the digester so that the gas naturally flows into the storage tank via gravity.

2. Using Biogas Safely

To use biogas safely, you must regulate the pressure and ensure it is stored in suitable containers. Gas burners, generators, and stoves designed to burn biogas can be connected to the storage system using a pipeline. It's essential to use gas regulators to prevent over-pressurization, which can cause system damage or leaks.

Troubleshooting and Maintenance

Biogas systems require ongoing maintenance to keep them running efficiently. Regularly monitor the system for any signs of malfunction, and resolve potential issues early, preventing them from developing into major problems.

1. Troubleshooting Gas Leaks

Gas leaks are one of the most serious risks in a biogas system. Ensure that all connections between the digester, storage tank, and gas utilization devices are airtight. If you notice a gas smell, check the system for leaks and seal them immediately.

2. Addressing Improper Digestion

If the biogas production is slow or the system is not producing enough gas, it may be due to improper digestion. Check the temperature, moisture, and aeration levels inside the digester. Make sure the organic waste is being added in the correct proportions and that the system is being maintained at the right conditions for bacterial activity.

21. Hazardous Waste Disposal: Handling Batteries, Chemicals, and Electronics Safely

Living off the grid offers many rewards, but it also requires careful attention to the management of waste, especially hazardous waste. Managing hazardous materials such as batteries, chemicals, and electronics safely is crucial not only for the health of your family but also for the long-term preservation of the environment. Improper disposal of these materials can have devastating effects on the ecosystem, and it is essential to approach disposal with care and responsibility. This section will guide you through the identification, handling, disposal, and recycling of hazardous materials, as well as the environmental impacts of improper disposal and sustainable waste management solutions.

Identifying Hazardous Waste: Types of Waste and Potential Risks

Hazardous waste includes materials that are potentially dangerous to human health or the environment. These can range from household items like batteries and paints to larger, more industrial-sized waste materials. In the context of off-grid living, it is vital to recognize and properly dispose of these substances, as improper handling may result in the contamination of your water, soil, or air.

1. Batteries

Batteries, including those found in household electronics, vehicles, and backup power systems, can contain dangerous chemicals such as lead, lithium, cadmium, and mercury. When batteries are disposed of improperly, these chemicals can leach into the environment and contaminate water supplies, posing serious health risks to humans and wildlife. Common battery types that require special attention include:

- Lead-acid batteries (used in off-grid power systems, vehicles)
- Lithium-ion batteries (commonly found in electronics and solar storage systems)
- Rechargeable batteries (nickel-cadmium, lithium-ion, etc.)

2. Chemicals

Chemicals in household products such as paints, cleaning agents, solvents, pesticides, and fertilizers can pose a significant environmental threat. These substances are often toxic and, if not handled properly, can lead to contamination of soil and water. Some chemicals may also pose fire hazards or cause respiratory problems if inhaled.

3. Electronics (E-Waste)

E-waste, or electronic waste, includes discarded electronics like phones, computers, televisions, and appliances. These items often contain hazardous materials such as heavy metals (lead, mercury, cadmium) and plastics that, if not disposed of correctly, can pollute the environment and release toxic fumes.

Handling Hazardous Waste: Safe Storage, Transport, and Labeling Practices

Properly handling hazardous waste is crucial to avoid any harm to you, your family, and the environment. The first step in managing hazardous materials is understanding how to store, transport, and label them to prevent accidents and leaks.

1. Storage

Hazardous materials should always be stored in their original containers, if possible, and in a designated, secure area away from children, pets, and food storage areas. Containers should be tightly sealed to prevent leaks or spills. Storage areas should be cool, dry, and well-ventilated. Flammable materials should be stored away from heat sources, and chemicals that could react with one another should be kept separately.

2. Transport

When transporting hazardous waste for disposal or recycling, use appropriate containers designed to hold the materials securely. If transporting large quantities, use leak-proof containers and avoid overloading the vehicle. Always ensure that you comply with local regulations when transporting hazardous materials.

3. Labeling

Proper labeling is essential for safety. Clearly mark containers with labels that identify the contents and include any hazard warnings. Many local regulations require specific labels, so be sure to check what is legally required in your area.

Disposal Methods: Safely Disposing of Hazardous Materials

There are several options for safely disposing of hazardous materials. Never dispose of them in regular trash or pour them down drains, as this can lead to contamination of groundwater and soil.

1. Recycling Centers

Many areas have recycling centers that accept hazardous waste, including batteries and electronics. Recycling centers can safely process these materials and ensure that hazardous components are properly handled and that reusable materials are recovered.

2. Special Disposal Services

Some communities offer special disposal services for hazardous waste. These services often include periodic drop-off events where you can bring your hazardous materials to a centralized location for safe disposal. Make sure to take advantage of these services to ensure proper disposal.

3. Eco-Friendly Disposal Methods

In addition to traditional methods, consider eco-friendly alternatives for disposing of hazardous materials. Some companies specialize in environmentally conscious disposal of chemicals and electronics, using methods that minimize environmental impact. For example, some bioremediation techniques can be used to break down chemical waste without causing harm to the surrounding environment.

Recycling Batteries and Electronics: Safe Disposal of Lithium-Ion Batteries, Lead-Acid Batteries, and E-Waste

Batteries and electronics often contain valuable and reusable materials. Recycling these items not only prevents hazardous materials from entering the environment but also allows you to recover valuable components like metals and plastics.

1. Lithium-Ion Batteries

Lithium-ion batteries are commonly used in off-grid power systems, electric vehicles, and electronics. These batteries contain lithium, cobalt, and nickel, which can be toxic if not disposed of correctly. Many recycling centers accept lithium-ion batteries, where they are processed to recover these valuable materials. Before disposal, remove the batteries from any devices and ensure that they are in a safe, non-leaking condition.

2. Lead-Acid Batteries

Lead-acid batteries, often used in solar storage systems or vehicles, can be recycled to recover lead and sulfuric acid. Many auto shops and recycling centers offer a take-back program for used lead-acid batteries. Make sure to store the battery upright and avoid puncturing or damaging the casing to prevent leaks.

3. E-Waste Recycling

E-waste recycling involves the collection, sorting, and safe disposal of old electronics like phones, computers, and televisions. Many electronics retailers offer e-waste recycling programs, where you can drop off your unwanted electronics. These items are then sent to specialized recycling facilities where hazardous materials are safely processed, and reusable materials are recovered.

Legal Requirements and Regulations: Understanding Local and National Regulations for Hazardous Waste Disposal

It is essential to become well-acquainted with both local and national regulations surrounding hazardous waste disposal. These laws ensure that hazardous materials are disposed of safely and prevent harm to both the environment and human health.

Many local authorities have strict rules regarding the disposal of items like batteries, electronics, and chemicals. These regulations may include designated disposal centers, the types of materials that can be disposed of, and penalties for improper disposal. Check with your local environmental protection agency (EPA) or local government offices for more information on the specific rules in your area.

Environmental Impact: The Long-Term Effects of Improper Disposal

Improper disposal of hazardous waste can have far-reaching effects on the environment. When materials like batteries and electronics are disposed of in landfills or incinerated, toxic chemicals can leach into the soil and water. These chemicals can contaminate groundwater, harm wildlife, and pose serious health risks to humans.

Improper disposal of chemical products, such as pesticides or solvents, can lead to soil and water contamination. Over time, this contamination can affect local ecosystems and agriculture, making it important to manage these materials responsibly.

By following proper disposal methods and recycling whenever possible, you can significantly reduce the environmental impact of hazardous waste and contribute to a healthier planet.

22. Household Waste Management & Off-Grid Recycling

Living off the grid means becoming more self-sufficient and minimizing waste. Proper waste management and recycling are essential to reducing the environmental impact of your household. By incorporating sustainable practices into your daily routine, you can reduce the amount of waste you produce and increase the amount of waste you recycle.

1. Waste Reduction Strategies

One of the most effective ways to manage waste is to reduce the amount of waste you create in the first place. This can be achieved through sustainable practices such as conscious purchasing decisions, minimizing packaging waste, and choosing reusable products. By thinking critically about the items you buy, you can reduce the amount of waste that ends up in landfills.

2. Recycling Systems

Design a home recycling system for items such as paper, plastics, glass, and metals. Setting up dedicated bins for each material will make it easier to sort recyclables. For off-grid homes, you may need to take the recyclables to a local recycling center or find creative ways to reuse materials.

3. Composting

Composting is a sustainable way to turn food scraps and yard waste into nutrient-rich soil for your garden. Set up a compost bin or pile in a designated area of your yard. By composting organic waste, you not only reduce the amount of waste that goes to a landfill but also create valuable soil for gardening.

4. Repurposing and Upcycling

Repurposing and upcycling are creative ways to reuse materials instead of throwing them away. Whether you're turning old wood into furniture or using scrap metal for a garden trellis, there are countless ways to repurpose materials that would otherwise be discarded. Look around your home for items that can be transformed into something new and functional.

5. Waste Sorting and Collection

To reduce waste and improve recycling, set up dedicated bins for compost, recycling, and trash. Having clearly labeled bins for each type of waste will make it easier to sort materials and ensure that recyclable items are properly processed.

6. Sustainable Disposal Methods

For items that cannot be recycled or repurposed, work with local waste disposal services that prioritize eco-friendly solutions. If local services are unavailable, look into building your own solutions for dealing with waste that cannot be reused or recycled. Consider DIY incineration options or exploring other alternative waste disposal methods.

By incorporating these strategies into your off-grid lifestyle, you can contribute to a cleaner, more sustainable environment and reduce your reliance on external waste management services.

23. Waste & Recycling Solutions Projects

Project: Build a Composting Toilet System

In this project, you will design and construct a self-contained, dry composting toilet for your off-grid home. This system will allow you to process human waste without relying on water, septic systems, or municipal services. By using a combination of sawdust, wood chips, or other carbon-rich materials, you'll create an environmentally friendly waste management solution.

Materials Needed:

- Large container or bucket for the base (preferably a 5-gallon or larger container)
- Lid for the container (a tightly fitting lid to prevent odors)
- Compostable material (sawdust, peat moss, wood chips, or coconut coir)
- Toilet seat (can be a repurposed seat or a purpose-built composting toilet seat)
- Ventilation pipe (PVC or flexible tubing)
- Small fan (optional, to assist with ventilation)
- Drainage materials (optional, for urine diversion if desired)
- Tools for cutting and assembling (saw, drill, screws, etc.)

Step 1: Plan and Design Your Toilet System

Before you begin constructing your composting toilet, you need to plan the system's layout and decide on the materials.

- Choose a location for the toilet that ensures good ventilation and is away from food-growing areas to avoid contamination.
- Design the system to be self-contained, meaning the waste will be collected in a container that you can easily empty and compost.
- Consider the size of the container based on how many people will be using it and how frequently it needs to be emptied.

Step 2: Prepare the Composting Chamber

The composting chamber is where the waste will be collected.

- Start by selecting a sturdy container or bucket that can hold waste. A 5-gallon bucket or larger is a common size for a home system.
- Ensure the container is made of durable plastic or metal that can hold up over time and prevent leakage.
- Drill or cut a hole in the lid for a ventilation pipe. This hole should be large enough to fit the pipe securely.
- Attach a ventilation pipe to the lid, leading to the outside of the structure. This will help promote

airflow and reduce odors. A small fan can be added to the pipe if needed to encourage airflow.

Step 3: Create the Toilet Seat

Now, it's time to make the seat.

- You can either buy a pre-made composting toilet seat or repurpose an existing toilet seat.
- If you're making your own, secure the toilet seat to the top of the container, ensuring it is stable and comfortable to sit on. You can use wood or plastic to make a simple frame that supports the seat and aligns with the container's opening.
- Make sure the seat is easily removable so you can clean the container as needed.

Step 4: Set Up the Composting Materials

For the composting process to work efficiently, you'll need to add a carbon-rich material to each use.

- The composting process relies on balancing the nitrogen in human waste with carbon to create compost.
- Common materials used for this include sawdust, wood chips, peat moss, or coconut coir.
- After each use, add a generous amount of your chosen material to cover the waste. This will help absorb moisture, reduce odors, and promote decomposition.

Step 5: Install Urine Diversion (Optional)

If you want to make your system more efficient, you can add an option for urine diversion.

- Urine is high in nitrogen and can create odors or make composting slower if left mixed with solid waste.
- You can either use a separate container for urine or install a simple diversion system with a small pipe that channels urine into a different container for disposal or irrigation.
- Be sure to regularly empty the urine container to prevent it from overflowing.

Step 6: Test Your System

Once the system is assembled, it's time to test it.

- Place the composting toilet in its designated location and make sure the ventilation pipe is working properly.
- After each use, remember to add composting material and ensure that it is being decomposed properly.
- Check for any leaks, odors, or issues with the composting process, and make adjustments as needed.

Step 7: Maintain the System

The composting toilet requires regular maintenance to ensure it functions efficiently.

- Periodically check the composting chamber for signs of overfilling and empty it when necessary. You can process the compost by allowing it to mature for a few months before using it as fertilizer.
- Clean the container, lid, and toilet seat regularly to maintain hygiene and prevent the buildup of bacteria or mold.
- If odors become problematic, ensure that the ventilation is working well and that there's enough carbon material to absorb moisture.

With your composting toilet system in place, you've successfully created an eco-friendly solution for off-grid sanitation. This system helps you save water, reduce waste, and contribute to a more sustainable and self-sufficient lifestyle.

Project: Build a Greywater Recycling System

[Diagram: Graywater Sources flow into Biofilter Stage, then to Sand/Gravel Filter, then to Storage Tank, with Overflow and Irrigation Pipes outputs]

In this project, you will design and construct a simple greywater recycling system for irrigation purposes. This system will allow you to reuse water from household activities like showers, sinks, and laundry to irrigate your garden or landscape, reducing your water usage and improving sustainability.

Materials Needed:
- Piping (PVC or polyethylene tubing)
- Diverter valve or tee fitting for greywater diversion
- Sand for filtration
- Activated charcoal for filtration
- Mesh screens or fabric for pre-filtering larger particles
- Storage container or tank (large plastic or metal container)
- Gravel or small stones for filtration layers
- Bucket or basin for overflow collection (optional)
- Tools (saw for cutting pipes, drill for holes, etc.)
- Optional: small pump (for pumped systems)

Step 1: Plan Your System Design

Start by planning the layout of your greywater recycling system. Identify the sources of greywater in your home (showers, sinks, or laundry). Choose a location for the storage tank, preferably near the garden or irrigation area where you'll use the water. Consider the flow of the water, ensuring that you can divert it easily from the household plumbing to the greywater system. Draw a simple diagram of your system to map out the components like piping, filters, and storage.

Step 2: Set Up the Greywater Diversion

Install a diverter valve or tee fitting in the pipe leading from the greywater source (e.g., your shower or sink). This will allow you to divert the water from your household plumbing into the greywater system. The valve should be easy to operate so that you can switch between using the water in the system and using the regular plumbing.

If you're installing the system on a sink or shower, ensure that the diverter is easily accessible, and the greywater pipe leads to your filtration and storage area.

Step 3: Build the Filtration System

The key to reusing greywater safely is filtering out contaminants like soap, oils, and debris. Build a basic filtration system using sand, charcoal, and mesh to capture solids and chemicals.

- Start with a container that will act as the filter. You can use a large bucket or a plastic container for this.

103

- Add a layer of gravel or small stones at the bottom to help with drainage and prevent the finer filter materials from getting clogged.
- Place a layer of sand on top of the gravel. Sand will trap larger particles and help with basic filtration.
- Add a layer of activated charcoal to further filter out chemicals and impurities from the water. Charcoal is great for removing odors and organic material.
- Use mesh screens or fabric to filter out larger particles like hair or food scraps before the water enters the filtration container.

Step 4: Install the Storage Tank

Once the water passes through the filtration system, it should be stored in a tank or container. The storage tank should be large enough to hold the amount of greywater produced in a day, depending on how many people are using the system.

- Choose a durable container made of plastic or metal to prevent leaks and contamination. If you're planning to use the water for irrigation, make sure the tank is opaque to avoid algae growth.
- Connect the outlet from your filtration system to the storage tank, ensuring that the water is flowing freely and that there is no risk of contamination.
- If needed, you can add an overflow container or basin to catch any excess water if the storage tank becomes full.

Step 5: Set Up the Irrigation System

Now that the greywater is collected and stored, it's time to set up the irrigation system to use the water in your garden. You can either use a gravity-fed system (if your tank is elevated) or install a small pump to move the water.

- For a gravity-fed system, ensure that the tank is placed at a higher elevation than the irrigation area. The water will flow naturally through the pipes and into the garden.
- For a pumped system, install a pump that will move the water from the storage tank to the irrigation system. Connect the pump to the piping leading to the garden, and ensure that it is positioned to prevent backflow or contamination.

Step 6: Test the System

Once your system is assembled, it's time to test it. Turn on the diverter valve to start diverting greywater from your household plumbing into the system. Watch the water flow through the pipes, filters, and into the storage tank. Check for any leaks or blockages in the system.

Observe the water's flow through the filters and ensure that the filtration is working properly to remove debris and contaminants. If you're using a pump, test it to make sure it moves water efficiently to your irrigation system.

Step 7: Regular Maintenance

For your greywater system to function effectively, it requires regular maintenance. Check the filtration system every few weeks to clean out the sand and charcoal and replace the mesh filters. Ensure the storage tank remains free of algae and that the overflow basin is functioning as expected.

Clean the pump (if using) and inspect the entire system for leaks or blockages. Also, check the diverter valve to ensure that it is still operating smoothly.

With your greywater system in place, you've successfully created a simple greywater recycling system <u>for irrigation</u>. This system will not only save water but also contribute to a more sustainable lifestyle.

Project: Build a Small-Scale Biogas Digester for Home Use

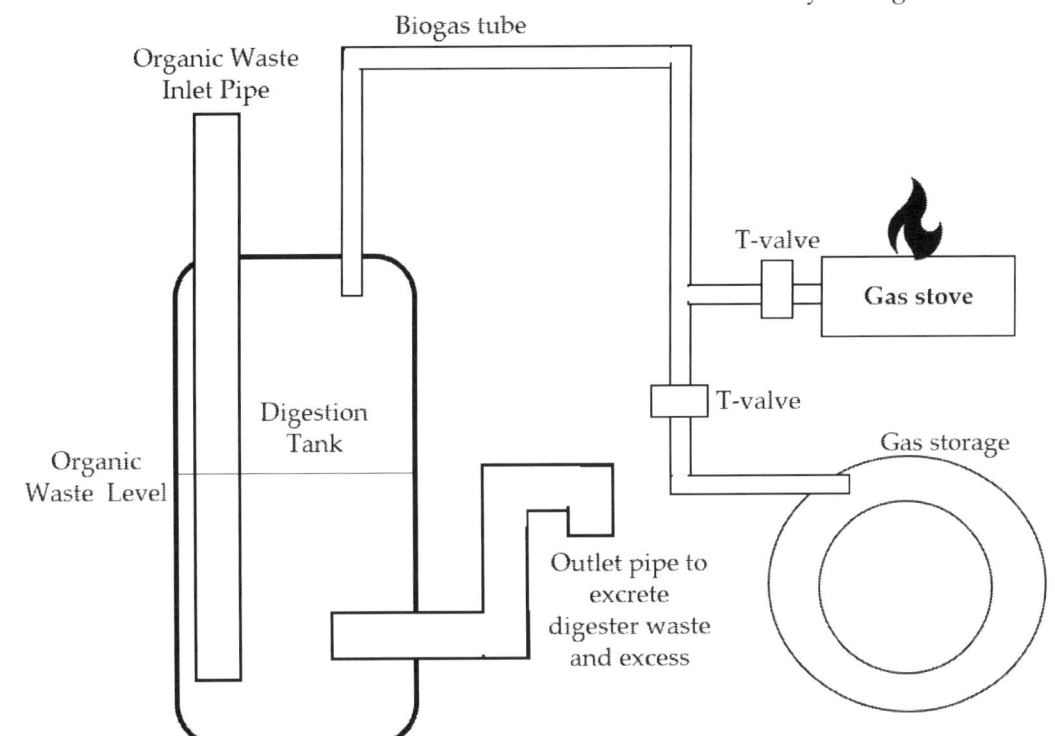

In this project, you will design and build a small-scale biogas digester that can convert household organic waste into usable biogas. This system will provide a renewable energy source for cooking or heating and reduce waste in your home. By following these steps, you will create a simple yet functional biogas system using affordable materials.

Materials Needed:
- Plastic container (large bucket or barrel) for the digester
- PVC pipes (for gas inlet and outlet)
- Flexible rubber hose (for gas collection)
- Sand, gravel, and charcoal for filtration layers
- PVC valve or valve connector
- Flexible plastic or rubber gas storage bag
- A biogas stove or small gas-powered generator (for biogas utilization)
- Tools: Saw, drill, scissors, wrench, and screwdriver
- Organic waste: Kitchen scraps, yard waste, or animal manure
- Water to adjust moisture levels

Step 1: Plan the System Layout

Before you begin construction, plan the design of your biogas system. Start by determining where you will place the digester, gas storage container, and gas utilization device (stove or generator). The digester should be placed in a location that is easy to access for waste input and maintenance, and it should have enough space around it for ventilation and safe gas storage. Sketch out the system, including all components like pipes, valves, and connections.

Step 2: Prepare the Digester Container

For the digester, you will use a plastic container, such as a large bucket or barrel. This container needs to be durable enough to withstand the pressure that will build up inside during gas production. Drill holes at the top of the container for the inlet pipe (where the organic waste will enter) and the gas outlet pipe (where the biogas will exit). Make sure these holes are tightly sealed once the pipes are in place to prevent gas leaks.

Step 3: Install the Inlet and Outlet Pipes

Cut PVC pipes to the appropriate lengths for the inlet and outlet. The inlet pipe should be long enough to allow for easy addition of organic waste. The outlet pipe will allow the gas to flow out of the digester and into the storage bag. Attach the pipes to the drilled holes using a sealant or waterproof adhesive to ensure a tight fit.

The inlet pipe should be positioned toward the bottom of the digester so the organic waste can sink, and the outlet pipe should be positioned near the top of the digester to allow for gas collection.

Step 4: Create the Gas Storage System

You will need a container to store the biogas once it's produced in the digester. This can be a flexible gas storage bag made from durable plastic or rubber. Connect one end of a flexible rubber hose to the gas outlet pipe on the digester, and attach the other end to the gas storage bag. Make sure the hose is secure and that there are no leaks.

The gas storage bag should be placed above the digester to allow the gas to flow naturally into the bag by gravity. Ensure that the storage bag can expand to accommodate the gas produced, and make sure it is airtight to prevent leaks.

Step 5: Set Up the Gas Utilization System

Once the biogas is collected in the storage bag, it can be used for cooking or heating. You can connect the gas storage system to a small biogas stove or a gas-powered generator. Attach the flexible gas hose from the storage bag to the gas input of your stove or generator. Ensure that the connection is secure, and use a regulator to control the flow of gas to the appliance.

Step 6: Add Organic Waste to the Digester

Now that the system is set up, begin adding organic waste to the digester. You can use food scraps from the kitchen, yard waste like leaves and grass clippings, or animal manure. Be sure to chop or shred larger items to help with the breakdown process. Avoid adding oils, grease, or chemical substances, as these can interfere with the digestion process.

Add water to the digester to help maintain the right moisture level for anaerobic bacteria to work. The ideal moisture level is around 60-70%. Stir the mixture gently to ensure that the waste is well distributed and that there is no clogging at the bottom.

Step 7: Monitor the Digester

After adding organic waste to the digester, check it regularly to ensure that the anaerobic digestion process is working. Over time, the organic waste will break down and produce methane gas. Monitor the temperature, moisture levels, and the flow of gas into the storage container. You should start seeing gas buildup after several days to a week, depending on the amount of organic material added and the temperature.

Step 8: Use the Biogas

As the gas builds up in the storage container, you can begin using it for cooking or heating. Open the valve connected to the storage bag and regulate the flow of gas to your stove or generator. The biogas should burn cleanly, and you will have a renewable source of energy for cooking or other household tasks.

Step 9: Maintain the System

Your biogas system will need regular maintenance to ensure it operates efficiently. Periodically check for leaks in the gas pipes and storage bag. If the gas production slows down, check the moisture levels and temperature in the digester. Add more organic waste as needed, and stir the contents to ensure proper digestion. You may also need to clean the filter or screen that prevents larger debris from clogging the pipes.

Step 10: Troubleshooting and Optimization

Over time, you may encounter issues such as a decrease in gas production or odor problems. If the system is producing little gas, check the temperature and moisture levels in the digester. You can also add more organic waste or adjust the balance of materials inside. If odors persist, inspect the seals and check for leaks, as improper sealing can lead to the release of unwanted gases.

Module E. Security & Property Defense

This module focuses on equipping you with the knowledge and tools necessary to safeguard your off-grid home and personal well-being from potential external threats. Whether you're concerned about intruders, natural disasters, or the challenges that come with living in a remote location, this module covers strategies for fortifying your property and implementing personal defense tactics. We'll explore ways to protect yourself and your family from harm, both inside your home and in the surrounding area, ensuring peace of mind and security in your off-grid environment.

24. Fortifying Your Property: Keeping External Threats at Bay

When living off-grid, one of the most important factors in ensuring your security and peace of mind is protecting your property from external threats. While remote locations offer many benefits, they can also leave you more vulnerable to intruders, wildlife, and even the unpredictable forces of nature. Fortifying your property involves creating multiple layers of defense—physical barriers, home fortification strategies, surveillance systems, and deterrents—all designed to keep unwanted visitors away while allowing you and your family to feel safe and secure.

Physical Barriers

A strong and well-planned physical barrier is often the first line of defense in keeping external threats at bay. The type of barrier you choose will depend on various factors such as the terrain, climate, and the level of security you require. Here's a closer look at the types of barriers that can be installed around your property.

Fencing: Types of Fencing (Wood, Metal, Electric)

Fencing serves as both a physical and psychological deterrent. It creates a boundary that is clear to intruders and can prevent casual entry onto your property. There are several types of fencing materials, each with its own benefits and considerations.

- **Wood Fencing**: Wood is a traditional fencing material that offers a natural aesthetic and privacy. However, it can be susceptible to weathering, rotting, and breakage. If you choose wood, make sure to maintain it regularly to ensure its longevity.
- **Metal Fencing**: Metal fences, such as chain link or wrought iron, provide durability and strength. They can withstand the elements and are difficult to breach. However, they offer less privacy than wood fences unless combined with other features like shrubbery or privacy slats.
- **Electric Fencing**: Electric fences are an effective way to deter intruders and animals without causing lasting harm. They provide a psychological barrier, as the risk of a mild shock often prevents people and animals from attempting to cross. Keep in mind that electric fences require maintenance to ensure they remain operational.

Each type of fence has its own strengths and weaknesses, so choosing the right one will depend on your specific needs and the level of protection you require.

Gates: How to Choose and Secure Gates for Better Entry Control

Gates are an important part of your perimeter defense system. They serve as controlled entry points to your property and should be designed to withstand external threats while allowing easy access for you and your family.

When selecting a gate, consider the following:

- **Material**: Like fencing, gates should be made from sturdy materials that will resist wear and tear. Metal gates, for example, are strong and difficult to tamper with, while wooden gates offer a more traditional aesthetic.
- **Locking Mechanism**: A high-quality lock is essential for keeping your gate secure. Consider a heavy-duty deadbolt lock or a combination of locks that will make it more difficult for intruders to gain access.
- **Automatic Gates**: In some cases, you may want to install an automatic gate that can be opened remotely. This is particularly useful for large

properties, allowing you to control access without leaving the security of your home.

Walls and Fortified Entrances: Designing and Building Physical Walls or Barriers Around Your Property

For an added level of security, consider installing walls around your property. Walls can be made from various materials, including brick, stone, or concrete, and they serve as a significant barrier to entry. Unlike fencing, which can be scaled or cut, walls provide a more formidable obstruction.

- **Height and Structure**: The height of your wall should be sufficient to prevent people from climbing over. A taller wall, combined with security features like barbed wire or spikes, can significantly reduce the likelihood of an intrusion.
- **Material Choice**: Concrete or stone walls are highly durable and difficult to breach. However, they may require more upfront investment in terms of both materials and labor. Choose a material that balances cost and effectiveness for your situation.

Landscaping for Defense: Using Natural Elements Like Thorny Bushes and Trees for Perimeter Defense

Landscaping can complement physical barriers by providing an additional layer of protection. Strategically placed trees, shrubs, and thorny plants can discourage intruders from attempting to approach your home.

- **Thorny Shrubs**: Planting thorny bushes like roses or hawthorn around your property's perimeter can create a natural barrier that is difficult and painful to navigate.
- **Trees**: Tall trees planted in strategic locations can obscure your home from view, offering both privacy and security. They can also act as a barrier, making it difficult for someone to approach your home undetected.
- **Dense Vegetation**: Dense shrubs and plants can create a thick wall of greenery that can prevent intruders from accessing your property. It's essential to maintain the vegetation to ensure it doesn't become overgrown and allow for easier entry.

Home Fortification

Once you have established external barriers, the next step is fortifying your home. The goal here is to make it difficult for anyone to gain unauthorized access to your living space. Here's how to enhance the security of your home.

Reinforcing Doors and Windows: Strengthening Doors and Windows with Security Bars, Locks, and Impact-Resistant Glass

The doors and windows of your home are the most vulnerable points of entry. It's crucial to fortify these entry points to make them resistant to forceful attempts to break in.

- **Security Bars**: Installing security bars on windows and doors can add an extra layer of protection. These bars should be strong and durable, ideally made from steel or another sturdy material.
- **Impact-Resistant Glass**: If you prefer to keep your windows clear, consider installing impact-resistant glass. This type of glass is designed to resist shattering, making it much harder for someone to break in.
- **Deadbolt Locks**: Invest in high-quality deadbolt locks for your doors. These locks are much more secure than standard latch locks, making it more difficult for intruders to pick or force open your doors.

Creating Safe Rooms: Designing a Secure Space Within Your Home for Emergencies

A safe room is a room within your home designed to protect you and your family in the event of an emergency. Safe rooms should be located in areas of your home that are easily accessible but not too obvious.

- **Room Location**: Ideally, a safe room should be in the center of your home, away from windows and doors. Basements or interior rooms work best.
- **Reinforced Walls**: To protect against forced entry, consider reinforcing the walls of the safe room with additional materials like concrete or steel.
- **Supplies**: Stock the safe room with essential supplies, including food, water, first aid kits, and communication devices like a phone or radio.

Hiding Entry Points: Concealing Windows and Doors to Make it Harder for Intruders to Access the Home

Concealing entry points can make it much harder for intruders to find a way into your home. By using creative design and landscaping, you can obscure your windows and doors, making them less obvious.

- **Window Treatments**: Use heavy curtains or blinds to prevent outsiders from seeing inside your home.
- **Camouflaged Doors**: In some cases, you may want to install concealed or camouflaged doors that blend in with the surroundings. This can be particularly useful for access points like emergency exits or hidden escape routes.

Surveillance Systems

Surveillance systems can provide real-time monitoring of your property and help deter intruders from attempting to gain access. Here's how to set up a surveillance system that's both efficient and reliable.

Security Cameras: Placement and Types of Cameras to Monitor Your Property

The key to an effective security camera system is placing cameras in strategic locations around your property. The goal is to monitor key areas while also ensuring that your cameras are well-hidden or discreet to avoid being tampered with.

- **Camera Types**: There are different types of security cameras, including wired, wireless, and solar-powered cameras. Choose cameras that suit your needs, whether it's constant monitoring or motion-activated recording.
- **Camera Placement**: Install cameras at entrances, gates, along the perimeter of your property, and in areas that are typically out of view. It's also wise to place cameras near high-value areas such as storage sheds or off-grid power systems.

Motion Detectors and Lights: How to Use Lighting to Deter Potential Threats

Motion detectors paired with lights are an effective deterrent against intruders. When someone approaches your property, the lights will automatically turn on, making it harder for them to approach without being noticed.

- **Placement**: Install motion sensors near gates, doors, and along pathways to detect movement. You can also use them around areas like tool sheds or other vulnerable points.
- **Lighting Options**: Use bright LED lights that will immediately alert you to movement on your property. Solar-powered lights are an excellent option for off-grid homes, as they require no wiring and are energy-efficient.

Remote Monitoring: Setting Up an Off-Grid Monitoring System Using Solar-Powered Cameras

For off-grid living, setting up a remote monitoring system that operates on solar power is an ideal solution. Solar-powered cameras can provide real-time surveillance without the need for a constant power source.

- **System Design**: Choose a solar-powered camera system with long battery life, high-resolution video, and remote access via smartphone or computer.
- **Backup Power**: Ensure your surveillance system has enough battery backup to function during cloudy days or in case of extended periods of low sunlight.

External Deterrents

In addition to physical barriers and surveillance, external deterrents play an essential role in keeping intruders at bay. These deterrents provide a psychological barrier, signaling that your property is protected and monitored.

Alarm Systems: Choosing and Installing Alarms That Notify You of Unauthorized Access

Alarm systems can notify you immediately if someone tries to enter your property. These alarms can be integrated into your security system and can be connected to your phone or a central monitoring service.

- **Types of Alarms**: There are various types of alarms, including motion detectors, door/window sensors, and break-glass alarms. Choose the ones that work best for your property layout.
- **Installation**: Install alarms on doors, windows, and any other entry points where an intruder may try to gain access.

Warning Signs: Using Signage to Discourage Intruders (e.g., "Protected by Surveillance")

Warning signs are a low-cost but effective way to discourage would-be intruders from attempting to enter your property. The presence of a sign indicating that the property is protected by cameras or alarm systems can deter criminals from approaching.

- **Sign Placement**: Place signs at visible entry points, including gates and entrances to your home.
- **Signage Options**: Use clear and noticeable signs that communicate the protection measures in place, such as "24/7 surveillance," "Alarm system in use," or "Protected by guard dogs."

Animal Deterrents: Using Guard Animals (Dogs, Geese) and Other Methods Like Motion-Triggered Noise Devices

Animals can be an excellent deterrent for keeping intruders away from your property. Guard dogs, in particular, are highly effective at alerting you to intruders and keeping people at a distance. Geese, too, can be surprisingly effective due to their loud calls and territorial nature.

- **Guard Dogs**: Train dogs to protect your property and alert you to any unusual activity. Their presence alone can be a strong deterrent for intruders.
- **Geese**: Consider raising geese as an alternative to dogs. Their loud honking can alert you to movement on your property.
- **Motion-Triggered Devices**: Use motion-activated noise devices like alarms or barking dog sounds to create the illusion of a protective animal presence.

Escape and Emergency Routes

In the event of an emergency, having multiple escape routes and an evacuation plan is crucial for the safety of everyone on your property.

Designating Multiple Exits: Planning Different Escape Routes for You and Your Family

Identify at least two or three escape routes from your home, ensuring that there are no obstacles or vulnerabilities along the way. These routes should lead to safe areas, such as a nearby forest, a neighbor's property, or an emergency shelter.

Concealed Paths: Designing Hidden Paths for Emergencies and Alternate Access Points

Hidden paths allow you to exit or enter your property without being noticed. These paths should be discreet and difficult for intruders to detect, giving you a means of escape if your main entry or exit points are compromised.

Safety Drills: Training Family Members for Potential Evacuation Scenarios

Regularly practice evacuation drills with your family members. These drills should include identifying escape routes, gathering necessary supplies (such as first aid kits and communication devices), and ensuring everyone knows what to do in case of an emergency.

25. Personal & Home Defense Strategies

Living off-grid offers you autonomy and freedom, but it also means taking responsibility for your safety and security. In a remote environment, you are more isolated, and that often comes with increased risks. Whether you're concerned about potential intruders, wildlife, or even unforeseen emergencies, having a solid home defense strategy can help you safeguard yourself, your family, and your property. This section will guide you through personal defense tools, training, legal considerations, home defense technology, and strategies to handle various threats that may arise.

Self-Defense Tools

In off-grid living, you need to be prepared for self-defense situations. Depending on your circumstances, you may choose from a variety of tools designed to protect you from threats. Whether you prefer non-lethal options or are considering firearms for home defense, knowing how to use these tools effectively and safely is key.

Non-lethal Weapons: Types of Non-lethal Weapons and Their Usage

Not all defense situations call for lethal force. Non-lethal weapons offer a way to protect yourself while minimizing harm. The right non-lethal weapon can provide you with an effective means to defend yourself and give you time to escape or secure help.

- **Pepper Spray**: One of the most popular non-lethal defense tools, pepper spray can incapacitate an attacker by causing temporary blindness, intense pain, and difficulty breathing. It's easy to use and can be carried in a pocket, purse, or on your keychain for quick access.
- **Stun Guns and Tasers**: Stun guns and Tasers are designed to incapacitate an attacker by delivering a high-voltage electric shock. The main advantage of these tools is that they don't require a high level of physical strength, making them accessible to people of all sizes. Stun guns are typically used in close-range encounters, while Tasers can incapacitate from a distance.
- **Batons**: A telescoping baton is a compact, durable weapon that can be carried easily. It is effective for close-range defense and can be used to disarm or disable an attacker. With proper training, a baton can be used effectively to protect yourself.

It's important to practice using these tools regularly to become proficient and ensure that you are able to use them effectively in stressful situations.

Firearms: Basic Knowledge of Firearms for Home Defense

Firearms are one of the most powerful self-defense tools, but they come with great responsibility. If you decide to use firearms for home defense, it's crucial to choose the right one, learn how to use it, and follow strict safety protocols.

- **Choosing the Right Firearm**: The best firearm for home defense depends on your personal needs and preferences. Shotguns are often favored for home defense because of their wide spread, making them effective even in close quarters. Handguns are more maneuverable and can be kept readily accessible. Rifles may offer longer range and higher accuracy, but they are less practical in confined spaces.
- **Safety Protocols**: Firearm safety is paramount. Always store firearms in a safe place, preferably locked away and separate from ammunition. Learn proper handling techniques, including how to load and unload, and familiarize yourself with safety mechanisms on your weapon. Practice shooting at a range to improve your skills and comfort level.

While firearms provide a strong line of defense, they also require respect and careful handling. Never point a firearm at someone unless you intend to use it, and always know what's behind your target to avoid accidental harm.

Improvised Weapons: Everyday Objects That Can Be Used for Self-Defense

Sometimes, you may find yourself in a situation where you don't have immediate access to your self-defense tools. In these moments, everyday objects can serve as effective weapons. From kitchen utensils to items found around your home, anything that can put distance between you and an attacker or incapacitate them can be useful.

- **Kitchen Knives**: A sharp kitchen knife is an effective tool for self-defense. If you're in a close-range encounter, you can use it to defend yourself or create an escape route. Always be mindful of how you handle and store knives to avoid accidental injury.
- **Fire Extinguisher**: A fire extinguisher can be used as a deterrent, both for its ability to cause confusion and disorient an attacker. The force of the spray is also strong enough to push someone away temporarily.
- **Heavy Objects**: Anything heavy and throwable, such as a vase, bottle, or chair, can be used to create distance and potentially incapacitate an attacker.

Being resourceful and thinking quickly in these situations can make all the difference. Learn to assess your surroundings and use whatever is at hand for your protection.

Training and Preparedness

No self-defense tool is effective if you don't know how to use it properly. Equally important is being mentally and physically prepared to handle high-pressure situations. This section covers the essential training and preparedness steps you should take to protect yourself and your family.

Personal Defense Training: How to Prepare Physically and Mentally for a Home Intrusion or External Threats

Personal defense training is not just about learning how to fight—it's about preparing yourself for any situation. This includes mental preparedness, physical conditioning, and building awareness of your surroundings.

- **Self-Defense Classes**: Take classes in martial arts or self-defense. These classes will teach you how to protect yourself, disarm an attacker, and use leverage to your advantage.
- **Situational Awareness**: Stay alert to your environment. Whether you're at home or in town, pay attention to who is around you and what is happening in your vicinity. Trust your instincts, and if something feels off, take precautions.
- **Physical Conditioning**: Building strength and endurance will improve your ability to respond to threats. Regular exercise can make a huge difference in a physical altercation, allowing you to move quickly and handle stress better.

Firearm Training: How to Safely Handle, Use, and Store Firearms for Self-Defense

If you choose to use firearms for home defense, it is absolutely critical to receive formal training in their safe handling, usage, and maintenance.

- **Training Classes**: Attend a firearm safety course where you'll learn how to operate firearms, how to aim, shoot, and reload safely, and how to handle a firearm in various situations. Many ranges also offer defensive shooting courses that simulate real-world threats.
- **Safe Storage**: Always store firearms in a locked safe that is inaccessible to children and unauthorized individuals. Keep ammunition separate from the firearm to ensure maximum safety.
- **Dry Fire Practice**: This technique involves practicing shooting without live ammunition. It allows you to develop muscle memory and improve accuracy.

First Aid Skills: Learning Essential First Aid Skills to Deal with Injuries During Emergencies

Whether it's a minor injury or something more serious, first aid skills are invaluable when it comes to home defense. You may need to address wounds during or after a defensive situation, or assist others who have been injured.

- **CPR and Basic First Aid**: Learn how to perform CPR, stop bleeding, treat burns, and deal with broken bones. Having these skills can make a significant difference in emergencies.
- **First Aid Kit**: Keep a well-stocked first aid kit in your home and car. It should include bandages, antiseptic wipes, gauze, pain relievers, and other essential medical supplies. Learn how to use each item effectively.

Defensive Tactics: Basic Combat Training for Close-Range Defense Situations

In close-quarters situations, understanding basic combat tactics can make a huge difference in your ability to defend yourself. Learn how to fight from a standing position, on the ground, and when you're armed.

- **Combative Techniques**: Take a basic self-defense or combat class. Techniques such as grappling, striking, and using leverage can help you defend yourself when there's no time to escape.
- **Weapon Defense**: Learn how to defend yourself if someone is attempting to take your weapon. This is an essential skill if you are armed and need to ensure that your attacker can't use your weapon against you.

Legal Considerations

Your right to self-defense is important, but it's also essential to understand the legal framework that governs the use of force, firearms, and surveillance in your area. Here's what you need to know.

Gun Laws: Understanding the Laws Regarding Firearms and Self-Defense in Your Area

Before acquiring firearms for home defense, make sure you're familiar with the laws surrounding gun ownership and self-defense in your jurisdiction. These laws vary widely from country to country and even from state to state.

- **Ownership Regulations**: Make sure you understand the legal requirements for purchasing, registering, and owning firearms in your area.
- **Use of Firearms in Defense**: Know when it's legally acceptable to use a firearm in self-defense. Some regions have strict laws regarding the use of deadly force, while others allow it under certain circumstances.

Use of Force: Learning When It's Legal and Ethical to Use Force in Self-Defense Situations

It's crucial to know the boundaries of self-defense. While you have the right to protect yourself and your family, there are limitations on when and how much force can be used.

- **Proportional Force**: The force you use should be proportional to the threat you face. You should only use deadly force if you believe your life is in immediate danger.
- **Castle Doctrine**: In some areas, the castle doctrine allows you to defend your home with deadly force if someone unlawfully enters. Be sure to understand the specifics of this law in your area.

Privacy and Surveillance: What You Can and Cannot Legally Monitor on Your Property

You have the right to monitor your property for security purposes, but there are legal limitations on surveillance. Know where you can place cameras and what you can monitor.

- **Camera Placement**: Ensure that cameras are focused on your property and not on neighboring properties, as this could violate privacy laws.
- **Audio Surveillance**: Many areas have strict regulations regarding audio recording, even if it's on your property. Make sure you're not inadvertently recording conversations without consent.

Community Support and Networking

In off-grid living, the safety of your property doesn't rely solely on individual efforts. Having a supportive community network can provide you with additional protection and resources in times of need.

Neighbor Watch Programs: Setting Up and Participating in Community Watch Programs

Creating a neighbor watch program can enhance security in your area. By keeping an eye out for suspicious activity and communicating regularly with neighbors, you can help protect your entire community.

- **Watch Group Meetings**: Organize regular meetings with your neighbors to discuss security concerns and share tips on how to improve safety in the area.
- **Shared Resources**: Pooling resources for surveillance or emergency

supplies can be beneficial. Neighbors can take turns monitoring the area or contribute to a shared emergency kit.

Bartering and Mutual Defense: How to Create a Network of People for Mutual Defense and Bartering in Emergencies

In off-grid living, having a network of people who can assist in emergencies is invaluable. This network can provide resources, defense, and support in times of crisis.

- **Bartering for Security**: Establish systems for bartering services and goods in exchange for security. For example, a neighbor may help with home defense while you provide them with food or tools.
- **Mutual Defense**: Work together with others to protect each other's properties. If there's a threat in the area, having a group of trusted individuals who can assist is crucial for ensuring everyone's safety.

Personal Protection Strategies

While physical barriers, weapons, and security systems are important, your awareness and ability to respond to threats are equally essential. Here's how to hone your personal protection strategies.

Situational Awareness: Developing Awareness of Your Surroundings

Situational awareness is one of the most important skills you can develop. By staying aware of your surroundings, you can detect threats before they become serious problems.

- **Being Observant**: Pay attention to the people around you and any unusual activity. When out in public or near your home, stay alert and avoid distractions, such as excessive phone use.
- **Trusting Your Instincts**: If something feels off, trust your gut. Don't hesitate to take precautions, such as moving to a safer location or contacting authorities.

Non-violent Conflict Resolution: Techniques for De-escalating Conflicts Before They Turn Violent

Not every conflict needs to escalate into violence. Learning how to de-escalate situations before they turn dangerous is a valuable skill.

- **Calm Communication**: Speak in a calm, steady voice, and use non-threatening body language to avoid provoking aggression.
- **Avoiding Provocation**: Recognize when someone is escalating a situation and attempt to remove yourself from it if possible. Sometimes, the best course of action is to walk away and diffuse the situation without confrontation.

Disguising Yourself: How to Blend into Your Surroundings

In some situations, blending into your environment is the best way to stay safe. Whether it's avoiding unwanted attention or escaping danger, being inconspicuous can be a lifesaver.

- **Clothing Choices**: Choose clothing that helps you blend into your surroundings rather than drawing attention to yourself. Avoid flashy or expensive-looking items.
- **Behavioral Cues**: Act calmly and confidently, avoiding actions that would make you stand out or look suspicious.

26. Security & Property Defense Projects

Project: Building Your Own Perimeter Fence

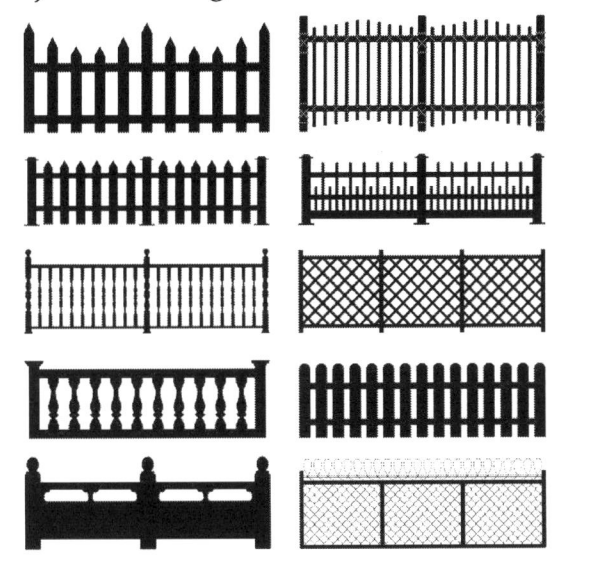

Step 1: Plan Your Fence Design
Start by assessing the size and layout of your property. Measure the perimeter to determine how much fencing material you will need. Decide on the type of fence you want based on your needs: a wooden fence for privacy, a metal fence for durability, or an electric fence for added security.

- For a wooden fence, you'll need wooden posts, boards, screws, and nails.
- For a metal fence, you'll need steel or wrought iron posts and panels.
- For an electric fence, you'll need electrified wire, insulators, a transformer, and grounding rods.

Step 2: Gather Materials
Once you've chosen the type of fence, gather all the materials required. Ensure that you have enough posts, fencing material, tools (like a post hole digger, hammer, and screwdriver), and safety gear (gloves, safety glasses).

Step 3: Mark the Fence Line
Using a measuring tape, mark where your fence will go around the perimeter of your property. Use stakes or spray paint to outline the fence path. This will help ensure your fence is placed in a straight line and within your property boundaries.

Step 4: Install the Posts
Dig holes for the fence posts, spaced according to the type of fence and local regulations. Typically, fence posts should be placed about 6-8 feet apart. Ensure the posts are securely anchored in the ground. You can use concrete or gravel to set the posts in place, especially for wooden or metal fences.

Step 5: Attach the Fence Material
Once the posts are set, attach the fence material to the posts. If using wooden boards, screw them into the posts. If using metal panels, fasten them with bolts. For electric fencing, install the wires according to the manufacturer's instructions, ensuring the wires are taut and elevated for proper function.

Step 6: Install Gates
Install gates at convenient entry points. Depending on your design, you may need to add latches or locks to secure the gates. Ensure that the gates open and close easily, and that they align properly with the fence to prevent unauthorized access.

Step 7: Check the Fence for Security
Once the fence is up, check for any gaps or weaknesses. Walk around the perimeter to make sure everything is tight and secure. If any sections are weak or loose, reinforce them.

Project: Reinforcing Your Home's Entry Points

Step 1: Assess Doors and Windows
Start by inspecting all doors and windows in your home. Look for any points that could be easily breached by an intruder. Check for weak or flimsy frames, locks, or glass. Ensure that your primary entry points are the most secure.

Step 2: Reinforce Doors
For each exterior door, consider adding security bars, a deadbolt lock, and an additional strike plate. Deadbolts are harder to break compared to regular door knobs, so installing one will increase the security of your entryways. If possible, install a security bar across the door or reinforce it with metal sheeting.

Step 3: Reinforce Windows
For windows, consider installing security bars or window locks that prevent them from being opened without a key. You can also replace regular glass with impact-resistant glass or add window film to reinforce the glass against shattering. Keep in mind that if you have a basement or lower-level windows, these should be particularly secure.

Step 4: Create a Safe Room
Designate a room in your home to serve as a safe room. This room should have minimal windows, a solid door, and be stocked with supplies such as water, food, first-aid kits, and a phone for communication. Install a reinforced door with a high-quality lock that can withstand forced entry. Consider adding a peephole or a small, fortified communication window.

Step 5: Conceal Entry Points
Try to make entry points less obvious to intruders. Conceal any windows or doors in areas that could be hidden from view, or install curtains or blinds that prevent outsiders from seeing inside. You can also use plants, screens, or trellises to obscure doors or lower-level windows.

Step 6: Install Motion-Activated Lights
Place motion-activated lights around your home, especially near doors and windows. These lights can startle an intruder and alert you to movement around your home. Position lights in places that give you clear sight lines while also providing illumination for your security cameras or spotlights.

Step 7: Regularly Inspect and Update Security
Make it a habit to check your doors and windows regularly for wear and tear. Over time, locks and reinforcement materials can degrade or loosen. Regular inspection allows you to fix issues before they become weaknesses in your home's security.

Project: Installing a Solar-Powered Home Defense System

Step 1: Plan Your Security Needs
Begin by assessing the areas around your home that are most vulnerable and require monitoring. This

might include the front and back doors, driveway, and any windows that are easy to access from the ground. Think about where you'd want to place security cameras and motion-activated lights. Make a list of these areas so you know exactly what equipment you'll need.

Step 2: Choose Your Solar-Powered Equipment

Once you've identified your security needs, choose the right equipment. Look for solar-powered security cameras with capabilities such as high-definition video resolution, night vision, and motion detection. For lights, consider using motion-activated solar lights, which will illuminate dark areas of your property when movement is detected. These solar lights are perfect for enhancing security and deterring intruders.

Step 3: Determine Where to Place Your Solar Panels

Solar panels need to be placed in an area that receives ample sunlight throughout the day to ensure they charge properly. Identify spots on your roof or along your property where sunlight hits the most, and ensure the solar panels are positioned to capture that sunlight. Avoid placing panels in areas where they may be obstructed by trees or other structures. The ideal placement would be a spot that receives direct sunlight for a majority of the day.

Step 4: Install the Solar Panels

Adhere to the manufacturer's instructions to mount the solar panels in the optimal location. Typically, solar panels are secured to a roof or a pole using mounting brackets. Use screws or other appropriate fasteners to secure them firmly in place. Ensure that the panels are angled correctly to receive the most sunlight possible, typically between 30-45 degrees depending on your location and the time of year.

Step 5: Mount the Cameras and Lights

Next, mount your solar-powered security cameras and motion lights in the predetermined locations. Secure the cameras in areas where they can cover large parts of the property, such as corners or higher elevations. Make sure the motion lights are positioned to illuminate areas like walkways or gates that you want to keep under surveillance. Use appropriate brackets and screws to ensure both cameras and lights are securely mounted.

Step 6: Connect the Solar Panels to the Cameras and Lights

Connect the solar panels to the cameras and lights using the provided cables. Depending on the system, you may need to connect the wiring to a central hub or directly to the devices. Ensure all cables are tightly secured and protected from the elements, especially if they run along the exterior of your home. Use cable clips or ties to keep everything organized and out of the way.

Step 7: Test the System

Once everything is connected and mounted, it's time to test your system. Turn on the security cameras and motion lights and walk around the areas they cover to check if they detect movement and activate as expected. Ensure that the cameras provide clear footage, both day and night. Adjust the settings on the cameras and lights for optimal performance, and make any necessary tweaks to the angles and sensitivity.

Step 8: Set Up Remote Monitoring

Many solar-powered security systems offer remote monitoring capabilities, which allow you to view footage from your cameras through a smartphone app or web browser. Set up this remote access, following the instructions provided with your system. This step will allow you to monitor your property even when you are away, ensuring that you always have access to real-time security information.

Step 9: Test the Entire System Regularly

Once installed and tested, it's important to regularly check the system's functionality. Test the cameras and lights periodically to ensure they continue to perform as expected. Clean the lenses of the cameras to ensure clear footage, and inspect the solar panels to ensure they are charging properly. If any issues arise, troubleshoot the system or consult the manual to resolve the problem.

Step 10: Review and Adjust

Over time, you may notice areas where the system could be enhanced. Consider adjusting the placement of cameras or adding additional lights to provide more coverage. Keep monitoring the system's performance, and make changes as necessary to keep your property secure.

Project: Building a Self-Defense Tool Kit

Step 1: Assess Your Personal Needs and Environment

Before you begin building your self-defense toolkit, take a moment to assess your needs based on your lifestyle and environment. Consider the areas you frequent and the potential threats you might face. For example, if you live in a rural area, animal encounters might be a concern, while if you live in an urban area, personal safety from human threats might be your primary focus. Reflect on whether you spend more time at home, in your car, or out in nature, and tailor your kit accordingly.

Step 2: Select Non-lethal Self-Defense Tools

Start with non-lethal tools that will allow you to defend yourself without causing long-term harm. Some of the most popular non-lethal weapons include pepper spray, personal alarms, and stun guns. Each tool has its specific purpose:

- Pepper Spray: Ideal for temporarily disabling an attacker by causing intense irritation to the eyes and respiratory system. Choose a compact, easy-to-carry model with a safety feature to prevent accidental discharge.
- Personal Alarms: These are small, portable devices that emit a loud, attention-grabbing sound when activated. They are ideal for alerting others in case of an emergency and can help scare off a potential attacker.
- Stun Guns: These work by delivering an electrical charge to the attacker, temporarily incapacitating them. Choose a model that is easy to operate and portable. It should also have a safety mechanism to prevent accidental activation.

Step 3: Add Improvised Self-Defense Tools

In addition to your primary self-defense tools, include everyday items that can be used for defense if necessary. The beauty of these tools is that they are discreet and multipurpose:

- Flashlight: A sturdy flashlight can temporarily blind an attacker when directed into their eyes. It can also be used to strike an attacker in an emergency. Opt for a tactical flashlight with a strong beam and durable construction.
- Keys: If you find yourself in close proximity to an attacker, your keys can be used as makeshift weapons. Hold a key between your fingers, and use it to strike pressure points or vulnerable areas of the body.
- Pen or Writing Instrument: A sturdy pen or similar object can be used for self-defense in a pinch. Hold it tightly in your hand and use it to target sensitive areas like the eyes, throat, or groin.

Step 4: Include a First Aid Kit

A self-defense toolkit is not complete without a small first aid kit to handle injuries that may occur during an altercation. Ensure your kit is compact and includes the following items:

- Bandages and gauze for wound dressing
- Antiseptic wipes to clean wounds
- Tweezers for removing splinters or debris
- Pain relievers such as ibuprofen or aspirin
- Antiseptic cream for infection prevention

Make sure to tailor the first aid kit to your specific needs and those of your family. If you are prone to specific health issues, consider adding medications or treatments you may require in an emergency.

Step 5: Choose a Sturdy and Portable Carrying Case

Once you've gathered the self-defense tools and first aid supplies, you'll need a compact and accessible way to carry them. Choose a carrying case that allows for quick access and can be easily carried in your bag, backpack, or car. Look for a case that is durable, waterproof, and has designated compartments to keep your items organized.

Step 6: Familiarize Yourself with Each Tool

It's not enough to simply have these tools in your kit; you need to be comfortable using them effectively. Spend time practicing with each item, so you're prepared if an emergency arises. Practice using pepper spray at a safe distance and learn how to properly hold and discharge it. If you have a stun gun, familiarize yourself with how it works and the areas on the body that will incapacitate the attacker most effectively. Practice using your flashlight as both a light source and a weapon in different scenarios.

Step 7: Learn Self-Defense Techniques

While tools can be extremely effective, learning basic self-defense techniques can further empower you to defend yourself in an emergency. Consider taking a self-defense class that teaches simple moves such as how to break free from holds, defend against grabs, and target sensitive areas of the body. Many classes also cover how to handle stressful situations and how to remain calm when facing a threat.

Step 8: Store Your Self-Defense Tool Kit in Strategic Locations

It's essential to keep your self-defense kit in locations where it can be easily accessed when needed. If you're at home, store your toolkit near your bedroom or front door. When traveling, keep it in your car or carry it with you in your bag or backpack. If you're outdoors or on a hike, make sure it's in a place where it's within reach at all times.

Step 9: Review and Update Your Kit Regularly

A self-defense toolkit is something that should be reviewed and updated regularly. Check the expiration dates on items such as pepper spray, medications, and first aid supplies. Make sure your stun gun or flashlight batteries are charged and operational. Swap out items as necessary and replace anything that has been used.

Step 10: Share Your Plan with Trusted Individuals

If you live with others, it's important to share your self-defense kit and plan with trusted family members or housemates. Ensure they know where the kit is stored and how to use the tools effectively. This is especially important if there are multiple people in your home who may need access to the kit in an emergency.

Project: Designing and Implementing an Emergency Evacuation Route

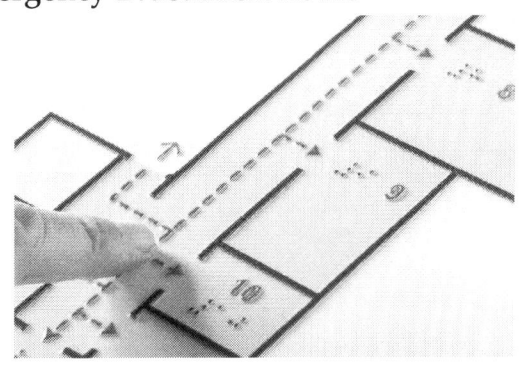

Step 1: Assess the Property Layout

Start by reviewing your property layout, both inside and outside the home. Identify all possible exits from the house such as doors, windows, and emergency exits. Map out the entire property, noting areas where you might need to escape in an emergency, such as rooms with limited access or upper floors that may require an alternative route. If you are living in a multi-story home, plan stairways, escape ladders, or any alternate exits from higher levels.

Step 2: Identify Primary and Secondary Escape Routes

For each room in the house, designate at least one primary and one secondary escape route. Primary routes should be the quickest and most accessible paths, such as doors. Secondary routes may be used in the event that the primary route is blocked. For example, if your main entry is blocked during a fire, windows or back doors should be your backup. Ensure that both routes are clear of obstructions and easy to open or unlock.

Step 3: Create Concealed Escape Paths

Next, think about areas that could be used to hide or disguise your escape routes. Hidden paths or concealed entry points can be crucial during a crisis, as they help protect against intruders. Design paths behind hedges, fences, or gardens that will provide safe and discreet routes. These routes can also be less obvious to an intruder who may be trying to follow you.

Step 4: Install Safety Features Along the Escape Routes

Ensure that your planned routes are safe to travel during an emergency. This might include installing exterior lighting along pathways, clearing away bushes, or ensuring there are no large obstacles that could cause tripping or block an exit. If you live in an area with snow or ice, consider adding de-icing solutions to prevent slipping. Make sure gates or doors along these routes can be opened easily and are not blocked by overgrown plants or equipment.

Step 5: Designate a Safe Assembly Point

Once everyone in the house has evacuated, they need a safe place to meet. This assembly point should be far enough away from the house to avoid any immediate danger, such as falling debris or fire. Choose a location that everyone can reach quickly, such as a neighbor's house, a nearby landmark, or even an open area on your property where you can gather and account for everyone.

Step 6: Install Emergency Signage

To make sure that everyone knows where to go, install clear emergency exit signs near exits, in hallways, or inside rooms. Use bright, reflective materials so that these signs are visible in low-light situations. You can also label escape routes or doors that lead to exits with simple, easy-to-understand symbols to help everyone navigate during an emergency.

Step 7: Practice the Escape Routes

Hold regular drills with your family to ensure that everyone knows the escape routes and how to reach the safe assembly point. Practice under different scenarios, such as fire, natural disaster, or intruder threats. Make sure that children and elderly family members can navigate these routes comfortably. Discuss the escape routes with everyone in your household so that each person understands where to go and what to do in case of an emergency.

Step 8: Evaluate and Adjust the Plan Regularly

After each drill, evaluate how well the escape routes worked. Were there any obstacles or difficulties in accessing certain areas? Did everyone reach the safe assembly point within a reasonable time? Make adjustments to your plan as needed. This may involve widening pathways, adding more visible signs, or creating alternative routes in areas that might be prone to flooding or other natural events.

Step 9: Review Safety Equipment

Ensure that you have any necessary equipment along these routes, such as fire extinguishers, flashlights, or first aid kits. Place these items near key exits or in a central location to be grabbed quickly. Test all safety equipment to make sure it is functional and ready for use.

Step 10: Share the Plan with Others

If you live with others or have visitors who might need to evacuate with you, ensure that everyone is aware of the emergency evacuation plan. Show them the designated escape routes and explain where the assembly point is located. If possible, distribute a printed copy of the emergency plan with clear instructions on how to evacuate. Make sure that everyone in the household understands the plan and is confident in using the routes during an emergency.

Module F. Sustainable Food & Self-Sufficiency

In an off-grid lifestyle, food self-sufficiency is one of the cornerstones of independence. Growing your own food, foraging, hunting, and fishing, combined with efficient food preservation and storage techniques, can ensure that your food supply remains reliable, sustainable, and free from external dependencies. This module will guide you through essential practices for cultivating a diverse and productive food system, from the basics of off-grid farming and gardening to more advanced methods like hydroponics and aquaponics. You will also explore sustainable techniques for harvesting wild food, preserving your harvests, and designing efficient food systems that thrive in off-grid environments. Whether you are just starting your journey toward food self-sufficiency or looking to improve your existing methods, this module will provide the tools and strategies you need to build a resilient and abundant food supply.

27. Growing Food: Off-Grid Farming & Gardening Techniques

Soil Preparation and Fertility

The importance of soil health in off-grid farming cannot be overstated. Your soil needs to be able to hold nutrients, water, and provide a stable environment for your plants to grow. Soil fertility is the lifeblood of your garden. It's what enables plants to access the necessary nutrients, oxygen, and water, ensuring that they can grow healthy and strong. But soil isn't something that just automatically happens; it requires your attention to maintain it year after year.

There are different types of soil, and each has its own properties that will influence what and how you plant. The most common types are loam, clay, and sand, and you should familiarize yourself with these before deciding how to amend and improve your soil.

Loam soil is the ideal soil type because it holds nutrients well, has good drainage, and is easy to work with. However, not all soil will be loam, so if you have sandy or clay-heavy soil, you'll need to take extra steps.

Clay soils tend to retain water and can become compacted, which restricts the plant roots' access to oxygen. Improving clay soil often involves adding organic matter to break it up and improve drainage.

Sandy soils, on the other hand, drain too quickly and don't hold nutrients well. If you have sandy soil, you'll need to improve it by adding organic matter to increase its nutrient-holding capacity.

Soil testing kits are an excellent first step in understanding your soil's health. These kits will help you determine its pH, nutrient levels, and overall fertility. Once you know what you're working with, you can begin improving it.

The next step is to add organic matter. This is where **composting**, **mulching**, and **natural fertilizers** come into play. Composting is a method of recycling plant and kitchen waste to create rich, organic soil that's full of nutrients. Composting works by breaking down organic materials, like food scraps, leaves, and grass clippings, into a rich soil amendment that improves soil texture, adds nutrients, and enhances water retention.

Mulch helps retain soil moisture, suppress weeds, and, over time, enriches the soil as it breaks down. Whether you use straw, leaves, or wood chips, mulching will help keep your soil temperature stable and protect it from the harsh elements.

Natural fertilizers, such as fish meal, bone meal, or well-rotted manure, can be applied to further boost soil fertility. These organic alternatives to synthetic fertilizers will not only help your plants grow strong but will also improve the long-term health of your soil.

Planting Techniques

When it comes to planting, you don't just throw seeds into the soil and hope for the best. Instead, you should carefully consider the best planting techniques to ensure your crops grow to their fullest potential. For a truly sustainable off-grid farm, companion planting and crop rotation should be staples in your gardening practices.

Companion planting is the practice of growing certain plants together because they benefit each other. For example, planting **tomatoes** alongside

basil can help deter pests and enhance flavor. Certain plants like **beans** can enrich the soil by fixing nitrogen, a nutrient that many plants need to grow. By planting complementary species next to one another, you can create a healthy, balanced ecosystem that minimizes the need for pesticides or fertilizers.

Crop rotation is a method that helps maintain soil health by planting different crops in the same space year after year. Different plants have different nutrient needs, and rotating your crops helps prevent soil depletion. It also helps disrupt pest cycles, reducing the chance of disease buildup in your soil. For example, one season you might plant legumes (which add nitrogen to the soil), followed by brassicas (which can take up the nitrogen left by the legumes).

Another essential aspect of planting is whether you should use **direct seeding** or **transplants**. For many crops, direct seeding works best because it encourages stronger root systems. However, for plants like **tomatoes** and **peppers**, it's better to start with transplants (young plants started indoors) to get a head start in the growing season.

Watering Systems

Watering your crops effectively is one of the most important aspects of gardening. Since you're living off-grid, it's essential to design a system that conserves water while still ensuring that your plants thrive. Traditional irrigation methods can be wasteful, but with the right techniques, you can minimize water use and maximize plant health.

Drip irrigation is one of the most efficient watering systems for off-grid gardening. This system delivers water directly to the base of the plants, minimizing evaporation and runoff. You can install a drip irrigation system with inexpensive materials like tubing, emitters, and a water source, such as a rainwater harvesting system.

Speaking of **rainwater harvesting**, this method collects rainwater from rooftops or other surfaces, stores it, and then channels it into your garden. Rainwater harvesting systems can be simple or complex depending on the amount of water you need. In dry climates, rainwater can be a lifeline for your crops.

Designing an efficient **irrigation system** is key to water conservation. You should consider the size of your garden, the type of crops you are growing, and your local climate. Even in a well-designed system, be sure to adjust the watering times for the seasons and check for leaks regularly.

Pest Control

Pests can wreak havoc on your garden, but as an off-grid gardener, you want to avoid using harmful chemicals. Fortunately, there are many **organic and natural pest management strategies** that can help protect your crops.

Beneficial insects, such as ladybugs, lacewings, and predatory beetles, are natural predators of common garden pests. By introducing these insects into your garden, you can naturally reduce pest populations without resorting to pesticides. Additionally, **planting pest-repellent plants**, like **marigolds** and **basil**, around your crops can help keep pests at bay.

Creating a **balanced ecosystem** in your garden is crucial. When you build a garden with multiple layers of plants—such as ground cover, flowers, herbs, and taller plants—you create a habitat that naturally supports biodiversity. This way, the garden can balance pests and predators, helping you avoid pest outbreaks.

Crop Selection

When choosing crops, it's vital to take into account your climate, soil, and how much sunlight your garden gets. Off-grid gardeners need to focus on crops that are easy to grow and can be harvested efficiently.

Selecting crops suited for your climate and soil is key to successful gardening. If you live in a cold climate, choose cold-hardy crops like **kale**, **spinach**, and **root vegetables**. In hot climates, consider heat-tolerant crops like **tomatoes**, **squash**, and **peppers**.

You'll also need to decide how to plant for **year-round food production**. This can involve **season extension techniques**, such as using row covers, cold frames, or greenhouses, to grow food through colder months. Additionally, incorporating **perennial crops**, like **asparagus**, **rhubarb**, and **fruit trees**, into your system can ensure that you have a steady food supply every year.

Integrating perennial and annual crops into your food system creates stability and diversity. Perennials provide reliable harvests, while annuals give you the flexibility to plant different crops every year.

28. Fundamentals for Raising Livestock Off the Grid

Choosing the Right Livestock for Off-Grid Living

Raising livestock on an off-grid homestead offers several benefits—sustainable food sources, waste management, and even labor. However, the first and most crucial step in building a successful off-grid livestock system is selecting the right animals for your environment. Different livestock has specific needs regarding climate, space, feed, and maintenance, so understanding these factors is key to long-term success.

Factors to Consider:
- **Climate:** The environment you live in directly impacts the animals you can raise. Cold climates may necessitate more shelter and winterizing strategies, while warmer climates might require ample shade and water. For example, chickens and ducks are hardy and adaptable to various climates, but goats and pigs, while resilient, may require more careful protection in extreme weather conditions.
- **Land Size:** Different animals require varying amounts of space. Larger animals, such as cattle and pigs, need ample grazing land, whereas smaller animals like chickens, rabbits, and goats are better suited for smaller areas, especially when rotated on pasture. When considering land size, keep in mind the long-term sustainability of the land and its ability to regenerate or provide continuous resources.
- **Feed Availability:** One of the most important factors in raising livestock off-grid is the availability of food. Can you produce enough feed on your land, or will you need to rely on outside sources? Animals like chickens and rabbits can thrive on foraged food and scraps, whereas cattle and pigs may require additional supplemental feed, particularly in winter.
- **Purpose:** Each animal serves a specific function, whether it's meat, eggs, milk, or wool. Some animals also provide other benefits, such as pest control (chickens for bugs), composting (pigs and chickens), or land maintenance (goats for clearing brush). It's important to select animals based on your homestead's goals.

Common Off-Grid Livestock:
- **Chickens:** Chickens are one of the most popular livestock for homesteads. They provide eggs, meat, and natural pest control. Chickens can be kept in smaller spaces, and their manure can be used to fertilize gardens. A simple chicken coop with a run is often all they need.
- **Ducks & Geese:** Ducks thrive in wet environments, so if you have ponds or streams, they may be ideal. They produce eggs, meat, and can help control pests like snails or slugs. Geese are great for larger properties, helping to clear grass and maintaining land health.
- **Rabbits:** Known for their high reproductive rates, rabbits are ideal for small-scale homesteads. They provide meat and fur, and are easy to breed. They can also be housed in small hutches or pens, making them ideal for limited spaces.
- **Goats:** Goats are incredibly versatile. They can be raised for milk, meat, or fiber (like wool). They also help with brush control, making them great for overgrown land. Goats require sturdy fences and a safe shelter to protect them from predators.
- **Sheep:** Like goats, sheep offer wool, meat, and milk. They are ideal for pasture management and are often raised in rotational grazing systems to prevent overgrazing.
- **Cattle:** Cattle require substantial land and resources but are invaluable for larger homesteads. They provide meat, milk, and manure for composting, and can even assist with draft work like pulling carts or plows. Cattle thrive in a rotational grazing system to prevent pasture degradation.
- **Pigs:** Pigs are excellent for waste management, land clearing, and providing meat. They thrive on food scraps, which makes them ideal for homesteads that generate a lot of kitchen waste. They do require strong fencing, as they are notorious for escaping.
- **Bees:** Bees are essential for pollination, which boosts crop production. In addition to helping your

garden thrive, bees provide honey, beeswax, and propolis, all of which can be used in various products or consumed.

Housing and Shelter Requirements

Once you've chosen the right animals, ensuring that they have adequate shelter is crucial for their well-being. Proper housing protects animals from extreme weather conditions, predators, and ensures that they remain healthy and productive.

Chicken Coops and Runs:
- Chickens require a coop that provides both shelter from the elements and space for laying eggs.
- The coop should allow for proper ventilation to prevent respiratory issues.
- Use predator-proofing materials such as hardware cloth, not chicken wire, and make sure that the doors and windows are secure.
- Consider a deep litter system to make use of the bedding for composting.

Rabbit Hutches:
- Rabbits should be kept in elevated hutches to prevent them from coming into contact with damp soil or predators.
- Insulate hutches during extreme cold weather to provide adequate protection.
- Some rabbits do well in colony-style pens with ample space for movement.

Goat and Sheep Shelters:
- Goats and sheep need shelter from harsh winds and extreme temperatures.
- Provide windbreaks using natural barriers or constructed walls.
- Ensure access to clean bedding, especially in cold weather.

Pig Pens:
- Pigs require sturdy pens to prevent escape, with high fencing and deep walls.
- They need access to wallows to cool down during hot weather, so a water source is essential.
- Provide adequate shelter during the winter months, such as a small barn or covered area.

Cattle Barns and Pastures:
- Cattle need a rotational grazing setup to avoid overgrazing and to allow the pasture to regenerate.
- In colder climates, cattle require some form of shelter, although they can tolerate exposure to mild weather conditions.

Beekeeping Hives:
- Beehives should be placed in sheltered areas with good airflow, away from direct sunlight.
- Regular inspection and maintenance are necessary to avoid diseases and pests.

Feeding and Nutrition for Sustainable Livestock Keeping

Sustainable livestock keeping relies on understanding and providing for the nutritional needs of your animals. The right balance of foraging, supplemental feeding, and DIY feed production ensures that your livestock remain healthy and productive throughout the year.

Foraging vs. Supplemental Feeding:

Foraging can be a highly sustainable method for feeding animals. Animals like goats, sheep, and cattle naturally graze, while chickens and ducks forage for insects and plants. However, supplemental feeding may be necessary, particularly in winter or when pasture availability is low.

DIY Animal Feed Production:

Growing your own animal feed is an excellent way to reduce reliance on external resources. Consider growing forage crops like clover, ryegrass, and alfalfa. Root crops such as carrots and turnips are excellent feed for rabbits and pigs. Additionally, fermenting grains like oats and barley improves their digestibility and provides additional probiotics for your animals.

Grass-Fed vs. Grain-Fed Approaches:

Grass-fed livestock are typically healthier, as they are consuming their natural diet. Grain-fed livestock, on the other hand, often grow faster and are sometimes easier to manage in confined spaces.

However, grass-fed animals require more land and space, which may not be feasible in smaller homesteads.

Breeding and Reproduction Management

Breeding livestock for self-sufficiency requires careful management. Understanding natural breeding cycles and implementing selective breeding practices ensures a steady supply of animals and promotes genetic diversity.

Understanding Natural Breeding Cycles:

Livestock such as goats, pigs, and cows have distinct breeding cycles. Cattle and pigs often breed year-round, while goats and sheep may be seasonal breeders. Understanding these cycles allows you to plan breeding schedules and optimize reproduction for your homestead needs.

Selective Breeding for Self-Sufficiency:

Selective breeding involves choosing animals based on desirable traits such as hardiness, disease resistance, and production (meat, milk, etc.). For example, breeding goats with higher milk yields or cattle with greater disease resistance can improve the overall health and productivity of your herd.

Artificial Insemination vs. Natural Breeding:

Natural breeding involves allowing animals to mate naturally, but artificial insemination is becoming more common, particularly in cattle and dairy animals. Artificial insemination allows for better control over genetics, improving traits like milk production and disease resistance.

Managing Birthing and Newborn Care:

Birthing and caring for newborn animals requires special attention. For example, chickens need warmth and food immediately after hatching, while goats and pigs require a clean and safe environment for birthing. Proper management ensures the survival of the offspring and the health of the mother.

Processing Livestock for Food and Other Uses

When you decide to raise livestock on your homestead, one of the primary goals is to utilize every part of the animal to maximize sustainability. Processing livestock for food and other uses—such as wool, leather, or byproducts like manure—is crucial for self-sufficiency. Every aspect, from the initial slaughter to the preservation of meat, is vital to ensure the best utilization of resources.

Egg Collection, Storage, and Incubation for Self-Sustaining Poultry Flocks

Egg production is a crucial source of food on an off-grid homestead. To maximize the benefits from your poultry, it's important to implement effective egg collection and storage methods.

- **Egg Collection:** Collect eggs daily to avoid damage or contamination. Use nesting boxes in your chicken coop, with clean bedding and privacy to encourage laying. Handle eggs gently and store them carefully.
- **Storage:** Eggs should be stored in a cool, dry place, ideally between 50-60°F (10-15°C). For long-term storage, consider coating eggs with mineral oil to extend freshness without refrigeration.
- **Incubation:** To maintain a self-sustaining flock, incubating eggs allows for continuous reproduction. Use an incubator or a broody hen, maintaining a temperature of 99.5°F (37.5°C) with proper humidity. Ensure proper turning of eggs for uniform development.

Milking Techniques and Processing Dairy Products

For goats, cows, and other dairy animals, milking is essential for providing food on a homestead.

- **Milking Techniques:** Milking should be done twice daily. Clean the teats and udder before milking to prevent contamination. Use clean, sanitized equipment to avoid infecting the milk.
- **Processing Dairy Products:** Once milk is collected, it can be used for cheese, butter, and yogurt. For cheese, pasteurize the milk before curdling. Make butter by churning cream until it separates. To make

yogurt, heat milk and add a starter culture, incubating it for a few hours.

Meat Processing and Preservation Techniques

Ethical and efficient slaughtering and processing of meat animals are key to utilizing them fully.

- **Ethical Slaughter Methods:** The method of slaughter should minimize pain and stress. A sharp knife is used for a quick, clean cut to the jugular vein, ensuring the animal is unconscious quickly.
- **Butchering Techniques:** Once slaughtered, the animal is butchered by removing organs and skin. For larger animals like cattle or pigs, the meat is broken down into cuts like roasts or steaks. Smaller animals like chickens or rabbits require simpler butchering methods.
- **Preserving Meat:** Meat preservation includes smoking, curing, and drying. Smoking adds flavor while preserving the meat. Curing involves salting or brining to draw out moisture, and drying removes moisture to prevent bacterial growth, making meat last longer.

Wool and Fiber Production

Raising sheep or fiber animals like alpacas can provide valuable materials for clothing and textiles.

- **Shearing:** Sheep are sheared annually in spring before the heat of summer. This is done with clippers or shears, and the fleece is carefully removed to avoid damage.
- **Cleaning:** Wool needs to be cleaned to remove dirt, lanolin, and grease. Wash it with warm water and a mild detergent.
- **Spinning and Weaving:** Once cleaned, wool is spun into yarn and woven or knitted into garments or other textiles. This adds self-sufficiency and versatility to your homestead.

Tanning Hides for Leather Production

Utilizing leather from animals like cows, goats, or pigs is a valuable skill.

- **Skinning and Preparing the Hide:** After slaughter, the hide is carefully removed and scraped to remove fat and muscle. It is then soaked in water to soften it.
- **Tanning Process:** Hides can be tanned using tannin-rich solutions like oak bark or by using chemical tanning agents. The hide is soaked in the solution, then worked by hand to soften it into usable leather.

Protecting Livestock from Predators and Harsh Weather

Predators and extreme weather conditions can pose significant threats to your livestock. Developing strategies to protect them ensures their survival and well-being.

Common Predators and Deterrents

Livestock are vulnerable to a variety of predators, including foxes, coyotes, raccoons, and even hawks or feral dogs. To protect your animals:

- **Secure Fencing:** Install strong fencing that is tall enough to deter predators from jumping over and buried deep enough to prevent digging under.
- **Electric Fencing:** Electric fences can be an effective deterrent for larger predators like coyotes or raccoons.
- **Guard Animals:** Consider using guardian animals such as dogs, donkeys, geese, or llamas. These animals can help protect your livestock by alerting you to predators or directly deterring them.

Weather Protection Strategies for Extreme Cold, Heat, and Storms

Extreme weather conditions, whether intense heat or freezing cold, require specialized protection for your animals.

- **Cold Weather Protection:** In colder climates, provide insulated shelters for animals like goats, sheep, and pigs. Use deep bedding, such as straw, to keep the animals warm. Windbreaks or barn structures protect against icy winds.
- **Heat Protection:** During hot weather, ensure that animals have plenty of shade and access to fresh water. Use fans or natural breezes to cool down animal pens, and ensure there is ample ventilation.
- **Storm Protection:** In areas prone to storms, create sturdy shelters and secure fencing. Keep animals in safe, enclosed areas during heavy winds or flooding, and ensure they have a dry, comfortable space.

Manure Management and Waste Utilization

Efficient manure management is essential for maintaining a clean, sustainable homestead and for improving soil health. Livestock manure can be a valuable resource for fertilization and biogas production.

Composting Livestock Manure for Soil Fertility

Manure can be composted to create nutrient-rich soil that can be used for gardening and crop production.
- **Composting Process:** Collect manure in compost bins or piles, and allow it to break down into humus over time. Regular turning of the pile ensures proper aeration, speeding up the composting process.
- **Using Manure in Gardens:** Once composted, manure can be spread over fields or gardens to enhance soil fertility, encouraging better plant growth.

Using Animal Waste for Biogas Production

Biogas is an efficient energy source that can be derived from livestock waste, primarily from cattle, pigs, and poultry.
- **Biogas System Setup:** Set up a sealed tank or digester where manure can be broken down anaerobically to produce methane gas. This gas can be used for cooking or heating.
- **System Maintenance:** Regularly maintain the biogas system by ensuring that the tank is kept sealed, and the waste is periodically added. This system helps reduce reliance on external fuel sources.

Deep Litter Systems for Chickens and Pigs

A deep litter system is an effective way to manage waste and produce compost.
- **System Setup:** Place bedding like straw or wood shavings in the animal pens. As animals defecate, the bedding absorbs the waste. Over time, the bedding breaks down, turning into compost.
- **Benefits:** This system reduces the need for frequent cleaning, helps maintain a clean environment for the animals, and produces valuable compost for gardens.

29. Hunting, Fishing, & Foraging

Living off-grid means that you need to rely on your ability to source food from nature. While growing crops and raising livestock are important components of self-sufficiency, hunting, fishing, and foraging provide valuable opportunities to diversify your food sources. These activities not only help you reduce reliance on store-bought food, but they also allow you to connect deeply with the natural environment around you.

In this section, we'll cover the basics of ethical and sustainable hunting, effective fishing techniques, and how to forage for wild plants, fruits, and mushrooms. Each of these skills can be invaluable when it comes to being self-sufficient in an off-grid lifestyle.

Hunting

Ethical and Sustainable Hunting Practices

When you hunt, it's crucial to approach it with respect for the environment and the animals you're hunting. Ethical hunting practices are centered around sustainability, meaning that you should hunt in a way that doesn't harm wildlife populations or ecosystems. Always be mindful of the long-term impact of your hunting practices.

One of the core principles of ethical hunting is only harvesting what you need, leaving enough animals to maintain the population. Overhunting depletes animal numbers and harms the local ecosystem. You should also strive to make the hunt as humane as possible, ensuring that the animal doesn't suffer unnecessarily.

Another important aspect is maintaining respect for the animals you hunt. Many hunters take steps to use as much of the animal as possible, not just the meat. This is part of a sustainable approach to hunting, where every part of the animal is put to good use, whether for food, tools, or other materials.

Required Licenses, Regulations, and Legal Considerations

Before you set out to hunt, it's essential to understand the legal framework surrounding hunting in your area. Hunting regulations can vary widely depending on your location, so always make sure you have the proper licenses and permits. These regulations are in place to ensure that hunting remains sustainable, protecting both wildlife and the environment.

Check local laws to find out what species are allowed to be hunted, the seasons during which hunting is permitted, and any restrictions on hunting methods. Many areas have specific rules for hunting certain species, such as bag limits (the number of animals you're allowed to hunt per season) or restrictions on hunting specific age classes or genders of animals (e.g., no hunting of female deer during certain times).

It's important to be knowledgeable about local wildlife conservation efforts and to participate in sustainable practices that align with these efforts.

Tools for Hunting

There are a variety of tools you can use for hunting, each with its pros and cons. Your choice of tools will depend on the animals you're hunting, your skill level, and your preference for hunting methods.

- **Firearms**: Rifles and shotguns are commonly used for hunting larger animals like deer, elk, or wild boar. Rifles offer accuracy and range, while shotguns are often used for bird hunting. If you choose firearms, ensure that you're familiar with safe handling, use, and cleaning practices.
- **Bows**: Bow hunting is a more challenging and rewarding skill. It requires precision and patience, as you must get much closer to your target. Archery is often used for hunting deer and other game animals, and it requires a bow, arrows, and knowledge of proper technique and safety.
- **Traps**: Trapping is another excellent way to secure food in an off-grid environment. Traps can be used for smaller animals like rabbits, squirrels, or even fish. There are various types of traps—box traps, snares, and deadfalls, to name a few—and each has its own method of setting and use.

Fishing

Types of Fishing Gear and Tools

Fishing can be a highly rewarding way to secure food, whether you're on a lake, river, or at sea. The type of gear you use will depend on the type of fishing you plan to do, the species you're targeting, and your environment.

- **Rods**: Fishing rods are one of the most common tools used for catching fish. Rods come in various lengths and strengths, so it's important to select one that suits your fishing style. For example, lightweight rods are ideal for catching smaller fish, while heavier rods are designed for larger species.
- **Nets**: Nets can be used for catching fish, especially in shallow water. There are several types of fishing nets, including cast nets, dip nets, and gill nets, each with different uses.
- **Traps**: Fishing traps are often used to catch fish in rivers and lakes. They can be left overnight and checked in the morning. These traps are often set up with bait to attract fish and can be highly effective for larger catches.

Sustainable Fishing Methods

Just like with hunting, sustainability is key when it comes to fishing. Overfishing can deplete local fish populations and harm aquatic ecosystems. To fish sustainably, you must follow local fishing regulations, including respecting catch limits, avoiding fishing in restricted areas, and only fishing during open seasons.

- **Catch limits**: These are restrictions on the number of fish you can catch in a day or season. Make sure you are aware of these limits for each species to ensure you aren't overfishing.
- **Species selection**: Some fish species are endangered or protected, and it's important to only fish for species that are abundant and allowed for harvesting. Check your local fishery regulations for more information on which species are legally fishable.
- **Habitat protection**: Respect the ecosystems where you fish. Avoid disturbing fragile habitats like coral reefs or nesting sites. When fishing, always ensure you're using environmentally-friendly

practices to minimize your impact on the environment.

Essential Fishing Skills

Fishing is a skill that takes time to learn, and you'll need to practice to improve. Here are some key skills you should master:

- **Knot tying**: Learn how to tie essential fishing knots, such as the **improved clinch knot** and **loop knot**, to attach hooks, lures, and other tackle to your fishing line.
- **Casting**: Mastering the art of casting is essential to fishing. Practice casting to increase your accuracy and distance. Whether you're fishing with a rod or using a net, learning the correct casting technique is essential for success.
- **Cleaning fish**: Once you've caught fish, knowing how to clean and fillet them properly is important for both food safety and ease of cooking. This skill can take time to perfect, but learning to gut and clean fish will be invaluable.

Foraging

Identifying Edible Wild Plants, Fruits, and Mushrooms

Foraging is the practice of gathering wild plants, fruits, mushrooms, and herbs from the natural environment. This is a highly valuable skill to have, especially in an off-grid lifestyle. It provides access to a wide variety of free food, but it requires knowledge and caution.

Before foraging, it's important to familiarize yourself with the plants in your local environment. You should be able to identify edible and medicinal plants, but also recognize dangerous or poisonous species.

Start by learning to identify common wild edibles such as dandelion, chickweed, blackberries, and wild garlic. Also, learn to identify edible mushrooms like chanterelles and morels. Keep in mind that some plants and mushrooms can look similar to poisonous varieties, so take your time and always double-check before consuming anything you've foraged.

Ethical Foraging Practices

When foraging, it's important to follow ethical practices to ensure that you're not damaging the environment or depleting local resources. Foraging is not about taking as much as you can, but about maintaining a sustainable approach.

Never harvest all of a plant or fruit-bearing tree. Leave enough behind so the plant can regenerate and continue to provide food for future harvests. This applies to both fruits and herbs. It's also essential to avoid foraging from protected areas or endangered species.

Preparing and Preserving Foraged Foods for Long-Term Use

Once you've foraged for edible plants and mushrooms, you'll need to know how to prepare them for consumption and long-term storage. Some foraged foods can be eaten immediately, while others require cooking, drying, or fermenting.

For example, wild berries can be eaten fresh, but you can also make jams, jellies, and preserves to store them for winter. Wild mushrooms can be dried or preserved by pickling for long-term storage. Herbs can be dried and stored in airtight containers for later use in cooking.

30. Food Preservation and Storage

Canning

Canning is one of the oldest and most reliable methods for preserving food, allowing you to store fruits, vegetables, meats, and liquids for long periods. As an off-grid individual, learning how to can your own food is essential for maintaining food security and reducing dependence on external sources. In this section, we will explore the two most common canning methods: water bath canning and pressure canning, as well as the tools needed and step-by-step instructions for each process.

Canning involves sealing food in airtight containers and heating them to a temperature that destroys harmful microorganisms, preventing spoilage. The difference between water bath canning and pressure canning lies in the temperature required to safely preserve the food.

pH and When to Use Each Method

The acidity of the food plays a crucial role in determining which method you should use:

- **Water Bath Canning**: This method works best for high-acid foods, as the acidity prevents the growth of harmful bacteria. Foods with a pH level of 4.6 or lower are ideal for water bath canning. High-acid foods include fruits (like berries, peaches, and apples), tomatoes (although tomatoes can vary in acidity), pickles, jams, and jellies.
- **Pressure Canning**: Low-acid foods, with a pH above 4.6, need the higher temperatures that pressure canning provides. This includes vegetables, meats, poultry, and dairy. These foods require a temperature higher than what boiling water can reach, which is where the pressure canner comes into play. The pressure cooker increases the boiling point of water, ensuring that these foods are safely preserved.

When to Use Water Bath vs. Pressure Canning:

- Use **water bath canning** for acidic foods such as fruits, jams, jellies, and pickled products.
- Use **pressure canning** for low-acid foods such as meats, poultry, vegetables, and soups to prevent bacterial contamination.

Water Bath Canning Step-by-Step

Tools Needed for Water Bath Canning

To successfully complete the water bath canning process, you'll need the following tools:

- **Water Bath Canner**: A large pot designed to hold multiple jars of food. It should have a lid and be deep enough for water to cover the jars by at least an inch.
- **Canning Jars**: Use jars specifically designed for canning, available in various sizes. Ensure they have no cracks or chips.
- **Lids and Bands**: Each jar will need a flat lid and a metal band to seal it.
- **Jar Lifter**: This tool allows you to safely lift the hot jars from the boiling water.
- **Funnel**: To help fill jars without spilling.
- **Bubble Remover**: To remove air bubbles from the jars.
- **Timer**: To ensure you process the jars for the correct amount of time.
- **Clean Cloths**: For wiping the rims of jars before sealing.

Step-by-Step Water Bath Canning Process

1. Prepare Jars and Lids: Wash your jars, lids, and bands thoroughly with hot, soapy water and rinse. Place the jars in the canner and fill it with water. Bring the water to a simmer, not a boil, to sterilize the jars. While the jars are sterilizing, prepare your lids by simmering them in hot water (but do not boil them).
2. Prepare the Food: Wash, peel, and cut your produce according to your recipe. For fruits, you may need to prepare syrup, and for pickles, a brine solution. Fill the jars with the prepared food, leaving the appropriate amount of headspace (usually 1/4 to 1/2 inch) at the top.
3. Remove Air Bubbles: Use a bubble remover or a non-metallic utensil to gently stir the food in the jar and release any trapped air. This helps ensure that the jars seal properly.
4. Wipe the Rims: Use a clean cloth to wipe the rims of the jars to remove any food residue or moisture. This step ensures that the lids can form a proper seal.
5. Seal the Jars: Place the lids on the jars and screw on the bands until they are fingertip-tight. The bands should be snug but not overly tight, as air needs to escape during processing.
6. Process the Jars: Carefully lower the jars into the canner, making sure they are covered with at least an inch of water. Bring the water to a full boil, then start your timer. The processing time will vary based on the type of food and altitude, so refer to your recipe for the correct time.
7. Cool the Jars: After the processing time has elapsed, remove the jars from the canner using the jar lifter. Place them on a clean towel or cooling rack and let them cool completely. The lids will "pop" as they seal, which is a good sign that the food has been preserved correctly.
8. Store the Jars: Once the jars are cool, check the seals by pressing down on the center of the lid. If it doesn't pop back, the jar is sealed. Store the jars in a cool, dark, and dry place.

Pressure Canning Step-by-Step

Pressure canning requires a different set of tools due to the higher temperatures involved:

- **Pressure Canner**: A large pot with a locking lid designed to build pressure. The canner should have a gauge to monitor the pressure inside.
- **Canning Jars and Lids**: Similar to water bath canning, you'll need sterilized jars and lids.
- **Jar Lifter**: To remove hot jars from the pressure canner.
- **Funnel and Ladle**: For safely filling the jars with hot food.
- **Bubble Remover**: To release air bubbles from the jars.
- **Timer**: To ensure accurate processing times.
- **Clean Cloths**: For wiping jar rims.

Step-by-Step Pressure Canning Process

1. Prepare Jars and Lids: Wash jars, lids, and bands thoroughly. Place the jars in the canner with hot water to sterilize them. Prepare the lids by simmering them in hot water. For meats, soups, and other low-acid foods, have them prepared in advance.
2. Fill the Jars: Pour the prepared food (such as meat, broth, or vegetables) into the jars, leaving the correct amount of headspace. Use a funnel and ladle to ensure the food is filled correctly. Be careful not to overfill the jars.
3. Remove Air Bubbles: Use a non-metallic utensil to stir the food gently in the jar and remove any trapped air bubbles.
4. Wipe the Rims: Use a clean cloth to wipe the rims of the jars to ensure no food residue prevents a proper seal.
5. Seal the Jars: Place the sterilized lids on the jars and screw on the bands until they are fingertip-tight.
6. Load the Pressure Canner: Add the jars to the pressure canner, ensuring they are not overcrowded. Adhere to the manufacturer's instructions for adding water to the canner. Seal the canner according to its design and ensure the lid is securely fastened.
7. Begin Heating the Canner: Turn the heat on and allow the canner to come to pressure. Watch for the pressure gauge to indicate the correct pressure for your altitude and the type of food being processed.
8. Process the Jars: Maintain the correct pressure for the specified time based on the type of food and altitude. Check the pressure throughout the process and adjust the heat if necessary.
9. Cool the Jars: After the processing time has passed, turn off the heat and allow the pressure canner to cool naturally. Once the pressure has completely subsided, carefully open the canner. Using the jar lifter, remove the jars and place them on a cooling rack.
10. Check the Seals and Store: Allow the jars to cool completely. Check the seals by pressing down in the center of the lid. If the lid doesn't pop back, the jar has sealed properly. Store sealed jars in a cool, dark, and dry place.

Certainly! Adding proper storage times for the different types of canned foods is crucial to ensure the best quality and safety of your preserved goods. Below is a guide on storing times for both water bath and pressure-canned foods.

Water Bath Canning Storage Times

Water bath canning is ideal for high-acid foods like fruits, jams, jellies, pickles, and tomatoes. These foods can be stored for a considerable amount of time if processed and stored correctly. Here's an overview of how long you can expect different items to last:

- **Fruits (e.g., peaches, apples, berries)**: Stored in a cool, dark place, water-bath canned fruits can last for up to **12 to 18 months**. The quality may degrade after a year, so it's best to consume them within the first year for optimal taste and texture.
- **Jams and Jellies**: These are typically good for **12 to 18 months**. Over time, the flavor may diminish, but they will still be safe to consume beyond this period as long as the seal remains intact.
- **Pickles (including cucumbers and other vegetables)**: Pickles can be stored for **6 to 12 months**. While they may still be safe to eat beyond this time, the texture might become softer, and the flavor may change.
- **Tomatoes**: When canning tomatoes using water bath canning, they can be stored for **12 to 18 months**. However, tomatoes can lose their flavor and color over time, so it's best to use them within a year.

Storage Tips:

- Always label your jars with the date they were canned.
- Store jars in a cool, dry, dark location, ideally between **50°F and 70°F** (10°C and 21°C).
- Inspect the seals regularly to ensure they remain intact.

Pressure Canning Storage Times

Pressure canning is suitable for low-acid foods, such as meats, poultry, vegetables, and soups. These types of food require higher temperatures, which are achieved through the pressure canner, to prevent the growth of botulinum bacteria.

- **Meats (beef, pork, chicken, etc.)**: Pressure-canned meats can last for up to **2 to 5 years**. The shelf life may vary depending on storage conditions, but they will remain safe and nutritious within this time frame as long as the seal is intact.
- **Vegetables (e.g., green beans, carrots, peas)**: These can be stored for **12 to 18 months**. Vegetables tend to lose some texture and flavor over time but should remain safe to eat for up to **2 years** if stored properly.
- **Soups and Stews**: Pressure-canned soups can last **1 to 2 years**, although the flavor may degrade slightly over time. The ingredients used in the soup (such as meat or beans) may affect how long the soup will maintain its best quality.
- **Stocks and Broths**: Homemade stocks or broths canned in a pressure canner typically last for **1 to 2 years**. If they are not consumed within this time, they may start to lose their flavor and nutritional content.

Storage Tips:

- As with water bath canning, always label jars with the date of canning.
- Store jars in a cool, dark, and dry place. **Avoid direct sunlight** and keep jars away from heat sources.
- Ensure jars are stored at temperatures **between 50°F and 70°F** (10°C to 21°C).
- If you notice any bulging lids, leaking, or other signs of spoilage, discard the jars immediately.

General Storage Guidelines for Canned Goods:

- **Do not store canned goods in humid areas** or near heat sources, as this can affect the integrity of the seal and lead to spoilage.
- **Rotate your stock**: Use the first-in, first-out (FIFO) method. Always consume the oldest items first and ensure that your pantry is replenished with new canned goods regularly.
- **Inspect cans before use**: Check for any signs of spoilage, including rust, leaks, or bulging lids. If any signs of spoilage are present, discard the item. Never eat food from a jar with a broken seal.

Drying and Dehydration

When you are living off the grid or looking to preserve food for the long term, drying and dehydration are excellent methods for preserving the bounty of your harvest. This method doesn't rely on refrigeration and allows you to store a variety of food for extended periods, which is especially important for sustainability and self-sufficiency. Here, you will learn how to dry foods effectively using different techniques, how to choose the right foods for dehydration, and how to properly store your dried foods to ensure they maintain their flavor, nutritional value, and safety.

Sun Drying vs. Dehydrators vs. Freeze-Drying

Each drying method has its own advantages and challenges. Depending on your resources, time, and the climate in which you live, one method might be more practical than another.

Sun Drying

Sun drying is the most traditional method of dehydrating food. It has been used for thousands of years and remains popular because it requires little energy input. It's especially effective in areas with dry, hot weather.

How it Works: Sun drying uses the natural heat and wind to dry foods. The food is placed on drying racks or trays and exposed to direct sunlight. As the sun's heat evaporates the moisture from the food, it becomes dehydrated and preserved.

Advantages:

- **Low cost**: Once you have the proper racks, no energy is required.
- **No electricity needed**: A great option for off-grid living.
- **Flavor retention**: Sun-dried foods often retain their natural flavor, especially fruits.

Challenges:

- **Climate dependent**: Sun drying is effective only in areas where you can rely on consistent sun and low humidity.
- **Long drying times**: It can take several days to fully dry foods, depending on the conditions.
- **Exposure to pests**: Foods may attract insects, birds, and other pests during the drying process.

Best for:

- Fruits (e.g., tomatoes, apricots, apples)
- Herbs (e.g., basil, oregano, thyme)
- Certain vegetables (e.g., bell peppers, beans)

Dehydrators

A food dehydrator is an electrically powered appliance that uses heat and airflow to remove moisture from food. It is more controlled than sun drying and works in almost any climate.

How it Works: Dehydrators work by circulating warm air around food placed in trays. The temperature is generally controlled to prevent the food from cooking while drying. It's typically set to a lower temperature (between 95°F to 145°F or 35°C to 63°C), ensuring that the food's nutritional content is preserved.

Advantages:

- **Consistent results**: You can dry foods at any time of the year, regardless of the weather.
- **Faster than sun drying**: Dehydrators can dry food much faster—usually within hours to a day.
- **Better pest control**: Since the process occurs indoors, there are fewer worries about insects.

Challenges:

- **Energy usage**: Dehydrators require electricity, which could be an issue in an off-grid setting unless you have a backup energy solution.
- **Initial cost**: Dehydrators can be expensive, though they're generally cost-effective in the long run.

Best for:

- A wide range of fruits, vegetables, and meats
- Making jerky or drying fish

Freeze-Drying

Freeze-drying is a more complex and expensive process than sun drying or using a dehydrator. It involves freezing the food and then removing the moisture using a vacuum process. Freeze-dried food retains nearly all of its flavor, texture, and nutrients, which is why it's often used in military and long-term storage food.

How it Works: The food is first frozen and then placed in a vacuum chamber. The pressure is lowered, causing the ice to sublimate (change directly from solid to gas) and leaving the food dry.

Advantages:

- **Preserves the highest nutritional value**: Freeze-dried foods retain almost all of their vitamins and nutrients.
- **Better texture**: Unlike dehydrated food, freeze-dried foods rehydrate better, keeping much of their original texture.
- **Longer shelf life**: Freeze-dried foods can last for 20 to 30 years when stored properly.

Challenges:

- **Expensive**: Freeze-drying equipment is costly, and the process requires more energy than other methods.
- **Complex**: The equipment required to freeze-dry food is more technical than dehydrators or sun drying.

Best for:

- Long-term food storage
- Preparing for emergencies or disaster preparedness
- Vegetables, fruits, full meals

Choosing the Right Foods for Drying and Preserving

While almost any food can be dried, not all foods are suitable for all drying methods. Some foods dehydrate better than others, and some may require special treatment before drying.

Fruits:

- **Best for Drying**: Apples, apricots, bananas, peaches, pears, plums, and tomatoes. These fruits have high natural sugars and acids that make them ideal for drying.

- **Treatments**: Many fruits, especially apples, should be dipped in an ascorbic acid solution or a mild salt solution to prevent browning during drying.

Vegetables:
- **Best for Drying**: Peas, beans, carrots, corn, spinach, and broccoli.
- **Pre-Treatments**: Many vegetables require blanching before drying to stop enzyme activity and preserve color and flavor.

Herbs:
- Best for Drying: Basil, oregano, thyme, rosemary, dill, and sage. Herbs generally dry well without any special treatment.

Meats and Fish:
- **Best for Drying**: Beef, venison, chicken, and fish. These can be turned into jerky using a dehydrator or oven.
- **Pre-Treatments**: Marinating meats before drying improves flavor and tenderness. Always ensure proper safety handling and drying to avoid foodborne illnesses.

Best Practices for Drying Food
- **Freshness**: Use fresh, ripe produce for the best quality. Avoid any overripe or damaged food.
- **Cutting Size**: Cut fruits and vegetables into uniform pieces for even drying.
- **Layering**: Avoid overcrowding trays to ensure that air circulates freely around the food.

Step-by-Step Drying Process

1. **Prepare the Food**: Wash, peel, chop, and pre-treat fruits and vegetables as necessary.
2. **Arrange on Trays**: Place the prepared food in a single layer on your dehydrator trays, allowing space between each piece for air circulation.
3. **Set the Temperature**: Refer to the dehydrator's manual for the correct temperature setting for each type of food.
4. **Dry the Food**: Turn on the dehydrator and let the food dry for several hours. The time will depend on the type of food, the thickness of the slices, and the humidity level.
5. **Test for Dryness**: Once drying is complete, check if the food is fully dehydrated by breaking it in half. If there's any moisture left inside, it needs to dry longer.
6. **Cool and Pack**: Allow the dried food to cool to room temperature before packing it away in containers.
7. **Store**: Place the dried food in airtight, moisture-proof containers such as Mylar bags, vacuum-sealed bags, or glass jars with tight-fitting lids.

Proper Storage of Dehydrated Foods

Proper storage is essential to maintain the quality of dehydrated foods and prevent spoilage.

- **Airtight Containers**: Use sealed glass jars, vacuum-sealed bags, or Mylar bags with oxygen absorbers to keep air and moisture out.
- **Cool, Dark, and Dry**: Store dehydrated foods in a cool, dark, and dry environment. Ideal storage temperatures range from **50°F to 70°F** (10°C to 21°C).
- **Avoid Direct Sunlight**: Exposure to light can degrade the nutritional content of dried foods.
- **Use Within Recommended Time**: Most dried foods should be used within **6 months to 1 year** for optimal flavor, texture, and nutrients.

Cold Storage

Cold storage is a critical aspect of off-grid living, especially when you rely on your harvest for sustenance during colder months. Without refrigeration, you need to understand how to preserve and store your food effectively. A well-maintained root cellar, along with other alternative cooling methods, can keep food fresh for months without electricity, and cold storage is essential for ensuring your off-grid food supply lasts.

Building and Maintaining a Root Cellar

A root cellar is a time-tested method for storing root vegetables and other perishables in a cool, stable environment. Unlike a conventional refrigerator, a root cellar uses the earth's natural insulation to maintain a consistent temperature that's ideal for preserving food.

Choosing the Location

When choosing a location for your root cellar, it's important to consider factors such as the depth, accessibility, and the local climate. A root cellar needs to be located in an area where temperatures stay cool, usually between 32°F and 40°F (0°C and 4°C), which is perfect for long-term storage of root vegetables like carrots, potatoes, onions, and apples.

You'll want to find a spot that has natural cooling, such as the north side of a hill or an area with minimal exposure to direct sunlight. If you are digging a root cellar, choose a location that's away from large trees whose roots might affect the structure. Ideally, the cellar should be at least partially underground, as this helps maintain the cooler temperatures.

Designing Your Root Cellar

There are several designs for building a root cellar, depending on your space, materials, and skill level. Here are the most common types:

- **Underground Root Cellar**: This type is dug into the earth to take advantage of the natural insulation and temperature regulation. The deeper the cellar, the more stable the temperature will be. A minimum depth of 3 feet (91 cm) is ideal for keeping a consistent temperature year-round. The walls and ceiling can be made from stone, brick, or cement blocks. Concrete is often used to line the cellar to prevent moisture from seeping in.
- **Hill Root Cellar**: For homes built on a slope, a hill root cellar is a great option. It requires less digging because part of the cellar is already under the earth. This type of design uses the hill's natural earth insulation to regulate temperature and humidity.
- **Crawl Space Root Cellar**: If you have an existing crawl space under your house, it can be transformed into a root cellar by improving ventilation and insulation. This option requires less construction but may have limitations in terms of space and temperature control.

Key Components

- **Ventilation**: Proper airflow is essential to prevent mold and mildew in your root cellar. Install vents at both the top and bottom of the structure to encourage airflow. A simple vent system using PVC pipes is a cost-effective option.
- **Temperature Control**: Ensure that the cellar remains at a cool temperature by digging deep enough and using insulating materials. The soil's natural insulation will help keep the temperature stable.
- **Shelving and Storage**: Wooden shelving works well in root cellars for organizing your food. It allows for air circulation around each item, preventing spoilage. Consider placing your root vegetables in burlap sacks or wooden crates to allow for ventilation.

Maintaining Your Root Cellar

Once your root cellar is built, it's important to check it regularly to ensure the conditions are ideal for food preservation. You should aim to maintain consistent temperature and humidity levels, as these are critical to keeping your food fresh. Use a thermometer and hygrometer to monitor temperature and humidity levels. If necessary, you can use a fan or humidifier to adjust conditions as required.

Alternative Cool Storage Methods

While root cellars are ideal for off-grid living, there are other cool storage methods that can help you keep food fresh without electricity. These methods are especially helpful in warmer climates or if you do not have the space to build a root cellar.

Coolers

For a simple off-grid cooling option, coolers can be a great choice for short-term storage. You can use ice or cold packs to maintain a cool environment inside a sealed cooler. While not as long-lasting as a root cellar, coolers can be useful for storing perishable items, such as dairy products or meat, for several days at a time.

Underground Storage

Underground storage is another alternative method that mimics the natural cooling effect of a root cellar but on a smaller scale. You can dig a hole in the ground, line it with plastic or another waterproof material, and place food inside. This works especially well in cool, damp environments and is an excellent method for preserving vegetables like potatoes and carrots.

Underground storage is not as regulated or sustainable as a root cellar, but it can be a good temporary solution.

Iceboxes and Ice Houses

In cooler climates, iceboxes or ice houses are still used in some off-grid areas for storing perishable items. This method involves collecting and storing ice during the colder months and using it to keep food cool through the warmer months. An ice house needs to be insulated and kept cool to be effective, and it can provide an excellent solution for larger quantities of food storage.

Best Crops for Long-Term Cold Storage

Certain crops are particularly well-suited for long-term cold storage, making them ideal for off-grid living. These crops are durable, nutrient-rich, and can last for months when stored correctly.

Potatoes

Potatoes are one of the best crops for cold storage. When stored in a root cellar or cool, dark place, they can last for several months. Store them in burlap sacks or wooden crates to allow for air circulation. Keep them away from sunlight to prevent sprouting.

Carrots

Carrots also store well in cold environments. Place them in containers filled with sand, and store them in a cool, humid location. Carrots will last for several months in optimal storage conditions.

Onions

Onions can last for months in cool storage. Store them in mesh bags or braided bunches and hang them in a well-ventilated area. They require good air circulation to prevent rot.

Cabbage

Cabbage can be stored in cold storage for several months, but it requires humidity control. It's best stored in a root cellar with consistent temperature and humidity levels to prevent wilting or decay.

Apples

Certain apple varieties, like Granny Smith and Fuji, store well in cold storage. Store apples in a cool, dark place and separate them from other produce to prevent them from releasing gases that speed up ripening.

Fermentation

Fermentation is one of the oldest methods of food preservation, allowing you to transform fresh foods into more flavorful, longer-lasting, and sometimes more nutritious items. By utilizing the power of beneficial bacteria, yeasts, and molds, fermentation can significantly enhance your food supply and contribute to a self-sufficient off-grid lifestyle.

What is Fermentation?

At its core, fermentation is the process of breaking down carbohydrates into alcohols, acids, or gases through the action of microorganisms. These microorganisms, such as bacteria, yeast, and fungi, convert sugars in the food into useful byproducts that enhance the flavor, texture, and shelf life of the food. The most common examples of fermented foods include sauerkraut, kimchi, kefir, yogurt, and kombucha.

Fermentation not only preserves the food but also provides unique probiotic benefits that can help with digestion, gut health, and immunity. During fermentation, the food develops flavors that vary from tangy and sour to mildly sweet, depending on the specific bacteria or yeast used and the conditions of the process.

What Foods Can Be Fermented?

Fermentation can be applied to a wide range of foods, from vegetables and fruits to dairy and grains. Common foods that are fermented include:

- **Vegetables**: Cabbage (for sauerkraut and kimchi), cucumbers (for pickles), carrots, beets, and radishes.
- **Dairy**: Milk (to make yogurt, kefir, and cheese).
- **Fruits**: Apples (for cider), grapes (for wine), and other fruits for vinegar.
- **Grains**: Barley (for beer), wheat (for sourdough bread), and rice (for sake).
- **Beverages**: Kombucha and ginger beer.

Fermentation can be used to preserve foods for long periods, reduce waste, and create delicious, health-boosting items for your off-grid pantry.

Step-by-Step Process of Fermentation

1. Choose Your Food: Start by selecting the type of food you want to ferment. Vegetables like cabbage or cucumbers are popular choices because they're simple and have high water content.
2. Prepare the Ingredients: Wash your vegetables thoroughly and cut them into the desired shapes—shredding cabbage for sauerkraut or slicing cucumbers for pickles.
3. Salting: For most vegetable ferments, you'll need to salt the food to encourage the growth of beneficial bacteria. The salt helps draw out water and creates the brine necessary for fermentation. A general guideline is to use 2-3% salt by weight of the vegetables.
4. Pack the Jars: Place the salted vegetables into clean jars, packing them tightly to ensure they're submerged in their own juices or the brine. The vegetables should be fully submerged to prevent exposure to air, which could cause mold growth.
5. Cover and Wait: Seal the jars and leave them at room temperature to ferment. The time it takes depends on the temperature and the specific recipe. For example, sauerkraut may take anywhere from 3 to 10 days to ferment, while kimchi can take longer.
6. Taste and Monitor: During the fermentation process, taste the food daily to see if it has reached the desired level of sourness. Monitor the jars to make sure there is no mold or unusual odor. Once it has fermented to your liking, it can be stored in the fridge or another cool area.

Storage Time for Fermented Foods

Fermented foods can last from several weeks to several months, depending on the food type and how well it has been stored. Once the fermentation process is complete, it's best to store the jars in a cool, dark place like a root cellar or fridge. The cooler temperature will slow down the fermentation process and allow the food to last longer. For instance, sauerkraut stored in a refrigerator can last up to 6 months or longer, while kimchi may last around 1 to 3 months.

Pickling

Pickling is another preservation technique similar to fermentation but differs in its method and the ingredients used. While fermentation relies on the natural growth of bacteria to preserve the food, pickling involves immersing food in an acid solution, typically vinegar or brine, to create a preserved product. Fermented foods often have a tangy or sour flavor, whereas pickled foods can have a sharp acidity but lack the depth of flavors that develop from fermentation.

What Foods Can Be Pickled?

Pickling is commonly used for vegetables, fruits, and even some meats. Common pickled foods include:

- **Vegetables**: Cucumbers (pickles), carrots, peppers, onions, cauliflower, and green beans.
- **Fruits**: Apples, peaches, and pears.
- **Meats**: Fish or eggs in some regional recipes.

Step-by-Step Process of Pickling

1. Choose Your Produce: Select fresh, firm vegetables or fruits. Pickles are typically made from cucumbers, but many other vegetables like peppers, carrots, and green beans also work well.
2. Prepare the Produce: Wash the vegetables or fruits thoroughly. You can slice them into desired shapes or leave them whole, depending on your preference.
3. Make the Brine: Combine water, vinegar, salt, and any additional seasonings like sugar, garlic, or mustard seeds in a pot. Bring the brine to a boil and let it simmer for a few minutes. The ratio of vinegar to water is typically 2:1 for a strong pickle, but this can be adjusted depending on taste.
4. Pack the Jars: Once the brine is prepared, place your produce into sterilized jars, leaving some room at the top. Pour the hot brine over the food until it is fully submerged.
5. Seal and Store: Once the jars are packed, seal them tightly and let them cool to room temperature. Store them in a cool, dark place like a pantry or cellar. Once opened, pickles can be stored in the refrigerator for up to several weeks.

Storage Time for Pickled Foods

Pickled foods can last several months when stored in a cool, dark place. The acidity of the vinegar helps preserve the food, making it last longer than fresh produce. Properly sealed jars can last up to a year or more. However, once opened, they should be stored in the refrigerator and used within a few weeks.

Vacuum Sealing

Vacuum sealing involves placing food in a plastic bag or container and using a vacuum machine to remove the air. This process creates a tight seal that prevents oxygen from entering, which slows down the degradation of the food.

What Foods Can Be Vacuum Sealed?

Many types of food can be vacuum sealed, including:
- **Meats**: Vacuum sealing preserves the freshness of meat and can extend its shelf life for months or even years when stored properly.
- **Vegetables and Fruits**: For long-term storage, vacuum-sealed vegetables and fruits can be frozen and retain their flavor, texture, and nutrients.
- **Grains and Beans**: Grains like rice and beans last longer when vacuum-sealed, making them ideal for off-grid food storage.

Step-by-Step Process of Vacuum Sealing

1. Prepare the Food: Wash, peel, or cut your food as needed. For meats, ensure they are cut into portions. For vegetables and fruits, blanch them before sealing if you plan to freeze them.
2. Place the Food in a Vacuum Bag: Insert your prepared food into a vacuum-seal bag. Leave enough space at the top of the bag to allow the machine to create a proper seal.
3. Seal the Bag: Use a vacuum sealer machine to remove air from the bag. Follow the manufacturer's instructions for your specific machine. Once the air is removed, the machine will heat-seal the bag to create an airtight seal.
4. Store the Vacuum-Sealed Food: Once sealed, store the bags in a cool, dry place, preferably a pantry or cellar. For long-term storage, freezing the sealed bags will help preserve the food's freshness for extended periods.

Storage Time for Vacuum-Sealed Foods

Vacuum-sealed food lasts much longer than food stored in regular bags. For example, vacuum-sealed meat can last up to 2 to 3 years when frozen, and vacuum-sealed vegetables can last for up to a year in the freezer. If stored in the pantry, dried goods can last up to 6 months or longer.

Combining Vacuum Sealing with Drying

Vacuum sealing dried foods like herbs or tomatoes can extend their shelf life by preventing moisture from getting back into the food. By combining vacuum sealing with drying, you effectively protect your preserved foods from air, moisture, and contaminants.

Smoking

Smoking is an age-old method of food preservation that uses smoke from burning or smoldering materials, such as wood or charcoal, to cure and preserve meat, fish, cheese, and even vegetables. The process not only helps in extending the shelf life of food but also imparts a distinctive smoky flavor that enhances the food's taste.

The two primary methods of smoking are hot smoking and cold smoking. Hot smoking involves cooking the food at temperatures between 130-180°F (54-82°C) during the smoking process. Cold smoking, on the other hand, occurs at temperatures below 90°F (32°C) and does not cook the food, but rather preserves it through the application of smoke. Hot smoking is typically used for meat, fish, and poultry, while cold smoking is more common for cheeses, sausages, and some vegetables.

For What Food?

Smoking is commonly used to preserve and enhance the flavor of foods such as:
- **Meat**: Beef, pork, lamb, and poultry are often smoked to enhance their flavor and ensure long-term storage. The process helps tenderize tougher cuts and adds rich, deep flavors to the meat.

- **Fish**: Fish like salmon, trout, and mackerel are ideal for smoking. Smoked fish is a staple in many cultures and offers a protein-packed food source.
- **Cheese**: Hard cheeses, such as cheddar and gouda, are often smoked to create a distinct flavor. Cold smoking is usually preferred for this.
- **Vegetables**: Smoking vegetables like peppers, tomatoes, and eggplants adds flavor while also preserving them for long periods.

Step-by-Step Process

1. Prepare the Food: Start by selecting the right cut of meat, fish, or vegetables for smoking. For meat, you might want to brine or marinate it before smoking to enhance the flavor and moisture content.
2. Set Up the Smoker: Choose your smoker, whether it's a traditional wood smoker, electric smoker, or even a DIY setup. If you're using wood, opt for hardwoods like hickory, oak, or maple, as they impart the best flavors.
3. Prepare the Wood Chips: Soak wood chips in water for about 30 minutes before adding them to the smoker to prevent them from burning too quickly. Alternatively, you can use wood chunks if you prefer longer smoking times.
4. Preheat the Smoker: Preheat your smoker to the appropriate temperature based on the food you're smoking. If you're hot smoking meat or fish, aim for a temperature range of 130-180°F (54-82°C).
5. Place the Food in the Smoker: Once the smoker is ready, arrange your food inside, ensuring that it's not crowded, and air can circulate around it. For meat, it is recommended to smoke at lower temperatures for longer periods to allow the smoke to fully penetrate the food.
6. Monitor the Smoking Process: Maintain a consistent temperature in your smoker throughout the process. Check the internal temperature of the food periodically to ensure it's cooking evenly and safely. For meat, use a meat thermometer to check for the correct internal temperature based on the type of meat you're smoking.
7. Smoking Time: Depending on the type of food and the size of the cuts, smoking times can vary. Generally, it will take anywhere from 2-6 hours for meats, while smaller items like fish or vegetables might take 1-2 hours. You can also opt for a longer smoking process for more intense flavor, up to 12 hours for certain cuts of meat.
8. Cooling and Storage: Once the food is done, allow it to cool before storing it. If you're planning to keep it for long-term use, vacuum-seal the smoked foods or store them in airtight containers to protect them from moisture and contaminants.

Storage Time

When smoked properly, food can be stored for a long time. However, storage conditions play a big role in ensuring the preservation. Smoked meats and fish, if vacuum-sealed and stored in a cool, dry place, can last up to 6 months. If refrigerated, the food will last around 1-2 weeks, and smoked cheeses can last up to 4 weeks. For long-term storage, freezing is a great option and can keep smoked food preserved for up to a year.

Salting

Salting is one of the oldest food preservation methods that involves using salt to draw moisture out of foods, creating an environment that inhibits bacterial growth and spoilage. Salt works as a preservative by reducing the water activity in the food, which is where bacteria and mold grow. This method is commonly used for preserving meats, fish, and even some vegetables.

For What Food?

- **Meats**: Salting is ideal for preserving meats like pork, beef, and poultry. Salt-cured meats like ham, bacon, and jerky are staples in many off-grid food storage systems.
- **Fish**: Fish, especially those that are caught in abundance, can be preserved using salt. Salted fish can last for months when stored in proper conditions.
- **Vegetables**: Some vegetables, like cucumbers and tomatoes, are also salted to create salted products, such as pickles.

Step-by-Step Process

1. Select Your Food: Choose your meat, fish, or vegetables to be salted. For meats and fish, it's

best to use fresh cuts to ensure optimal preservation.
2. Prepare the Salt: Use non-iodized salt, as iodine can affect the preservation process. You will need coarse sea salt or kosher salt for effective salting.
3. Apply the Salt: Generously coat your food with salt. For meats and fish, rub the salt directly onto the surface, ensuring it is fully covered. For vegetables, you can either pack them in salt or immerse them in a brine solution.
4. Storage: After salting, you'll need to allow the food to sit for a while. Store the salted food in a cool, dry place. For meats and fish, this can be in a container or wrapped in cloth to keep air out.
5. Drying the Food: Depending on the type of food, you may need to dry the salted food in a cool, dry location or even in a dehydrator to further reduce moisture. This step is particularly important for meats and fish.

Storage Time

Salted food, when stored in a cool, dry environment and properly sealed, can last from a few weeks to several months. When stored in an airtight container, salted meats and fish can last for up to 6 months in a cool storage area, while salted vegetables should be used within 2-3 months. For long-term storage, vacuum-sealing salted products is an excellent option.

31. Hydroponics & Aquaponics: Advanced Off-Grid Food Production

Growing your own food off-grid requires resourcefulness and innovation. One of the most effective ways to achieve this in small spaces or challenging environments is through hydroponics and aquaponics. These advanced methods of food production allow you to grow vegetables, herbs, and even fish without relying on traditional soil-based agriculture. Whether you're looking to maximize space or conserve water, both hydroponics and aquaponics offer powerful solutions for sustainable food production.

Hydroponics Systems

Hydroponics is the practice of growing plants without soil, using a nutrient-rich water solution to deliver essential nutrients directly to the roots. This method offers greater control over plant growth and can be highly efficient in terms of water and space usage.

Basics of Hydroponics

In hydroponics, plants are suspended in a medium (such as clay pellets, perlite, or vermiculite) that supports their roots while allowing them to absorb water and nutrients. Unlike traditional soil-based farming, the absence of soil means you can control the nutrients plants receive more precisely, promoting faster and more abundant growth.

How Hydroponics Works:

- **Water and Nutrient Solution**: Plants are grown in a water-based solution containing all the nutrients they need, such as nitrogen, phosphorus, potassium, calcium, and trace minerals.
- **Support Medium**: While there is no soil, the roots need support to anchor themselves and access the nutrients in the water. Materials like perlite or coconut coir are commonly used.
- **Water Circulation**: The water solution is typically circulated to provide continuous access to nutrients, ensuring the plants are well-fed and hydrated.

Types of Hydroponic Systems

There are several types of hydroponic systems, each with its own set of benefits and challenges. The system you choose will depend on the space available, the crops you want to grow, and your level of experience.

1. Wick System

    This is the simplest form of hydroponics. The plants are placed in containers, and a wick draws the nutrient solution from a reservoir to the plant roots. It's a passive system, meaning there's no need for pumps or electricity. However, it's best for smaller plants that don't require a lot of water.

2. Deep Water Culture (DWC)

    In this system, plant roots are submerged directly in a nutrient-rich water solution. An air pump provides oxygen to the roots, ensuring they stay healthy and thrive. DWC systems are great for growing leafy greens like lettuce and herbs.

3. Nutrient Film Technique (NFT)

NFT systems use a thin film of nutrient solution that constantly flows over the roots of the plants. This system is more efficient in terms of water use and is often used for growing high-density crops like herbs or small vegetables.

4. Aeroponics
   In aeroponics, plant roots are suspended in the air and misted with a nutrient solution. This method uses less water than other systems and encourages faster growth because the roots have access to more oxygen. However, aeroponics requires careful attention to misting frequency to prevent the roots from drying out.

Benefits, Challenges, and Applications of Hydroponics for Off-Grid Living

Benefits:

- **Water Efficiency**: Hydroponics uses up to 90% less water compared to traditional soil-based farming, which is especially valuable in off-grid settings with limited water sources.
- **Space Saving**: Since plants don't require soil, you can grow them in stacked or vertical systems, maximizing your use of available space.
- **Faster Growth**: With optimal conditions, plants grown hydroponically often grow faster than those grown in soil.

Challenges:

- **Setup Cost**: Initial setup for hydroponics systems can be expensive, particularly if you choose advanced systems like aeroponics or NFT.
- **Maintenance**: Hydroponic systems require regular monitoring of nutrient levels, water pH, and the overall health of the plants.
- **Electricity Dependency**: Many hydroponic systems require pumps or lights, which means access to reliable power is necessary.

Applications: Hydroponics is perfect for small-scale food production, especially in areas where the soil is not fertile or where water conservation is a priority. It's also ideal for urban environments where space is limited.

Aquaponics Systems

Aquaponics combines hydroponics with aquaculture, creating a symbiotic relationship between fish and plants. The fish produce waste that serves as a natural fertilizer for the plants, while the plants help filter the water, which is then returned to the fish tanks.

Introduction to Aquaponics

Aquaponics operates on the principle of sustainability. By integrating fish farming and hydroponics, you can create a closed-loop ecosystem that benefits both the plants and the fish. The system typically involves a fish tank, grow beds for plants, and a filtration system to clean the water before it circulates back to the fish tank.

How Aquaponics Works:

- **Fish Waste**: Fish waste contains nutrients that are vital for plant growth. As the fish swim in the tank, they produce ammonia-rich waste.
- **Biofiltration**: The waste is filtered by biofilters, breaking down the ammonia into nitrates, which are then absorbed by the plants.
- **Plant Filtration**: The plants act as a natural filtration system, purifying the water and helping to maintain a healthy environment for the fish.
- **Recirculating Water**: Water is pumped from the fish tank to the grow beds, where plants take up the nutrients. The clean water is then returned to the fish tank.

What You Need to Set Up a Small-Scale Aquaponics System

To set up a small-scale aquaponics system, you'll need several components:

- **Fish Tank**: This is where the fish will live. Choose a tank size based on the number of fish you plan to have and the size of the grow beds.
- **Grow Beds**: These are where your plants will be grown. The size of the grow beds should be proportional to the size of the fish tank. Gravel or hydroton is often used in grow beds as a medium for plant roots.
- **Water Pump**: A water pump circulates the water between the fish tank and the grow beds.
- **Air Pump**: An air pump is used to ensure there's enough oxygen in the water for both the fish and the plants.
- **Filtration System**: This helps break down fish waste into nutrients that the plants can use.

Choosing Fish and Plants Suitable for Aquaponics

Fish: Choose fish that are hardy, fast-growing, and thrive in a closed-system environment. Some popular choices include tilapia, goldfish, and catfish. Ensure the water temperature and pH are suitable for the fish species you select.

Plants: Not all plants are ideal for aquaponics, but many leafy greens, herbs, and vegetables grow well. Consider plants like lettuce, spinach, basil, and kale, which thrive in an aquaponic environment.

Benefits and Challenges of Aquaponics

Benefits:
- **Sustainability**: Aquaponics systems create a closed-loop system where both fish and plants are cared for without external fertilizers or pesticides.
- **Water Efficiency**: Like hydroponics, aquaponics systems use minimal water, making them suitable for off-grid living.
- **Multi-Use Production**: Aquaponics provides both protein (from fish) and fresh produce, increasing food security in off-grid systems.

Challenges:
- **Initial Cost**: Aquaponics systems can be more expensive to set up than traditional gardening or hydroponics due to the need for tanks, pumps, and filtration systems.
- **System Complexity**: Aquaponics systems require regular monitoring and management of both plant and fish health.
- **Energy Requirements**: Some components, such as the water pump and air pump, require electricity to run.

Designing and Sizing Systems

Designing and sizing your hydroponics or aquaponics system is crucial for maximizing efficiency and minimizing waste.

Calculating Water and Nutrient Requirements

To size your system properly, calculate how much water your plants and fish will need. For hydroponics, the amount of water required depends on the type of plants and the size of your system. For aquaponics, you must balance the water requirements of both the plants and the fish.

Space Considerations

Before designing your system, assess the space you have available. Hydroponics and aquaponics systems can be scaled to fit various spaces, from small indoor units to large outdoor setups. Consider the amount of sunlight, shade, and temperature variation in your space when deciding where to place your system.

Cost and Resource Efficiency

Hydroponics and aquaponics systems can be expensive to set up, but they provide long-term sustainability. Look for ways to reduce costs, such as using second-hand equipment or building DIY systems. Consider using solar energy to power your pumps and lights to make the system more resource-efficient.

Sustainability and Maintenance

Sustainability is key to the success of your hydroponics or aquaponics system. Regular maintenance is necessary to keep the system running smoothly and efficiently.

Minimizing Energy and Water Consumption

Hydroponics and aquaponics are already more efficient than traditional farming, but you can further minimize your environmental impact. Use low-energy pumps, timers to control water flow, and water-saving techniques like rainwater harvesting.

Regular System Checks

Monitor your system regularly to ensure it's functioning properly. Check water levels, pH, nutrient concentrations, and the health of your plants and fish. Keeping an eye on these factors ensures you catch any problems early and keep your system running efficiently.

32. Sustainable Food & Self-Sufficiency Projects

Project: Building a Composting System for Soil Health

Materials Needed:
- Compost bin or designated compost area (a simple wooden frame or plastic bin)
- Kitchen scraps (vegetable peels, coffee grounds, eggshells, etc.)
- Yard waste (leaves, grass clippings, small twigs, etc.)
- Shovel or pitchfork for turning the compost

What To Compost
- Vegetables
- Houseplants
- Yard trimmings
- Coffee, tea
- Fruits
- Nut shells
- Eggshells
- Paper napkins
- Paper scraps and cardboard

What Not To Compost
- Dairy products
- Fats and oils
- Eggs, meat or fish bones and scraps
- Pet waste
- Diseased plants
- Produce stickers
- Medication
- Cigarettes
- Broken glass

COMPOST

Objective: Create a composting system to improve soil fertility, reduce waste, and create nutrient-rich compost for your garden.

- Water source for maintaining moisture
- Compost thermometer (optional but helpful)
- Gloves (for handling compost)

Step 1: Choose Your Composting Method

You have a few options for composting depending on your space and needs:

1. **Compost Bin**: A small, contained bin made of wood, plastic, or metal that holds compost materials. This is ideal for smaller spaces.
2. **Compost Pile**: A pile of compost materials placed in a corner of your garden. This is best for larger properties.
3. **Worm Composting (Vermiculture)**: If you're interested in speedier composting, use worms to help break down organic matter. This method requires a specific bin or container and red wiggler worms.

Choose the one that best fits your space and lifestyle.

Step 2: Set Up Your Compost Bin or Pile

If you're using a bin, place it in a well-drained area, preferably on bare soil to allow microbes to enter the compost. For a compost pile, select a shady, dry area in your yard away from plants that you don't want to risk contaminating.

Step 3: Layering the Compost

Start by layering brown and green materials to encourage aeration and proper decomposition.

Brown Materials: These are dry, carbon-rich materials that help balance the high nitrogen of green materials. Examples include:

1.1. Dry leaves
1.2. Shredded newspaper or cardboard
1.3. Small twigs and straw
1.4. Wood chips or sawdust (ensure they're untreated)

Green Materials: These are moist, nitrogen-rich materials that break down quickly and help fuel the composting process. Examples include:

1.5. Fruit and vegetable scraps (peels, cores, etc.)
1.6. Coffee grounds
1.7. Grass clippings (if not treated with pesticides)
1.8. Eggshells

Alternate layers of brown and green materials, ensuring that each layer is about 3-4 inches thick.

Step 4: Maintain the Correct Moisture

Composting materials need moisture to decompose. If the pile or bin feels dry, sprinkle it with water to keep it damp—like a wrung-out sponge. You don't want it soaking wet, as too much water can cause the compost to become soggy and smelly.

Step 5: Turn the Pile Regularly

Turn the compost pile or stir the contents of the bin every 1-2 weeks with a shovel or pitchfork. This helps mix the materials, aerates the pile, and speeds up the decomposition process. If you're using a bin, you can simply rotate the bin every week or so, if it's designed to spin.

Step 6: Monitor Temperature and Decomposition

The compost pile should reach temperatures of 130-160°F (54-71°C) in the center to break down properly. If you have a compost thermometer, insert it into the center of the pile to monitor the temperature. If it's too cool, add more green materials to increase the heat.

Step 7: Add Compostable Materials Gradually

As your compost pile grows, continue adding materials to it. It's best to add a variety of green and brown materials, but make sure not to overload it with one type. Too much of one kind (like leaves or food scraps) can slow the decomposition process.

Step 8: Harvesting the Finished Compost

Your compost is ready to use when it's dark, crumbly, and smells earthy—this can take anywhere from 3 months to a year, depending on the materials and conditions. Use a screen or sift to remove any large, undecomposed bits. The final product should be fine and rich in nutrients.

Step 9: Apply Your Compost to Your Garden

Once ready, spread the finished compost over your garden beds or mix it into your soil. Compost improves soil structure, provides essential nutrients, and helps retain moisture in the soil.

Step 10: Ongoing Maintenance

Keep your compost pile or bin active by continuing to add kitchen scraps, yard waste, and other compostable materials. Regular turning and moisture management will ensure that you have a steady supply of fresh compost for your garden.

Project: Designing a Simple Drip Irrigation System for Your Garden

Objective: Set up a basic and efficient drip irrigation system to ensure your garden receives consistent, water-saving irrigation. This system is ideal for reducing water waste and providing precise watering to your plants.

Materials Needed:
- Drip tubing (typically 1/2 inch in diameter)
- Drip emitters or drip stakes
- Stakes or hold-down clips to secure the tubing
- Hose bib adapter (for connecting to the water source)
- Filter to remove debris
- Pressure regulator (to ensure consistent water flow)
- Connector fittings and end caps (for joining and sealing)
- A pair of scissors or pipe cutters
- Optional: Timer to automate watering schedule

Step 1: Plan Your System Layout

1. **Assess your garden**: Take measurements of your garden to determine how much tubing you will need. Identify the areas where you want to water plants—such as rows or beds of vegetables, flowers, or trees.
2. **Design your irrigation layout**: Sketch out your garden and mark where you'll run the tubing. Be sure to account for access to your water source, usually a hose bib. Plan for the tubing to follow a layout that ensures water reaches each plant or area of your garden evenly.

Step 2: Set Up Your Water Source Connection

1. **Connect to the water source**: Attach the hose bib adapter to your water source (faucet). You'll then connect the adapter to the main tubing that will feed your system. Ensure the connection is secure to prevent leaks.
2. **Install the filter**: Connect a filter to the hose bib adapter. This will prevent debris from clogging the tubing and emitters. Most filters are simple to install by screwing them onto the adapter before connecting the main tubing.

Step 3: Install the Pressure Regulator

1. **Attach the pressure regulator**: Connect the pressure regulator after the filter. The pressure regulator is necessary to reduce the water pressure and ensure that the drip emitters work correctly. If your water pressure is too high, it could cause the emitters to spill water too quickly, wasting water and energy.
2. **Check the recommended pressure**: Drip irrigation systems typically operate well between 20-30 psi (pounds per square inch). Make sure to check the pressure rating for your system, and adjust accordingly.

Step 4: Lay Out the Drip Tubing

1. **Lay down the tubing**: Start at the water source and unroll the tubing along the row or bed of plants. Use your design to guide the placement, ensuring you can reach each plant with the tubing.
2. **Secure the tubing**: Once the tubing is in place, use the stakes or hold-down clips to secure the tubing to the ground. This helps keep the tubing in place and prevents it from moving around, which could disrupt the water flow to your plants.

Step 5: Install Emitters or Stakes

1. **Select appropriate emitters**: Based on your plant types, choose the right emitters or stakes. There are several options:
    1.1. **Inline emitters**: These are built into the tubing and deliver a consistent flow of water along the length of the tubing.
    1.2. **Drip stakes**: These are small stakes that attach to the tubing and direct water straight to the base of individual plants.
2. **Install emitters**: Use a punch tool to make small holes in the tubing where you want the emitters or stakes. Insert the emitters into the holes and ensure they fit securely. For rows of plants, evenly space the emitters to ensure all plants are watered.

Step 6: Set Up Water Flow and Adjust

1. **Turn on the water**: Open the valve and allow the water to flow through the system. Check for leaks or blockages and fix any issues you come across.
2. **Test the emitters**: Ensure that each emitter is providing a steady flow of water. You can test by checking for even distribution—each plant should be receiving water at the base.
3. **Adjust the flow rate**: If needed, adjust the emitters or pressure to ensure a balanced and sufficient watering schedule for all your plants.

Step 7: Add Optional Features

1. **Install a timer**: If you'd like to automate the watering process, connect a timer to your system. This allows you to set up a consistent watering schedule without the need to manually turn the water on and off each time.
2. **Rain sensors**: You can also add a rain sensor to stop the system from running if there is sufficient rainfall. This helps conserve water when it's not needed.

Step 8: Monitor and Maintain the System

1. **Inspect regularly**: Regularly check the system for clogs, leaks, or issues with the emitters. Over time, the filter may need cleaning, and emitters may need replacing.
2. **Adjust for seasonal changes**: As the seasons change, you may need to adjust the watering schedule or change the emitters based on the plant's needs.
3. **Clean the system**: At the end of the growing season, flush the system with clean water to prevent debris buildup and maintain the system's efficiency.

Step 9: Enjoy Your Efficient Irrigation System

With your drip irrigation system installed, your garden will receive consistent watering, and you can rest easy knowing that water is being used efficiently. Regular maintenance will ensure that the system continues to function effectively season after season.

Project: Setting Up a Crop Rotation Plan for Your Garden

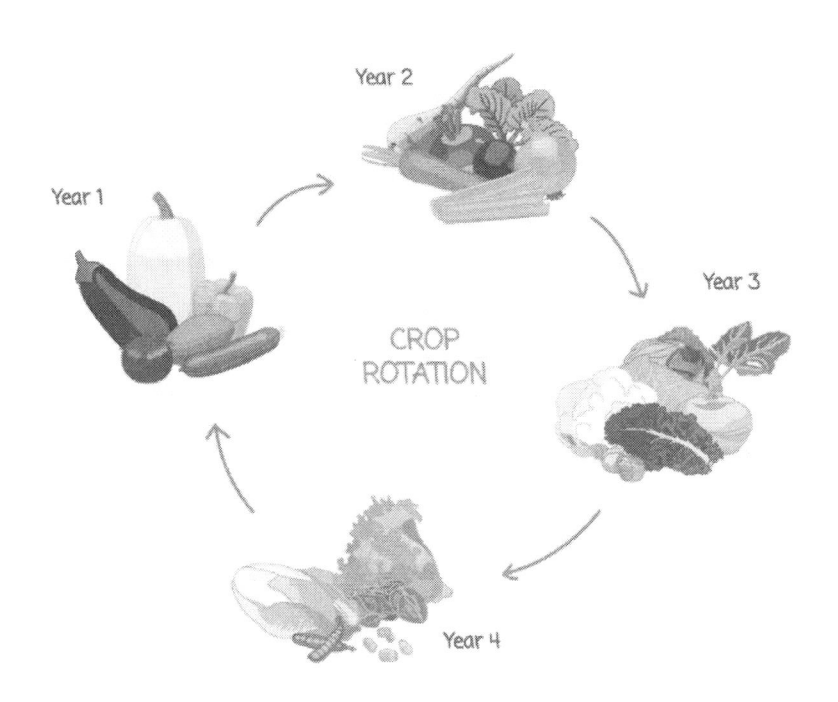

Objective: Design a crop rotation plan to optimize soil health, prevent pests and diseases, and improve the productivity of your garden over time.

Materials Needed:
- A notebook or gardening journal
- Measuring tape or garden grid for marking plots
- Seed packets or plant varieties
- A calendar or planner
- Marker pens or colored pencils
- Compost or organic soil amendments (optional)

Step 1: Assess Your Garden Space

1. **Survey your garden**: Measure the total area of your garden and divide it into smaller sections or planting beds. These sections will allow you to easily rotate crops each season.
2. **Note sunlight exposure and soil conditions**: Assess the amount of sunlight each section receives and take note of any areas that tend to be wetter or drier. This will help determine what types of crops should go where.

Step 2: Choose Your Crops

1. **Select your crops**: Choose a mix of vegetables that grow well in your region and climate. Consider the growth cycle of each crop, its nutrient requirements, and its susceptibility to pests and diseases.
2. **Group crops by family**: Organize your crops into different plant families, such as:
   2.1. Leafy greens (e.g., lettuce, spinach, kale)
   2.2. Root vegetables (e.g., carrots, beets, potatoes)
   2.3. Legumes (e.g., beans, peas)
   2.4. Fruit-bearing plants (e.g., tomatoes, peppers, cucumbers)
   2.5. Brassicas (e.g., broccoli, cabbage, cauliflower)

Step 3: Plan the Rotation Schedule

1. **Rotate based on plant family**: Ensure that plants from the same family do not follow each other in the same section of your garden, as this can lead to soil depletion and pest buildup. For example, after growing tomatoes (a fruit-bearing plant), plant beans (a legume) in that bed the following season.
2. **Create a cycle**: Plan your crop rotation over a 3-4 year cycle, allowing each bed to host a variety of crop families. This helps balance the nutrients in the soil and reduces the risk of disease and pest buildup. A sample rotation might look like this:
Year 1: Leafy greens, Brassicas
Year 2: Root vegetables, Fruit-bearing plants

Year 3: Legumes, Leafy greens
Year 4: Brassicas, Root vegetables

Step 4: Map Out the Rotation on Paper

1. **Draw a garden layout**: On a piece of paper, sketch your garden space and mark out the sections or beds. Label each section clearly with the corresponding crops that will be grown in each one.
2. **Use colors to differentiate**: Color code each section based on the plant family or crop group. This will help visually track the crop rotation and ensure that plants from the same family do not repeat in the same section.

Step 5: Implement the Crop Rotation

1. **Prepare the soil for planting**: Before planting each season, amend the soil with compost or other organic materials to maintain soil fertility. Ensure the bed you are planting in is free from weeds and debris.
2. **Plant according to your plan**: Follow your crop rotation schedule by planting the appropriate crops in each section. Keep track of when you plant and what is planted in each section to help with future rotations.
3. **Maintain soil health**: Throughout the growing season, monitor soil moisture, pests, and plant health. Apply organic fertilizers or compost as needed and remove any pests or weeds that might affect your crops.

Step 6: Record Your Results

1. **Track your garden performance**: At the end of each growing season, document how well the crops performed in each section. Take notes on which plants grew well, which faced challenges, and any pest or disease issues encountered.
2. **Adjust the plan**: Use your observations to refine your crop rotation plan for the following year.

Adjust plant families as needed based on crop performance and garden conditions.

Step 7: Review and Repeat

1. **Monitor long-term soil health**: After a few seasons, assess the overall health of your soil and the diversity of pests. If necessary, adjust your crop rotation strategy to further improve soil health or manage specific pest issues.
2. **Adapt for future growth**: As your gardening skills improve and you learn more about your plants and soil, continue adapting your crop rotation plan to meet the evolving needs of your garden.

Project: DIY Greenhouse Building

Building a greenhouse is an essential project for off-grid food production, as it extends the growing season, protects plants from extreme weather conditions, and allows you to grow a variety of crops year-round. This project will walk you through the process of building a simple and effective greenhouse using basic materials and tools.

Step 1: Plan the Greenhouse Design

1. **Choose the greenhouse size:** The size of the greenhouse will depend on the available space, the number of plants you want to grow, and the

materials at hand. A small greenhouse can be as simple as 6x8 feet, while larger greenhouses can be 10x20 feet or more.
2. **Decide on the shape:** The most common greenhouse designs are A-frame, hoop house, and gable roof. Each has its advantages:
   2.1. **A-frame** is great for snow shedding.
   2.2. **Hoop house** is easy to build and inexpensive.
   2.3. **Gable roof** offers more headroom and better light distribution.
3. **Select the location:** Choose a sunny spot with good drainage. Avoid areas with heavy winds or flood-prone land. The greenhouse should face south (in the Northern Hemisphere) to maximize sunlight exposure.

Step 2: Gather Materials and Tools
1. **Wooden posts or PVC pipes** for the frame
2. **Plastic or polycarbonate sheets** for covering
3. **Metal brackets and screws** to secure the frame
4. **Rebar or wooden stakes** for anchoring the structure to the ground
5. **Twine or wire** for securing plastic or fabric coverings
6. **Tools required:**
   6.1. Measuring tape and level
   6.2. Hammer or drill
   6.3. Saw (if using wood)
   6.4. Shovel (for digging holes)
   6.5. Gloves and safety glasses

Step 3: Prepare the Ground
1. **Clear the area:** Remove any grass, rocks, or debris from the chosen location to create a flat and even surface.
2. **Level the ground:** Use a shovel and rake to make sure the area is level. This is important for stability and water drainage.
3. **Mark the perimeter:** Use stakes and string to mark the dimensions of the greenhouse, ensuring you have the correct shape and size.

Step 4: Build the Frame
1. **Construct the base:**
   1.1. If using wood, cut four pieces to form the base of the greenhouse. These will act as the perimeter frame.
   1.2. If using PVC, create four rectangular or square sections for the base.
2. **Attach vertical posts:**
   2.1. For a wooden frame, cut vertical posts to the desired height of the greenhouse (typically around 6 to 7 feet tall).
   2.2. For a PVC frame, cut PVC pipes to the correct height and attach them to the corners of the base using elbows or connectors.
3. **Support arches (for hoop house design):**
   3.1. If building a hoop house, bend PVC pipes or metal rebar into arches and secure them to the base at equal intervals.
   3.2. If using wood, construct A-frame or gable roof structures by joining wooden beams or slats at the top.

Step 5: Install the Roof and Walls
1. **Attach the roof beams:**
   1.1. For gable or A-frame designs, connect horizontal beams at the peak of the frame.
   1.2. For hoop houses, secure additional horizontal supports along the top.
2. **Cover the structure:**
   2.1. Lay polycarbonate or plastic sheets over the frame, ensuring there are no gaps. Secure them with wire or twine, stretching tightly for durability.
   2.2. Overlap the edges and seal them to prevent drafts and moisture from escaping.
3. **Secure the sides:**
   3.1. If using plastic sheeting, wrap it tightly around the sides of the frame and secure it with twine or metal clips.
   3.2. Ensure the plastic overlaps the base to avoid water seeping in.

Step 6: Create Ventilation and Access Points
1. **Install vents:**
   1.1. Install vents at the top and bottom of the greenhouse to allow hot air to escape and encourage airflow. These can be simple flaps made of plastic or wood.
   1.2. For hoop houses, you can add vent panels to the ends or along the sides.
2. **Create a door:**
   2.1. Construct a simple door frame using wooden slats or PVC, and attach it to one side of the greenhouse using hinges.
   2.2. Use a latch or lock to secure the door and prevent it from swinging open unexpectedly.

Step 7: Add Flooring and Additional Features
1. **Prepare the floor:**
   1.1. If you plan to add plants directly to the ground, remove any weeds and lay down a layer of mulch or cardboard to prevent regrowth.
   1.2. Alternatively, you can place gravel or wood chips to create a solid, permeable surface for water drainage.
2. **Add shelves or tables:**
   2.1. If you want to maximize growing space, consider adding simple wooden shelves or tables along the sides for plants.
   2.2. These can be made from recycled wood or sturdy planks.
3. **Install irrigation system (optional):**
   3.1. Set up a basic irrigation system using PVC pipes, drip lines, or soaker hoses to water plants efficiently. This will help save time and conserve water.

Step 8: Final Inspection and Testing
1. **Check the structure:**
   1.1. Ensure all connections are secure, and that the frame is sturdy and well-anchored.
   1.2. Test the door to make sure it opens and closes smoothly.
2. **Inspect the covering:**
   2.1. Make sure there are no tears or holes in the plastic or polycarbonate.
   2.2. Adjust the tension if necessary to ensure tight coverage.
3. **Test ventilation:**
   3.1. Open the vents to ensure proper airflow, and check for any blockages or obstructions.
4. **Ensure proper drainage:**
   4.1. Make sure water flows away from the greenhouse, and adjust the ground slope if necessary to prevent pooling inside.

Step 9: Ready for Use
1. **Introduce plants gradually:**
   1.1. Start by placing a few plants in your new greenhouse to test temperature, humidity, and overall growing conditions.
   1.2. Monitor the environment and adjust ventilation and watering as needed.
2. **Expand as needed:**
   2.1. As you get comfortable with your greenhouse, add more plants or even a small raised bed to grow additional crops.

Project: Build Your Own Fishing Gear

Materials Needed:
- Natural cordage or strong twine (hemp, jute, or plant fibers)
- Knife or sharp tool (for shaping hooks)
- Bone, small wood, or metal scraps (for hook material)
- Small stones or pebbles (for weight)
- Branch or strong stick (for rod)
- Scissors or utility knife (for cutting and shaping)
- Water container (to test the gear)

Step 1: Gather Materials

Start by gathering all the materials needed to create your fishing gear. The most important parts of the gear are the fishing line, hook, and rod.

- If you don't have commercial fishing line, you can make your own by collecting strong plant fibers like hemp, yucca, or flax. Alternatively, you can use twine or cordage if available.
- Choose a strong bone, small branch, or metal scrap to form your hook.
- Gather some small, rounded stones that can serve as weights.

Step 2: Making the Fishing Line

If you're using natural fibers, you'll need to create your own fishing line.

- Take your chosen material and twist or braid it to create a long line. Aim for a length of at least 10-12 feet (3-4 meters) to ensure you have enough line to reach the water.
- If you're using twine, you can skip the twisting and go straight to measuring and cutting it to the desired length.

Step 3: Creating the Hook

Now, you'll need to create a hook to catch your fish. Hooks are essential to hook the fish securely, so it's important to ensure they're strong and sharp.

- **Bone Hook:** If you're using a bone, carefully carve it into a hook shape. Use a sharp stone or knife to smooth the bone and shape one end into a curved hook.
- **Wood Hook:** If you're using a small branch, cut a piece that's around 4-6 inches long and carve the sharp tip into a hook shape.
- **Metal Hook:** Alternatively, you can use metal scraps. Shape a small piece of metal into a hook using pliers or a sharp tool.

Your hook should have a pointed end to pierce the fish and a small eye at the base to attach it to the line.

Step 4: Attaching the Hook to the Line

Once you have your hook ready, it's time to attach it to your fishing line.

- Take the end of the fishing line and thread it through the eye of the hook.
- Tie a knot that will securely hold the hook in place, such as the improved clinch knot or a simple overhand knot. Ensure the knot is tight so that the hook doesn't slip off the line.

Step 5: Adding Weight to Your Line

Adding weight to your fishing line can help sink your hook and keep it at the right depth in the water.

- Find a few small, smooth stones or pebbles. The weight of the stones will depend on the size of your line and hook, but aim for enough weight to sink the hook.
- Attach the stones to the line above the hook by tying them securely with the same knot as you did

with the hook. Space the stones a few feet apart from each other and the hook to ensure proper distribution of weight.

Step 6: Making the Fishing Rod (Optional)

While a fishing rod isn't strictly necessary, it can help improve your casting distance and ease of use.

• Find a flexible stick or branch that is about 4-6 feet (1.2-1.8 meters) long. A willow or bamboo branch works well for this purpose.
• Strip any leaves or unnecessary parts of the branch to create a smooth, even rod. If the branch is flexible enough, it can act as a simple fishing rod for casting and reeling.
• You can tie the end of your fishing line to the tip of the rod for a simple fishing setup. If you prefer, you can create a reel by wrapping the line around a small spool, but this is optional.

Step 7: Testing Your Fishing Gear

Once everything is ready, it's time to test your fishing gear.

• Head to a location with a body of water such as a pond, lake, or river.
• Cast your line into the water. Let it sink to the bottom and wait for a fish to bite.
• Keep an eye on the line for any movement. When a fish bites, gently reel it in by slowly pulling the line in, keeping tension on the hook.

Step 8: Using the Gear to Catch Fish

If your fishing gear is working as expected, you can begin catching fish!

• When you catch a fish, remove it from the hook carefully. Handle the fish gently to avoid injuring it, or you can prepare it for cooking.
• Clean and gut the fish before cooking, ensuring you remove all inedible parts. You can preserve the fish by drying it, smoking it, or eating it fresh, depending on your needs.

Project: Build a Homemade Trap for Sustainable Hunting

Materials Needed:
• Sturdy twine or rope (preferably natural fiber like hemp or jute)
• Wooden sticks or bamboo poles (around 2-4 feet in length)
• Sharp knife or utility blade (for cutting and shaping)
• Heavy rocks or other objects for weights
• Nails, screws, or a hammer (optional for extra sturdiness)
• A small piece of netting or mesh (optional, for additional trapping material)
• A trigger mechanism (can be made from a small stick or a piece of flexible wood)

Step 1: Gather Materials

Collect all the necessary materials for building your trap. You will need sturdy twine or rope, wooden sticks or bamboo poles, and tools for cutting and shaping the wood. Having a sharp knife or utility blade is important for cutting the twine and shaping the trigger mechanism. You may also need nails or screws for reinforcement.

Step 2: Create the Trap Frame

The trap frame is the structure that will hold your trap in place and allow it to catch animals effectively.

- Start by taking two pieces of wood or bamboo, each approximately 2-4 feet long. These will serve as the sides of the trap.
- Attach them together at a 45-degree angle to create a triangular or A-frame structure. You can use nails, screws, or lash the pieces together tightly with twine. If using twine, tie the knots securely to ensure stability.
- If you prefer, you can create a more simple rectangular frame by cutting the wooden sticks to the desired length and joining them at right angles. The frame doesn't need to be too large; it should be big enough to catch a small to medium-sized animal.

Step 3: Design the Trigger Mechanism

The trigger mechanism is what will release the trap once an animal interacts with it. You can build a simple yet effective trigger with the materials you have.

- Cut a small, flexible stick about 6-8 inches long. This stick will be used as the triggering mechanism.
- Position the stick across the top of the trap frame. You'll need to create a notch or tension point at each end of the stick to ensure it stays in place until triggered.
- Attach a piece of twine to the trigger stick. This piece of twine will connect to the trap door or netting that will close over the animal once it's caught.

Step 4: Construct the Trap Door

The trap door is the part that will drop down once the trigger mechanism is activated. This door will trap the animal inside.

- Cut another piece of wood or bamboo to serve as the door. This should be large enough to completely cover the entrance of the trap frame.
- Attach the door to one side of the trap frame using twine or rope. It should swing freely, but be held in place by the trigger mechanism until the animal disturbs it.
- You can use a rock or heavy object on the opposite side of the door to create enough weight to pull the door down once the trigger is activated. This ensures the trap will close securely when the animal is inside.

Step 5: Set the Trap

Now that the trap frame, trigger mechanism, and door are complete, it's time to set the trap.

- Position the trap in an area where you have seen signs of animal activity, such as trails, feeding areas, or near burrows.
- Tie the free end of the twine from the trigger stick to the door or netting, ensuring that it will hold the door in place.
- Place the door in an upright position and balance it against the trigger stick. Ensure that the twine remains taut, and the door stays in place until triggered.

Step 6: Place Bait (Optional)

While your trap can work without bait, adding bait can increase the chances of catching an animal. You can use food, such as small pieces of fruits, grains, or even scraps from previous meals. The scent will attract animals to investigate the trap.

- Place the bait inside the trap, directly underneath the door, so the animal must interact with the trigger mechanism to reach it.
- Be careful not to overdo the bait, as too much may scare animals away or alert them to the trap's presence.

Step 7: Wait and Monitor

Once the trap is set, give it time to work. Animals may take some time to wander into the trap, so patience is key.

- Regularly check the trap to see if it has been triggered.
- If you have caught an animal, carefully open the trap door, ensuring that you handle the animal gently. If you're using the trap for survival or sustainable hunting, ensure that you handle and process the animal humanely.

Step 8: Reset the Trap

If the trap was triggered without catching anything, or if you successfully caught an animal and released it, reset the trap for another round.

- Simply lift the door, reattach the trigger mechanism, and adjust the bait as needed. Ensure that the trap is stable and ready for the next animal.

Project: Create a Fishing Line and Hook from Natural Materials

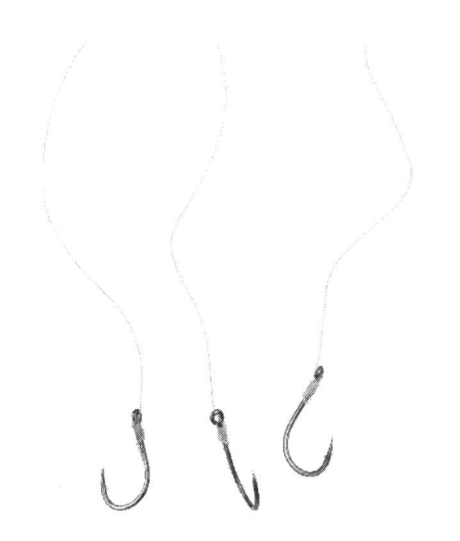

Materials Needed:
- Strong, flexible natural cordage (e.g., twine, hemp, or plant fibers)
- Sharp objects for making hooks (such as bones, small rocks, or metal scraps)
- Needle or a sharp object to pierce through natural cordage
- Small branches or wood pieces for making a rudimentary fishing rod (optional)
- Rocks or stones for weight (if needed)

Step 1: Gather Your Materials

Start by collecting the natural materials you'll need for your fishing line and hook. If you don't have access to pre-made cordage, you can make your own from plant fibers. Plants like hemp, flax, or yucca are ideal choices for making durable cordage.

- For cordage, look for fibrous plants or vines that can be stripped down into threads and twisted or braided together.
- If you can find a sharp bone or strong piece of wood, these can be shaped into your hook.
- Alternatively, use small metal scraps or stones as tools to create a hook-like shape.

Step 2: Making the Fishing Line

If you don't have access to a pre-made fishing line, you can create one using your natural cordage. The line needs to be strong enough to hold a fish but thin enough to cast effectively.

- Take the cordage and determine how long you want your fishing line to be.
- If using plant fibers or vines, twist or braid the fibers together to form a line that is around 10 to 12 feet long.
- Ensure the line is smooth and even throughout.

Step 3: Making the Fishing Hook

Making your own fishing hook is a rewarding task, especially when working with natural materials.

- If you are using bone, carefully shape it into a curved form with a sharp point. The hook should resemble a small "J" shape.
- If you have access to metal, use it to form the hook in the same shape.
- Alternatively, use small sticks or stones. Sharpen one end to create a hook-like point.
- Test the sharpness of your hook by running it over a piece of cloth to see if it catches.

Step 4: Attaching the Hook to the Line

Once you have your hook ready, you need to attach it securely to your fishing line.

- Thread the end of the cordage through the eye of your hook.
- Tie a knot such as the improved clinch knot, which is commonly used in fishing, or simply tie a secure loop.
- Make sure the hook is firmly attached to the line.

Step 5: Adding Weight (Optional)

In some cases, you may need to add weight to your fishing line so that it sinks properly in the water.

- Look for small, heavy stones or pebbles that can serve as makeshift weights.
- Tie one or more stones to the line a few feet above the hook. This will help keep your hook submerged in the water and increase your chances of catching a fish.

Step 6: Setting Up Your Fishing Rod (Optional)

While not necessary, you can make a simple fishing rod to help cast the line further.

- Find a flexible, sturdy stick that is around 4 to 6 feet long.
- Tie your fishing line to the end of the stick.
- If desired, you can create a basic reel by wrapping the remaining fishing line around a small spool, a stick, or a rock. This will allow you to wind and release the line as needed.

Step 7: Testing Your Fishing Line and Hook

Once your fishing line, hook, and rod (if used) are ready, it's time to test them.

- Choose a location with a water source such as a stream, river, or pond.
- Cast your line into the water and allow it to sink. Keep an eye on the line to notice when a fish bites.
- Reel in your line slowly, keeping tension on it to avoid losing your catch.

Step 8: Using Your Fishing Line to Catch Food

Now that you've set up your fishing line and hook, it's time to put your skills to use.

- Find a location with abundant fish and patiently wait for them to bite.
- If you're successful, gently remove the hook from the fish and prepare it as needed.

By using natural materials, you've created a sustainable, self-sufficient method for catching fish. This not only helps you provide food for yourself, but it also encourages resourcefulness and an understanding of how to use available materials in your environment.

Project: Map and Document Edible Plants in Your Area

Materials Needed:
- Notebook or digital device for documentation
- Camera or smartphone for photos (optional)
- Field guide for local plants (book or app)
- GPS or map tool to record locations (smartphone with GPS or paper map)
- Pen or pencil
- Plastic bags or containers for collecting samples (optional)
- First aid kit (just in case)

Step 1: Research Local Edible Plants

Before heading out, start by researching which plants are edible in your area. This can include fruits, vegetables, herbs, and wild plants. Use online resources, local plant guides, or apps dedicated to identifying local flora. Be sure to also check for any poisonous plants that could be harmful if misidentified.

- Research both cultivated plants (like wild greens or fruit trees) and native species that are commonly found in the area.
- Use local knowledge and talk to foragers, gardeners, or local experts who can help with identification.

Step 2: Prepare Your Documentation Tools

Make sure you have the necessary materials for mapping and documenting your findings.

- Bring along a notebook or a digital device (like a tablet or smartphone) to record the plant details.
- If you prefer, use a plant identification app or a simple note-taking app that allows you to log text, images, and locations.
- Have a camera or smartphone on hand to take photos of the plants and any distinctive features to help with later identification.
- Use GPS or mapping tools to note the exact locations of the plants.

Step 3: Plan Your First Exploration Trip

Choose a location where you can start your search for edible plants. This could be a park, nature trail, field, or your own backyard. Walk through the area and start observing your surroundings.

- Pick an area with diverse ecosystems—forests, meadows, and gardens are good places to find a variety of plants.
- Focus on areas with a mix of sunlight and shade, as different plants thrive in various conditions.

Step 4: Identify and Record Each Edible Plant

Start by identifying edible plants that you come across. For each plant, document key information.

- Record the plant's name (common and scientific, if possible), its location, and any distinctive features (color, size, shape, leaves, flowers, etc.).
- Take clear photographs of the plant from different angles to aid future identification.
- Note any additional information, such as when the plant blooms or bears fruit, what part of the plant is edible (roots, leaves, flowers, etc.), and whether it's commonly found in your area.
- Use a plant guide or an app to cross-reference and confirm your identification.

Step 5: Document Edible Parts and Uses

As you identify each plant, document the edible parts and any known culinary or medicinal uses.

- For each plant, write down what parts are edible (roots, stems, leaves, flowers, fruit, etc.).
- Record any preparation methods: can it be eaten raw, or should it be cooked? Does it need to be boiled or dried before consumption?
- If the plant has medicinal uses, make a note of those as well.

Step 6: Record the Plant's Growing Conditions

Understanding where and how a plant grows can help you return to find it later or cultivate it yourself. Note down the growing conditions for each plant.

- Record the type of soil the plant grows in (sandy, clay, loamy), the amount of sunlight it requires (full sun, partial shade, etc.), and its proximity to water (riverbanks, damp areas, dry areas).
- Also, take note of whether the plant grows in clusters or individually, and if it's native or invasive to the area.

Step 7: Map the Locations

Using a map or GPS tool, mark the exact locations of the plants you've found. This will help you track them in the future.

- If using a smartphone, mark the locations on a map app or take a screenshot to document each plant's location.
- If using paper maps, mark the location with a pen or pencil and create a reference legend for your notes.

Step 8: Collect Samples (Optional)

If you wish, collect small samples of plants for further examination or for later use.

- Use small bags or containers to collect leaves, flowers, or fruits.
- Make sure you leave enough of the plant in the wild to ensure it continues to thrive and propagate.
- If necessary, take additional photographs of the plants to help with identification back at home.

Step 9: Verify and Double-Check

After identifying and documenting the plants, go back and cross-check your findings. This can be done by researching plant identification online or consulting with local experts.

- If in doubt, seek assistance from local gardening groups, foraging communities, or botanical gardens.
- Cross-reference plant guides or apps to verify the plants' safety and edibility.

Step 10: Share and Contribute

After completing your documentation, consider sharing your findings with local community groups, foraging organizations, or online forums.

- Create a digital or physical guide to share with others interested in local edible plants.
- Contribute to a local database or library of wild plants, helping others learn about sustainable food sources.

Project: Making Hardtack Survival Bread

Hardtack is a simple, long-lasting survival food, perfect for off-grid living or emergency preparedness. This type of bread has been used for centuries by sailors, soldiers, and travelers due to its minimal ingredients, easy preparation, and impressive shelf life. In this project, you'll learn how to make your own hardtack survival bread. The process is simple, and the bread is highly durable, capable of lasting for years if stored properly.

Ingredients You Will Need
- 2 cups of all-purpose flour (or whole wheat flour if you prefer)
- 1/2 to 3/4 cup of water
- 1/2 teaspoon of salt
- Optional: 1 tablespoon of sugar or honey for sweetness (this is optional but can enhance flavor)
- Optional: Spices or herbs for added flavor (e.g., garlic powder, rosemary, etc.)

Tools Required
- Mixing bowl
- Measuring cups and spoons
- Wooden spoon or a dough scraper
- Rolling pin or flat surface for rolling out the dough
- Knife or dough cutter for cutting the dough
- Baking sheet
- Fork (for poking holes in the dough)
- Oven or campfire (to bake the bread)
- Airtight container or storage bags for storing the bread

Step 1: Mix the Ingredients

In a large mixing bowl, combine the flour, salt, and optional sugar or honey. Stir well to distribute the dry ingredients evenly. Gradually add water to the dry ingredients, a little at a time, until a stiff dough forms. Use your hands or a spoon to bring the dough together. You may need to add more or less water, depending on the humidity of your environment and the type of flour you're using.

If you're adding spices or herbs for flavor, now is the time to mix them into the dough.

Step 2: Knead the Dough

Once the dough has come together, knead it for 3-5 minutes on a clean, flat surface. The goal is to make the dough firm and smooth, but not sticky. If necessary, sprinkle a little more flour to prevent sticking. The dough should feel like a dense bread dough, and you should be able to handle it without it falling apart or becoming too wet.

Step 3: Roll the Dough

Next, roll the dough out to about 1/2 inch (1.25 cm) thick on a lightly floured surface. It doesn't need to be perfect in shape, but try to keep the thickness consistent throughout. You can roll it into a rectangle, square, or whatever shape you prefer.

Step 4: Cut the Dough into Pieces

Using a knife or dough cutter, cut the rolled dough into squares or rectangles. Each piece should be roughly 3-4 inches (7.5-10 cm) in size, but you can adjust the size based on your needs. These pieces will become your individual hardtack crackers.

Step 5: Poke Holes in the Dough

Using a fork, poke several holes across the top of each piece of dough. This will allow the air to circulate during baking, which will help dry the hardtack out and prevent it from puffing up too much.

Step 6: Bake the Hardtack

Preheat your oven to 375°F (190°C). Place the dough pieces on a baking sheet, making sure there's some space between each piece for even baking. Bake for 30-45 minutes, or until the edges are lightly browned and the pieces feel hard to the touch. You want the hardtack to be firm and dry, so don't worry if the pieces seem a bit overbaked — they should be crisp and dry to the core.

If you're cooking over a campfire or using an alternative heat source, you can bake the hardtack in a cast-iron skillet or on a griddle over low heat. Be sure to turn the bread occasionally to ensure even cooking on all sides.

Step 7: Dry the Hardtack (Optional but Recommended)

After baking, remove the pieces from the oven and let them cool on a wire rack for a few minutes. To make the hardtack even harder (and increase its shelf life), you can dry it out further by placing it back into the oven at a low temperature (around 200°F or 93°C) for another 30-60 minutes. This extra drying process will remove any remaining moisture, making the hardtack last much longer.

Step 8: Store Your Hardtack

Once your hardtack has fully cooled, store it in an airtight container or vacuum-sealed bags. For maximum shelf life, keep it in a cool, dry place. If you need to store it for long-term survival use, you can also keep it in a sealed mylar bag with oxygen absorbers to extend its freshness.

Hardtack can last for several months or even years when stored correctly. The drier the hardtack, the longer it will last, so make sure you've removed as much moisture as possible during the baking and drying process.

Final Notes

• **Serving Suggestions**: Hardtack is very hard when fresh, but it softens when dipped in water, tea, or soup. You can also crumble it into a porridge or stew for added nutrition.

• **Variations**: You can add herbs, spices, or even some sugar for flavor, depending on your preferences or the type of survival situation you are preparing for.

Project: Build a Simple Hydroponic Wick System

Materials Needed

• A plastic container or tub (for nutrient solution)
• Hydroponic nutrients
• Wick material (e.g., cotton rope or nylon wick)
• Growing medium (perlite, vermiculite, or coconut coir)
• Net pots or plastic cups (for holding plants)
• pH testing kit and pH adjustment solutions (optional but recommended)
• Seeds or seedlings
• Drill and bit for holes (optional)

Step 1: Prepare the Container

• Start by selecting a plastic container to hold the nutrient solution. A shallow tub or plastic bin works well for small systems. The size of the container will depend on the scale of your system.
• Clean the container thoroughly to remove any debris or dirt.

Step 2: Set Up the Wick System

- Measure and cut your wick material to an appropriate length. The wick should be long enough to reach from the bottom of the nutrient solution to the bottom of the growing tray or container.
- Insert one end of the wick into the growing medium, ensuring it has good contact with the medium to absorb moisture. The other end should be submerged in the nutrient solution.

Step 3: Prepare the Growing Medium

- Place your chosen growing medium into the net pots or plastic cups. The medium should be light and well-draining (perlite, vermiculite, or coconut coir are good choices).
- Make sure there is enough medium to hold the plants securely and keep the roots steady.

Step 4: Fill the Container with Nutrient Solution

- Mix your hydroponic nutrients according to the manufacturer's instructions. Make sure the water temperature is optimal, typically between 65-75°F (18-24°C).
- Pour the nutrient solution into the container, ensuring that the wick is submerged in the liquid while leaving some room for the plant roots to grow in the medium.

Step 5: Plant Your Seedlings

- Once the wick is set up and the solution is ready, place your seedlings into the net pots. The plants should be placed so that their roots will grow through the holes in the net pots and make contact with both the growing medium and the wick.
- Ensure the seedlings are upright and stable in the pots.

Step 6: Monitor and Adjust pH Levels

- It's important to regularly check the pH of your nutrient solution to make sure it's within the proper range (5.5-6.5). Use a pH testing kit to monitor levels.
- If the pH is too high, use pH down solution; if it's too low, use pH up solution to adjust. Keep a close eye on these levels, especially during the first few weeks.

Step 7: Maintain the System

- Over the next few weeks, monitor the water levels in the container. Top off the nutrient solution as needed to ensure the wick stays submerged and the plants are receiving enough water.
- Clean the wick system every month to ensure there's no algae growth or debris blocking the flow of water.

Step 8: Harvest the Plants

- Once the plants reach maturity, it's time to start harvesting! Begin picking leaves or produce as they ripen, which will also encourage further growth.
- For leafy plants, you can trim the leaves and allow the plant to continue growing.

Project: Set Up a Home-Scale Aquaponics System

Materials Needed:

- A 20-gallon fish tank (or larger)
- Water pump (submersible)
- Plastic tubing (to connect pump to grow bed)
- Grow bed (plastic or wooden container)
- Grow media (expanded clay pellets or gravel)
- Fish (tilapia, goldfish, or other hardy species)
- Fish food
- Plants (lettuce, herbs, or other fast-growing vegetables)

- PVC pipes for the overflow system
- Water pH test kit and pH adjustment solution
- Net pots or trays for the plants
- Drill and bit (for holes in the grow bed)

Step 1: Set Up the Fish Tank
- Choose a location that receives good natural light or is suitable for keeping fish. Set up the tank and make sure it's level.
- Fill the tank with dechlorinated water. If using city water, let it sit for 24 hours or treat it to remove chlorine.
- Install the pump inside the tank, ensuring it's positioned to circulate water effectively.

Step 2: Install the Grow Bed
- Position the grow bed above the fish tank. It should sit securely and be able to drain back into the tank after water flows over the plants.
- Drill holes in the grow bed for the water inlet pipe and overflow system. The inlet pipe will carry the water from the tank to the grow bed, and the overflow pipe will return excess water to the tank.

Step 3: Install Water Circulation System
- Connect the pump to the inlet pipe using the plastic tubing. The pump should push water into the grow bed, where it will flow over the grow media.
- Set up the overflow system, ensuring that the water will flow back into the fish tank through the overflow pipe.

Step 4: Prepare the Grow Media
- Fill the grow bed with the chosen grow media (expanded clay pellets or gravel). The media should support the plant roots and allow for proper water flow.
- Make sure the media fills the bed without blocking the flow of water to the plants.

Step 5: Add Fish to the Tank
- Choose hardy fish species that will thrive in the environment, such as tilapia or goldfish.
- Introduce the fish to the tank once the water conditions are stable. Feed them fish food according to their needs.

Step 6: Plant Your Crops
- Once the fish are settled in and the water circulation system is running smoothly, it's time to plant.
- Use net pots or trays to hold your plants. Choose fast-growing vegetables like lettuce or herbs. Place the plants in the grow media, ensuring their roots make contact with the water.

Step 7: Monitor Water Quality
- Regularly check the water's pH, ammonia, nitrate, and nitrite levels. Ideal pH levels should be between 6.5 and 7.5 for both the fish and plants.
- Adjust the water as necessary using pH up or down solutions. Monitor for any signs of nutrient deficiencies or imbalances.

Step 8: Maintain the System
- Monitor the fish health and feed them regularly.
- Clean the filters in the pump as needed to prevent blockages.
- Check plant growth and remove any dead or decaying matter from the system.
- If any problems arise, such as algae growth or low nutrient levels, take corrective action by adjusting the water, nutrient levels, or system components.

Step 9: Harvest Plants and Fish
- Once the plants are mature, begin harvesting. For leafy plants, trim the leaves and allow the plant to continue growing.
- If you wish to harvest fish, wait until they reach the appropriate size and maturity, then carefully remove them from the tank.

Module G. First Aid & Medical Emergency Preparedness

In any off-grid living situation, being prepared for medical emergencies is essential. Whether you're managing your homestead, working in the wilderness, or simply living away from traditional healthcare access, having the knowledge and tools to respond to injuries and illnesses can make all the difference. This module will guide you through key aspects of emergency medical preparedness, first aid, and trauma care, specifically tailored for off-grid environments. You'll learn how to manage common medical emergencies, assess and treat injuries, and provide life-saving interventions when professional help is unavailable. We'll also cover critical skills like basic life support, infection control, and psychological first aid, ensuring you're ready to handle emergencies calmly and effectively. Whether you're a seasoned off-gridder or just starting, mastering these skills will empower you to take control of your health and safety in any situation.

33. Emergency Medical Preparedness & First Aid

In an off-grid environment, you're often your first responder. Knowing how to handle medical emergencies could mean the difference between life and death. Whether it's treating a wound, performing CPR, or preventing infection, this section gives you the practical skills and tools you need to act swiftly and effectively. From building your first aid kit to understanding the basics of life support, you'll learn to manage injuries and medical situations with confidence—no matter where you are.

Understanding Basic First Aid

First aid is the initial care you give to someone who is injured or ill. Knowing what to do in these critical first moments is essential to preventing a situation from worsening before professional help arrives—or in the absence of help at all. You must be prepared to act quickly and correctly, assessing the severity of the situation and deciding what kind of action is needed.

Start by learning the common life-threatening conditions: severe bleeding, difficulty breathing, and unconsciousness. You can save a life by knowing how to identify and act on these signs early. Some basic first aid skills you need to know include CPR, controlling bleeding, and dealing with choking. These simple, effective steps can make all the difference.

Basic First Aid Skills

• **CPR (Cardiopulmonary Resuscitation):** CPR is performed when a person's heart stops beating or they stop breathing. The main actions in CPR are chest compressions and rescue breaths. Follow these steps:
1. Place the person on a firm surface, ideally a flat floor.
2. Place your hands, one on top of the other, in the center of their chest.
3. Push down hard and fast at a rate of 100-120 compressions per minute.
4. After every 30 compressions, give 2 rescue breaths.

• **Choking:** If someone is choking and cannot breathe, act immediately:
1. If the person is conscious, perform the Heimlich maneuver by giving abdominal thrusts.
2. If they lose consciousness, begin CPR immediately.

• **Bleeding Control**
• Apply direct pressure to the wound using a clean cloth or bandage. If bleeding is severe, use a tourniquet above the wound to stop the flow.
• Elevating the limb can also help reduce bleeding.
• Apply pressure for at least 10 minutes to allow clotting.

First Aid Kit Essentials

A well-equipped first aid kit is the foundation of any off-grid emergency plan. It's crucial to have the right tools on hand to treat common injuries, illnesses, and conditions. Here's what you'll need:

• **Bandages and Gauze Pads:** For covering wounds and preventing infection.

- **Antiseptic Solutions:** To clean and disinfect cuts and scrapes.
- **Sterile Gloves:** To protect yourself when providing first aid.
- **Pain Relievers:** Aspirin, ibuprofen, and acetaminophen for pain and inflammation.
- **Thermometers:** To check for fever.
- **Scissors and Tweezers:** For cutting bandages and removing foreign objects like splinters.
- **CPR Mask:** To protect both you and the person you're helping when performing CPR.

Customizing Your First Aid Kit

Customize your first aid kit based on where you live and the environment you're in. For instance, if you live in an area where you are likely to experience bites or stings, include an anti-venom kit or antihistamines. If you're living off-grid and away from hospitals, consider adding more advanced tools such as a splint or clotting agents.

Organizing and Maintaining Your First Aid Kit

Keep your first aid kit easily accessible, such as in a labeled, waterproof container. Regularly check the kit to ensure all items are present, clean, and not expired. Replace any used or expired items promptly.

Assessing Medical Emergencies

Knowing how to assess the severity of an injury or illness is key to effective first aid. If you arrive at an emergency scene, you must quickly determine whether it's life-threatening and decide if the situation requires professional medical help.

- **Breathing:** If the person isn't breathing or breathing abnormally, begin CPR.
- **Bleeding:** If there's excessive bleeding, apply pressure immediately.
- **Consciousness:** If someone is unconscious and unresponsive, check their airway and breathing.

When in doubt, call for help or transport the person to a medical facility. Time is critical.

Understanding Vital Signs

Vital signs give you insight into a person's condition. These include:

- **Heart rate:** Normal is about 60-100 beats per minute.
- **Breathing rate:** Normal is 12-20 breaths per minute.
- **Blood pressure:** Ideal is around 120/80 mmHg.
- **Temperature:** Normal is about 98.6°F (37°C).

Knowing these can help you assess the severity of the situation. A rapidly increasing or decreasing heart rate, for instance, may indicate shock or internal bleeding.

Handling Common Medical Emergencies

Burns, Cuts, and Scrapes: When someone sustains a burn or cut, you should:

- Clean the wound immediately, apply sterile bandages, and elevate the injured area if necessary.
- For burns, cool the burn under running cold water for 10 minutes and cover with a sterile bandage.

Sprains and Fractures:
- For sprains or strains, rest the injured area and apply ice to reduce swelling.
- For fractures, stabilize the injured area by using a splint made of sticks and cloth.

Heart Attacks and Strokes: For heart attack symptoms (chest pain, shortness of breath, nausea), encourage the person to remain calm and take aspirin if they are not allergic. For strokes (facial drooping, slurred speech, weakness on one side), time is critical. Call emergency services immediately.

Signs of Infection and Wound Care

Watch for signs of infection like redness, warmth, swelling, or pus in a wound. If you notice these signs, clean the area again and reapply fresh bandages. For deeper wounds, consider using antiseptic ointments or consulting a professional if available.

Infection Prevention and Control

In off-grid environments, infections can escalate quickly. To avoid this:
- Always clean your hands thoroughly before treating any wound.
- Sterilize your tools with boiling water or alcohol after each use.
- Use sterile bandages to cover wounds and change them regularly.

Basic Life Support (BLS)

Basic Life Support (BLS) includes techniques like CPR, using an Automated External Defibrillator (AED), and managing an airway. In an off-grid environment, it's essential to be familiar with BLS because you may need to act as the sole provider of care.

Psychological First Aid

In addition to physical care, emotional well-being is critical during and after an emergency. Psychological first aid includes providing comfort, ensuring safety, and listening actively to the person's needs. You may need to offer stress management techniques such as deep breathing to help the person relax.

Stress Management Techniques for Victims and Responders

In stressful situations, you need to manage both your own emotions and the person's. Some helpful techniques include:
- **Deep breathing exercises** to calm yourself and the person in distress.
- **Positive affirmations** to encourage a sense of control and safety.
- **Focusing on the present moment** to avoid overwhelming thoughts.

34. Trauma Care & Managing Medical Emergencies Off-Grid

Living off the grid offers great freedom and independence, but it also requires you to be self-reliant when it comes to handling emergencies. Trauma care and managing medical emergencies off-grid require more than just basic first aid. In a remote environment, the absence of immediate professional medical help means you must act decisively and efficiently. Whether it's a severe injury caused by a fall, an animal attack, or exposure to extreme temperatures, knowing how to properly assess and treat trauma can make a huge difference.

Understanding Trauma

Trauma is any physical or emotional injury caused by a sudden, external force. When you're off-grid, the risk of trauma is higher, as you may be more exposed to dangerous situations. Trauma can be classified into several categories: physical trauma, emotional trauma, and psychological trauma. For the purposes of managing medical emergencies, the focus is on physical trauma, which includes injuries such as cuts, fractures, burns, and more.

Types of Trauma

Blunt Trauma: This type of trauma occurs when a force impacts the body without breaking the skin. Examples include injuries from falls, car accidents, or blunt objects. The injury can lead to bruising, internal bleeding, or organ damage.

Penetrating Trauma: Penetrating trauma happens when an object enters the body, such as from stabbing, shooting, or an animal bite. This type of injury is often severe and may require immediate care to control bleeding and prevent infection.

Burns: Burns occur when the skin is exposed to heat, chemicals, electricity, or radiation. They can range from mild to severe, with third-degree burns being the most dangerous. Burns require careful treatment to prevent infection, manage pain, and promote healing.

Frostbite and Hypothermia: Cold-related trauma occurs when the body is exposed to extreme cold for extended periods. Frostbite affects extremities like fingers, toes, and ears, while hypothermia occurs when the body temperature falls below safe levels.

Initial Trauma Assessment and Prioritization (ABCs)

When you come across someone with a traumatic injury, your first priority is to assess their condition.

Follow the ABCs of trauma care to determine the most urgent needs:

1. **Airway**: Check if the person's airway is open. If the person is unconscious or has difficulty breathing, clear the airway by tilting the head back and lifting the chin.
2. **Breathing**: Assess the person's breathing. If they are not breathing or are gasping for air, begin CPR immediately.
3. **Circulation**: Check for signs of bleeding. If there's severe bleeding, apply direct pressure to the wound and elevate the injured area. If the person is in shock, maintain their body temperature and keep them calm while waiting for further medical assistance.

Managing Trauma Injuries

Once you've completed an initial assessment, it's important to take action to treat any injuries. Each type of injury requires specific care to prevent complications and ensure the best possible outcome.

Stopping Severe Bleeding

Severe bleeding is one of the most immediate threats in trauma care. Follow these steps to control bleeding:

1. **Direct Pressure**: Apply pressure to the wound using a clean cloth or bandage. Press firmly to help stop the bleeding.
2. **Tourniquets**: If bleeding is not controlled by direct pressure, apply a tourniquet above the wound. Secure it tightly and ensure the bleeding stops. Note the time the tourniquet was applied.
3. **Elevation**: If possible, elevate the injured limb above the level of the heart to reduce blood flow to the area.

Immobilizing Fractures and Dislocations

Fractures and dislocations can cause significant pain and, if not treated properly, may lead to long-term damage. Here's how to stabilize the injury:

1. **Splints**: Use available materials like sticks, cloth, or boards to create a splint. Secure the splint above and below the injury site, ensuring the bone is immobilized.
2. **Do Not Move the Limb**: Avoid moving the limb unnecessarily. Only move the person if the environment is unsafe (e.g., they're at risk of further injury from exposure or danger).
3. **Pain Management**: Offer pain relief, such as over-the-counter medications, if available. Use caution with pain management if you're unsure of the person's medical history.

Treating Burns, Frostbite, and Hypothermia

Burns, frostbite, and hypothermia all require specific care to prevent long-term damage and complications:

Burns:
1.1. Cool the burn under running cold water for at least 10 minutes.
1.2. Avoid breaking blisters or applying ointments unless advised by a medical professional.
1.3. Cover the burn with a clean, non-stick bandage.

Frostbite:
1.4. Gradually warm the affected area by immersing it in warm (not hot) water.
1.5. Do not rub the frostbitten area, as it may cause further tissue damage.

Hypothermia:
1.6. Move the person to a warmer area and remove any wet clothing.
1.7. Use warm blankets or clothing to rewarm them gradually.
1.8. Offer warm, non-alcoholic beverages to help raise body temperature.

Advanced Wound Care

When you're managing an off-grid trauma situation, it's important to have the skills to clean and dress wounds properly. This helps to prevent infection and promotes healing.

Cleaning Wounds:
1.1. Use clean water or saline solution to rinse out dirt and debris.

1.2. Avoid using alcohol or hydrogen peroxide, as these can damage healthy tissue.

Dressing Wounds:

1.3. Apply a sterile dressing or bandage to cover the wound.
1.4. Change the dressing regularly, especially if it becomes wet or dirty.

Infection Control:

1.5. Monitor the wound for signs of infection, including redness, warmth, swelling, or pus.
1.6. If signs of infection appear, clean the wound again and apply antiseptic ointment.

Treating Shock

Shock is a life-threatening condition that can occur after trauma, blood loss, or infection. It requires immediate attention to stabilize the person.

Causes of Shock:

1.1. Loss of blood, severe burns, or dehydration can all lead to shock.
1.2. Shock reduces the oxygen supply to vital organs, which can result in organ failure.

Managing Shock:

1.3. Keep the person warm by covering them with blankets to prevent further loss of body heat.
1.4. Keep them calm and still to prevent an increased heart rate or blood loss.
1.5. Elevate the person's legs if possible to improve blood flow to vital organs.

Monitoring Vital Signs:

1.6. Check the person's pulse, breathing, and skin color regularly.
1.7. If the person's condition worsens, seek professional help immediately.

Pain Management in Off-Grid Environments

Pain is a common symptom of trauma and injury. While you may not have access to the full range of medical pain relief options in an off-grid setting, there are methods to manage pain effectively.

Recognizing Pain:

1.1. Pain may present as discomfort, sharp pain, or even shock symptoms.
1.2. Be mindful of the person's facial expressions, body language, and verbal cues.

Using Natural Remedies:

1.3. Herbs like willow bark (a natural source of salicin) or ginger can provide pain relief.
1.4. Hot or cold compresses can also help reduce swelling and numb pain.

Medication:

1.5. If you have access to pain-relieving medication like ibuprofen or acetaminophen, use them according to the recommended dosage.

Transporting Injured Individuals

In off-grid situations, you may need to move someone who has been injured. This can be dangerous if not done properly. The goal is to move them safely without causing additional harm.

Lifting and Carrying:

1.1. If the injury is to the legs or feet, you may need to use a makeshift stretcher to lift the person.
1.2. If necessary, use a buddy system for support during the transport.

Makeshift Stretchers:

1.3. Use available materials like long boards, poles, or sturdy cloth to create a simple stretcher.
1.4. Ensure the person is securely tied down to prevent further injury during transport.

Communication:
- 1.5. If you're in a remote area and need outside help, use a satellite phone or radio to communicate your location and the severity of the injury.
- 1.6. Provide clear, concise information to emergency responders.

Off-Grid Medical Resources

When living off the grid, you may not have access to traditional healthcare services. However, there are alternative methods you can use to ensure you're prepared for medical emergencies.

1. **Alternative Medical Supplies**:
    - 1.1. Stock up on medical supplies such as bandages, antiseptic solutions, and over-the-counter medications.
    - 1.2. If you're running low on supplies, consider bartering or trading with others in the community.
2. **Using Natural Remedies**:
    - 2.1. Herbal medicine can help treat minor injuries and illnesses. Plants like aloe vera, lavender, and comfrey are great for soothing burns and cuts.
    - 2.2. Always ensure you're properly identifying plants to avoid toxic substances.
3. **Telemedicine and Remote Consultations**:
    - 3.1. Use telemedicine services to consult with healthcare professionals when necessary.
    - 3.2. You may be able to connect via satellite phones or radio to get advice from medical professionals.

Managing Medical Emergencies in Remote Areas

Living off the grid means being prepared for medical emergencies that could occur at any time. From extreme weather conditions to wildlife encounters, there are several factors to consider.

1. **Planning for Emergencies**:
    - 1.1. Create a first-aid plan for your family and community. Know how to handle common emergencies like bites, hypothermia, and heat exhaustion.
2. **Key Preparations**:
    - 2.1. Set up emergency shelters, emergency kits, and fire-starting tools.
    - 2.2. Develop emergency plans with neighbors to ensure mutual support when an emergency arises.

35. First Aid & Medical Emergency Preparedness Projects

Project: Create Your Own First Aid Kit

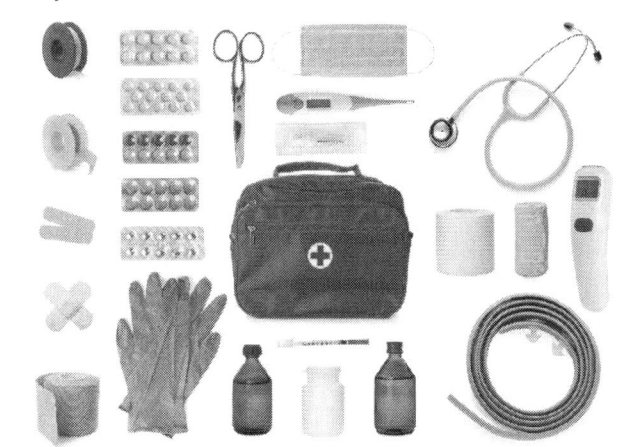

Materials Needed

- Empty, durable container (waterproof and easy to access)
- Adhesive bandages (various sizes)
- Gauze pads and rolls
- Medical tape
- Antiseptic wipes or solution
- Alcohol pads
- Tweezers
- Scissors
- Disposable gloves
- Thermometer
- Cotton balls or swabs
- Instant cold packs
- Eye wash or saline solution
- Burn cream or gel
- Pain relievers (e.g., aspirin, ibuprofen)

- Antihistamines
- Bandage for sprains or strains (elastic bandage)
- First-aid manual or printed instructions
- Hand sanitizer
- Emergency contact numbers
- Flashlight and spare batteries
- Pain relief ointment (e.g., Neosporin)
- Antiseptic cream
- Moleskin for blisters
- EpiPen (if necessary)

Step 1: Choose the Right Container
- Choose a waterproof, durable container with compartments, making it easy to organize the items. It should be large enough to fit all the necessary items but compact enough to store and transport easily. A plastic bin, tool box, or first aid case works well.

Step 2: Organize Your Kit
- Begin by organizing the kit into categories such as:
- Bleeding Control (bandages, gauze, medical tape)
- Pain and Injury Management (pain relievers, bandage for strains)
- Wound Care (antiseptic wipes, ointments, tweezers)
- Emergency Supplies (cold packs, flashlight)
- Personal Medications (antihistamines, EpiPen)
- Basic First-Aid Supplies (scissors, gloves, cotton balls)
- Reference Materials (first-aid manual, emergency contact list)

Step 3: Check for Expiry Dates
- Examine all medications, ointments, and bandages for expiration dates. Replace any expired items to ensure their effectiveness in an emergency. Regularly check your first aid kit to refresh supplies, especially medications and bandages.

Step 4: Label Your First Aid Kit
- Clearly label your first aid kit so that anyone can quickly identify it in an emergency. Include instructions for basic first aid, and place emergency numbers and a first-aid manual inside for reference.

Step 5: Personalize Your Kit

- Add any personal items you may need, such as an asthma inhaler, allergy medications, or a spare EpiPen if you or a family member has specific health conditions. Keep a small notebook for noting any changes in health or additional needs.

Step 6: Store Your Kit in a Readily Accessible Location
- Store your kit in a location that is easy to access in case of an emergency. Consider storing it in a central location at home, and also have a smaller version that you can take with you when traveling.

Project: Trauma Care Simulation Exercise

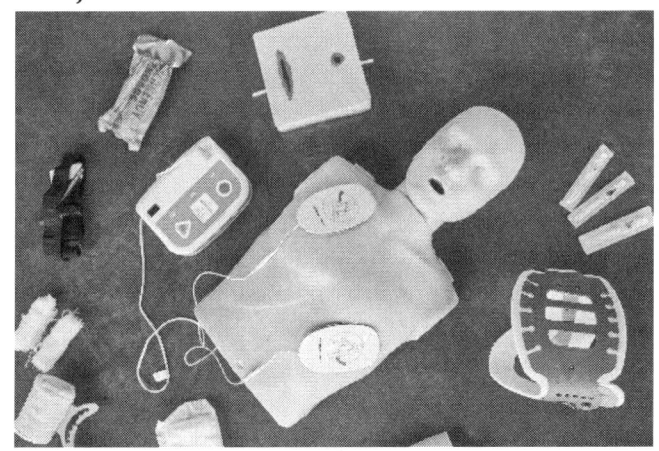

Materials Needed

- A willing volunteer (for the exercise)
- Bandages and gauze
- First aid kit (from previous project)
- Timer or stopwatch
- Simulation injuries (use makeup or fabric for bruises, cuts, and sprains)
- A safe area to conduct the simulation (preferably outdoors or in a large space)
- A blanket or makeshift stretcher (if needed)
- Water for washing injuries (for practice)

Step 1: Plan the Scenario
- Decide on the trauma scenario you want to simulate. This could include a sprained ankle, deep cut, or even a heatstroke situation. If you're practicing multiple skills, you could simulate a series of injuries (e.g., a fall that leads to a head

injury, bleeding, and dehydration). Write down the details to make the exercise feel realistic.

Step 2: Prepare the Area
- Create a safe space to perform the simulation. You will need space for movement and for setting up any tools. If the scenario involves moving the injured party, set up a safe path for transport. Clear any obstacles and make sure the area is easy to navigate.

Step 3: Apply the Simulation Injuries
- Using makeup or fabric, simulate the injuries on your volunteer. For example, use red makeup or fake blood to simulate cuts, bruises, and abrasions. For sprains or fractures, create realistic swelling with fabric and tape. Ensure that the volunteer is comfortable and aware of their role in the exercise.

Step 4: Perform Initial Assessment
- Start the exercise by assessing the "injured" volunteer. Check the injury using the ABCs (Airway, Breathing, Circulation). This will help you prioritize care based on the severity of the situation. Use the first-aid kit to clean, bandage, or immobilize the simulated injuries.

Step 5: Stop Severe Bleeding
- If the simulated injury involves bleeding, practice applying direct pressure with gauze or clean cloth. For severe bleeding, practice the application of a tourniquet, if necessary. Understand when and how to use each technique.

Step 6: Immobilize Fractures
- For simulated fractures or dislocations, practice immobilizing the injured limb using splints, bandages, or other available materials. This step should ensure that the injury is stabilized before moving the injured person.

Step 7: Apply the Right Pain Relief
- If the simulation involves pain, practice applying the right methods of pain relief. This could include applying a cold pack, giving pain-relief medications (if available), or immobilizing the injured area to minimize discomfort.

Step 8: Handle Shock
- If the simulated injury involves signs of shock (e.g., pale skin, rapid pulse, confusion), practice the steps for shock management. This could involve keeping the injured party warm, elevating their legs, and ensuring they are hydrated. Pay attention to monitoring vital signs and checking for signs of shock.

Step 9: Practice Communication for Off-Grid Settings
- In a real emergency, you may need to communicate with medical professionals or responders. Practice how you would notify them about the situation. Use a phone, radio, or other off-grid communication systems to simulate notifying emergency services.

Step 10: Simulate Transporting the Injured Person
- Practice how you would move an injured person from the scene of the injury to safety or a medical facility. Depending on the injury, you may need to improvise a stretcher or use a blanket to transport them.

Step 11: Evaluate Your Response
- After the simulation, evaluate your response and the care you provided. Consider the steps you took and identify any areas for improvement. How well did you assess the injury? How efficient was the application of first aid? Did you follow the appropriate trauma care protocols?

Step 12: Review and Reiterate
- To reinforce learning, repeat the scenario or switch to different types of injuries (e.g., heatstroke, burns, or sprains). Practicing these skills regularly will ensure that you are always prepared to handle medical emergencies effectively.

Module H. Cooking, Heating, Cooling & Lighting

Living off the grid means embracing self-sufficiency, and one of the most important aspects of that is learning how to cook, heat, cool, and light your home without relying on external power sources. This module explores the key systems and techniques you can implement to maintain comfort and functionality in your off-grid home. From alternative cooking methods like solar ovens and wood stoves to efficient heating and cooling systems that work without electricity, this section covers the essentials for off-grid living. Additionally, you'll learn how to power your home with off-grid lighting solutions, ensuring you can live sustainably while minimizing your environmental impact. Whether you're looking to preserve resources or simply seeking independence from grid systems, mastering these skills is essential for a successful off-grid lifestyle.

36. Cooking Without Power: Alternative Cooking Methods

When living off the grid, finding alternative cooking methods is key to self-sufficiency. In this section, we will explore four common techniques for cooking without relying on electricity: solar cooking, wood-fired cooking, fuel-based cooking, and the use of clay and stone ovens. Each of these methods has its unique benefits, challenges, and ideal applications, allowing you to choose the most suitable option based on your environment, resources, and cooking preferences.

Solar Cooking

What is it and how it works

Solar cooking harnesses the power of the sun to cook food. It relies on solar energy, captured by reflective panels and concentrated in a cooking chamber, to heat food without the need for a fire or stove. Solar cookers can range from simple devices like box cookers to advanced solar ovens. These systems use reflective materials (such as mirrors or aluminum foil) to direct sunlight into an insulated cooking area. The heat builds up inside the cooking chamber and is trapped, cooking food at temperatures that can range from 200°F to 350°F (93°C to 177°C), depending on the design and weather conditions.

When it is recommended

Solar cooking is ideal in sunny climates with minimal cloud cover, as it relies directly on sunlight. It's best suited for slower cooking methods, such as baking, roasting, or simmering, and is particularly effective for foods like stews, casseroles, and baked goods. Solar cookers are also an energy-efficient option when cooking for longer periods, as they require no fuel and produce no emissions. If you're in a region that experiences consistent sunshine, solar cooking can be a reliable and sustainable option for preparing meals.

Wood-Fired Cooking

What is it and how it works

Wood-fired cooking is one of the oldest and most traditional methods, where food is cooked using direct heat from a wood-burning stove or firepit. This method typically uses firewood as a fuel source, which is burned to generate heat. Wood-burning stoves and ovens, such as rocket stoves or traditional brick ovens, provide a steady and controllable heat source. They can be used for a variety of cooking techniques, from grilling to baking, and can also be used to boil water.

When it is recommended

Wood-fired cooking is highly effective in colder climates where a steady, reliable heat source is needed for both cooking and warmth. It is recommended for off-grid homes with easy access to firewood. It is particularly useful for larger, long-duration meals such as stews or roasts. Additionally, wood-fired cooking offers the benefit of creating warmth in your home during the cooking process, which can reduce heating costs in the winter.

Fuel-Based Cooking

What is it and how it works

Fuel-based cooking includes methods such as cooking with propane, butane, kerosene, or other fuel sources. These cooking methods typically use portable stoves or ovens that burn the fuel to create heat for cooking. They are versatile and can be used both indoors and outdoors. Propane stoves are the most common example of fuel-based cooking in off-grid settings. They provide a reliable source of heat and can quickly bring water to a boil or heat up a meal.

When it is recommended

Fuel-based cooking is recommended when wood is not readily available or in situations where solar or wood-fired cooking is not practical. It's ideal for areas where the weather may be too unpredictable for consistent solar cooking. Fuel-based cooking is also convenient for mobile setups, such as RVs or small cabins. However, it's important to ensure proper ventilation when using fuel-based cooking systems to avoid the buildup of harmful gases. Fuel-based stoves work well in colder climates where steady heat is necessary, but be mindful of the cost and storage of the fuel.

Clay and Stone Ovens

What is it and how they work

Clay and stone ovens, often called earth ovens or cob ovens, are traditional cooking methods that use thick earthen walls to retain and radiate heat. These ovens are typically built using clay, sand, and straw to form an insulated cooking chamber. Once the oven is heated by a fire (usually using wood), it can maintain a high temperature for an extended period, allowing for baking, roasting, and slow cooking. The heat stored in the stone or clay is evenly distributed and can cook a variety of dishes.

When it is recommended

Clay and stone ovens are ideal for long-term off-grid living, especially in warmer climates. They are most beneficial when you need a consistent heat source for baking bread, pizzas, or roasting large amounts of food. These ovens can also be used as a multi-purpose cooking tool, providing a reliable cooking source over extended periods. Since they require a significant upfront investment in materials and labor to construct, they are most suitable for off-grid homesteads where the oven can serve as a central cooking method over time.

37. Heating Systems: Warming Your Space Efficiently

In off-grid settings, managing your home's temperature efficiently is crucial for comfort and energy conservation. Several heating options can be used, depending on the resources available and your climate. This section will explore different heating methods, including wood stoves, rocket stoves, passive solar heating, thermal mass, and alternative heating options. Each of these methods comes with unique benefits, challenges, and practical considerations for different off-grid environments.

Wood Stoves

What is it and how it works

Wood stoves are the most common and effective way to heat an off-grid home. They burn firewood, providing both heat and, in some cases, the ability to cook food. The stove is typically vented through a chimney, allowing the smoke to escape while the heat radiates throughout the room. Modern wood stoves are highly efficient and designed to burn fuel slowly, producing a consistent and controllable heat source. Many wood stoves are equipped with a cooktop, allowing you to use them for cooking as well as heating.

When it is recommended

Wood stoves are recommended in cold climates, especially when access to electricity is limited or unreliable. They are ideal for larger off-grid homes or cabins that require consistent heat throughout the winter. Wood stoves are also beneficial for areas with access to firewood, as they provide both heating and cooking capabilities. Keep in mind that wood stoves require regular maintenance, such as cleaning the chimney and removing ash, to ensure they work efficiently.

Rocket Stoves

What is it and how it works

Rocket stoves are highly efficient wood-burning stoves that use a small, insulated combustion chamber to burn wood efficiently with minimal fuel. The design allows for extremely high combustion temperatures, which results in more complete burning and less smoke. Rocket stoves can be used for both heating and cooking, with many models designed to burn twigs and small pieces of wood, making them ideal for areas with limited access to larger firewood.

When it is recommended

Rocket stoves are best for smaller spaces or for areas where wood needs to be burned efficiently to reduce fuel consumption. They are particularly useful for cooking and heating in areas where resources are limited, as they use less fuel than traditional wood stoves. Rocket stoves work well in both rural and urban off-grid settings, as they can be used for both outdoor and indoor heating. They are also excellent for emergency heating or cooking during power outages.

Passive Solar Heating

What is it and how it works

Passive solar heating is a method of heating your home using the sun's natural energy. By positioning your home to take advantage of sunlight during the day, passive solar design captures and retains heat in your home. Large south-facing windows, thermal mass (such as concrete floors or brick walls), and strategically placed insulation can help store the sun's heat during the day, releasing it into the home during cooler evenings or overcast days.

When it is recommended

Passive solar heating is recommended for homes in climates where sunlight is consistent and strong, particularly in colder months. It works well in areas that receive ample sunlight throughout the year. The system is ideal for new homes or homes being designed with energy efficiency in mind, as it can significantly reduce heating costs. It is also a low-maintenance solution, as it relies on the natural heating ability of the sun rather than mechanical systems. Passive solar heating can be used in combination with other heating methods to optimize energy use.

Thermal Mass

What is it and how it works

Thermal mass refers to the ability of materials (such as concrete, brick, or stone) to absorb, store, and later release heat. In homes with high thermal mass, these materials can absorb excess heat from the sun during the day and release it at night, reducing the need for artificial heating. Thermal mass can be incorporated into floors, walls, or even ceilings to help regulate indoor temperatures.

When it is recommended

Thermal mass is best used in homes located in climates with temperature fluctuations between day and night. If you are in a region where it gets hot during the day and cools down significantly at night, thermal mass can help maintain a stable temperature. This method is best used in combination with passive solar heating to optimize energy efficiency. For homes that need minimal supplemental heating, thermal mass can be an essential component in keeping the space comfortable without additional energy input.

Alternative Heating (Compost Piles, Biogas, Propane and Kerosene)

Heating with Compost Piles or Biogas

Composting piles and biogas are alternative heating methods that utilize organic waste to generate heat. When organic material breaks down, it releases heat, which can be harnessed for heating purposes. In addition, biogas systems, which capture methane from decomposing waste, can be used for heating and cooking.

When it is recommended

Composting piles and biogas systems are most effective for off-grid homes located in areas with abundant organic waste, such as farms or rural properties. These methods are ideal for individuals who want to recycle organic waste while simultaneously heating their homes.

Propane and Kerosene Heaters

Propane and kerosene heaters are portable, efficient, and relatively easy to use. These heaters burn propane or kerosene to produce heat and can be used in a variety of settings, from small rooms to larger cabins.

When it is recommended

These heating methods are ideal in off-grid homes where space is limited, or for individuals who need supplemental heat in addition to wood-burning or passive solar systems. Propane and kerosene heaters are also effective for temporary heating solutions, such as in emergency situations or during power outages.

Heating Efficiency Considerations

Insulation and Weatherproofing Techniques

Proper insulation and weatherproofing can significantly reduce the amount of heat needed to keep your home warm. Using high-quality insulation in your walls, ceilings, and floors ensures that heat stays inside during the winter and remains cool in the summer. Weatherproofing your windows and doors also prevents drafts and helps to maintain consistent temperatures.

Managing Heat Loss through Windows and Doors

One of the biggest sources of heat loss is through windows and doors. By adding double-glazed windows or weatherstripping around doors, you can reduce drafts and prevent heat from escaping. In colder climates, heavy curtains or thermal blinds can also help retain warmth.

Temperature Regulation and Comfort

Maintaining a consistent temperature throughout your home is crucial for comfort. Use programmable thermostats or passive systems like thermal mass to help regulate the temperature in your home. A balanced approach to heating—using multiple methods, such as passive solar, wood stoves, and thermal mass—can keep your home warm without relying too heavily on any single method.

38. Natural Cooling: Keeping Your Home Comfortable

A well-designed off-grid home must remain comfortable even in the peak of summer without relying on energy-intensive air conditioning. Natural cooling methods use passive strategies, efficient materials, and smart ventilation to regulate indoor temperatures. By designing your home with airflow, thermal mass, and shading in mind, you can create a comfortable living space that remains cool in hot conditions.

Ventilation and Airflow

Air movement is one of the most effective ways to regulate temperature and humidity inside a home. Proper ventilation allows warm air to escape while drawing in cooler air, reducing heat buildup.

- **Cross ventilation**: Ensure windows or vents are placed on opposite sides of the home to create a continuous airflow path.
- **Stack effect**: Warm air rises, so high vents or windows allow hot air to escape while cooler air is drawn in through lower openings.
- **Operable vents and louvers**: Adjustable vents at different heights enable controlled airflow throughout the day.
- **Chimney ventilation**: A vented roof or attic space prevents heat from accumulating at the top of the house.

- **Clerestory windows**: Placed high on walls, these windows allow hot air to escape while bringing in diffused natural light.

Thermal Mass

Thermal mass refers to materials that absorb and store heat during the day and release it at night, helping regulate indoor temperatures.

- **Concrete floors and walls**: Thick concrete absorbs heat during the day and gradually releases it when temperatures drop.
- **Stone or brick interiors**: Dense materials like stone or adobe store heat energy and balance temperature fluctuations.
- **Water barrels or tanks**: Water has a high thermal capacity, making it an effective passive cooling element.
- **Earth-sheltered designs**: Walls partially built into hillsides or underground provide natural insulation against extreme heat.

To maximize effectiveness:

- **Position thermal mass strategically**: Place heat-absorbing materials where they receive direct sunlight during cooler seasons but remain shaded in summer.
- **Combine with shading**: Prevent excessive heating by blocking midday sun exposure.

Green Roofs and Walls

Green roofs and walls add a layer of insulation and reduce heat absorption by using plants to shield your home.

1. **Green roof construction**:
   1.1. Install a waterproof membrane to prevent leaks.
   1.2. Add a drainage layer to direct excess water away.
   1.3. Use lightweight soil and drought-resistant plants to minimize maintenance.
2. **Vertical gardens**:
   2.1. Attach trellises or hanging planters to exterior walls.
   2.2. Use fast-growing climbers like ivy or passionflower for shade.
   2.3. Install an irrigation system or rainwater catchment to sustain the plants.

Shading and Insulation

Blocking direct sunlight before it reaches your home significantly reduces heat buildup.

1. **Shading structures**:
   1.1. Install **awnings** or overhangs to block high-angle summer sun while allowing winter sunlight.
   1.2. Use **pergolas** with climbing plants to provide seasonal shade.
   1.3. Build **porches or verandas** to shade exterior walls.
2. **Natural shading materials**:
   2.1. **Bamboo mats or reed screens** filter sunlight while allowing airflow.
   2.2. **Hanging fabric shades** or **woven grass panels** cool windows and doorways.
3. **Window coverings**:
   3.1. Install **exterior shutters** or **reflective blinds** to prevent indoor overheating.
   3.2. Use **thermal curtains** to block radiant heat.

Evaporative Cooling

Evaporative cooling works by increasing humidity while promoting air movement, lowering temperatures through moisture evaporation.

1. **DIY swamp coolers**:
   1.1. Use a wet cloth draped over an open window to allow air to cool as it enters.
   1.2. Place a container of water with a fan blowing across it for localized cooling.
2. **Water features**:
   2.1. Position **shallow ponds** near windows or entrances to cool the air.
   2.2. Install **water misting systems** on porches or patios.

Energy-Efficient Cooling Solutions

Even off-grid, you can utilize minimal electricity to enhance cooling efficiency.

- **Solar-powered fans**: Small, solar-charged fans circulate air in rooms or ventilation shafts.
- **Cooling ponds**: Bodies of water near a home reduce ambient temperatures.
- **Radiant cooling panels**: Cooling surfaces that dissipate heat from interiors.

39. Off-Grid Lighting Solutions

Lighting plays a crucial role in off-grid living, ensuring safety and functionality at night while conserving energy. The right lighting setup depends on available resources, sustainability, and efficiency.

Solar-Powered Lighting

Solar-powered lights harness energy during the day and provide illumination at night.

1. **How to install solar lighting**:
    1.1. Position panels where they receive maximum sunlight exposure.
    1.2. Use **rechargeable batteries** to store energy.
    1.3. Select **motion-sensor** or **low-wattage LED** options for efficiency.
2. **Maintenance**:
    2.1. Keep solar panels clean and free of dust.
    2.2. Check battery life periodically.
    2.3. Replace LED bulbs when dimming occurs.

LED Lights

LEDs use very little energy while delivering bright, durable illumination

1. **Choosing LED bulbs**:
    1.1. Select **warm white (2700K-3000K)** for ambient lighting.
    1.2. Use **cool white (4000K-5000K)** for workspaces.
    1.3. Opt for **dimmable LEDs** where brightness control is needed.
2. **Placement considerations**:
    2.1. Install **task lighting** above kitchen counters or workstations.
    2.2. Use **motion-activated LEDs** in hallways or entryways.
    2.3. Place **outdoor LED floodlights** for security.

Oil and Gas Lighting

Oil lamps and gas lanterns provide reliable illumination when electricity is unavailable.

1. **Using and maintaining oil lamps**:
    1.1. Fill with **kerosene, lamp oil, or vegetable-based alternatives**.
    1.2. Trim wicks for steady, smoke-free burning.
    1.3. Store **fuel safely** away from heat sources.
2. **Gas lantern safety**:
    2.1. Use **pressurized propane or butane lanterns** for outdoor or emergency lighting.
    2.2. Ensure **adequate ventilation** when using indoors.

Candles

Candles provide a low-tech, functional light source.

1. **Making homemade candles**:
    1.1. Melt **beeswax, soy wax, or tallow**.
    1.2. Insert a **cotton wick** into a mold or jar.
    1.3. Pour wax and let it cool before trimming the wick.
2. **Candle safety**:
    2.1. Place candles in **stable, non-flammable holders**.
    2.2. Maintain a safe distance from flammable materials and provide adequate ventilation at all times.

Fire-Based Lighting

For outdoor settings or emergency use, torches and lanterns offer long-lasting light.

1. **Building a DIY torch**:
    1.1. Soak a **cloth-wrapped stick** in **animal fat or oil**.
    1.2. Secure the cloth with **wire or twine**.
    1.3. Light carefully and position in a safe area.
2. **Creating a lantern**:

2.1. Use a **glass jar with oil and a wick** for a simple lantern.
2.2. Position lanterns **away from wind** for consistent burning.

40. Cooking, Heating, Cooling & Lighting Projects

Project: Build a Solar Cooker

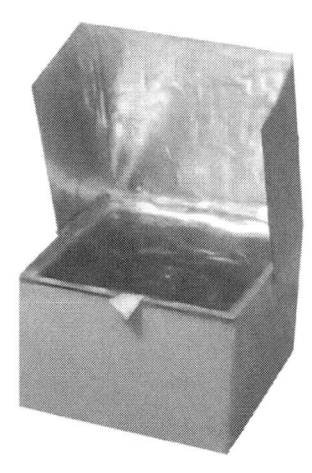

Materials Needed:
- 1 large cardboard box (sturdy, with flaps intact)
- 1 smaller cardboard box (fits inside the large one, leaving space around it)
- Aluminum foil
- Black non-toxic paint
- A piece of clear glass or plastic (large enough to cover the opening)
- Straw (or alternative insulation: shredded newspaper, Styrofoam peanuts, wool)
- Scissors or utility knife
- Ruler and marker
- Strong tape (duct tape or aluminum tape)
- Glue or spray adhesive

Step 1: Prepare the Outer Box
1. Select a large, sturdy cardboard box that will act as the outer shell of your solar cooker.
2. Ensure that the box has flaps intact to help with insulation and heat retention.
3. Check that the smaller box fits inside the larger one, leaving an even space (about 1–2 inches (2.5–5 cm)) on all sides.
4. Cut one side of the large box slightly shorter (about 1–2 inches (2.5–5 cm)) so that sunlight can enter more easily.

Step 2: Prepare the Inner Box
1. Take the smaller cardboard box and place it inside the larger one.
2. Make sure there is an even gap between the walls of both boxes for insulation.
3. If the inner box has flaps, trim them down or fold them so they don't interfere with the structure.
4. Leave the top open—this will be the cooking chamber.

Step 3: Insulate with Straw
1. Gather a good amount of straw or alternative insulation materials like shredded newspaper, Styrofoam, or wool.
2. Begin placing the straw in the gap between the inner and outer boxes.
3. **How to place the straw:**
   3.1. Gently push handfuls of straw into the gaps, starting from the bottom and working your way up the sides.
   3.2. Ensure that the straw is evenly distributed, with no large empty spaces.
   3.3. Lightly press the straw in place, but do not compact it too tightly—trapped air inside the straw helps with insulation.
   3.4. Continue adding straw around all sides until the entire gap is filled.
   3.5. If using shredded newspaper, crumple it slightly before stuffing it in to create air pockets.
4. Once fully insulated, use tape or a piece of cardboard to seal the gaps and keep the insulation secure.

Step 4: Paint the Interior with Black Paint

1. Apply a generous coat of **black, non-toxic paint** to the entire inside of the inner box.
2. Use a brush or roller to ensure an even coat on the bottom and side walls.
3. Let the paint dry completely before moving to the next step.

Step 5: Line the Inner Walls with Aluminum Foil

1. Cut sheets of aluminum foil slightly larger than each side of the inner box.
2. Apply a thin layer of glue or spray adhesive to one side of the box at a time.
3. Carefully press the aluminum foil onto the glued area, smoothing out wrinkles.
4. Repeat the process for all interior walls, ensuring full coverage.
5. If necessary, use tape to secure any loose edges.

Step 6: Create and Attach the Lid

1. Take a separate piece of cardboard large enough to cover the top of the cooker.
2. Cut out a **rectangular window** in the lid, leaving a 2-inch (5 cm) border around the edges.
3. Cover this window with **a sheet of clear glass or plastic**, taping or gluing it securely.
4. Attach one side of the lid to the cooker using tape, creating a flap that can be opened and closed easily.

Step 7: Position the Cooker

1. Place the solar cooker outdoors in direct sunlight.
2. Adjust the lid angle to reflect maximum sunlight into the cooking chamber.
3. Use a stick or small wooden piece to prop open the lid if needed.

Step 8: Cooking with Your Solar Cooker

1. Preheat the cooker by leaving it in the sun for 15–30 minutes.
2. Place a **dark-colored, heat-absorbent pot** with food inside the cooker.
3. Keep the glass or plastic lid closed to trap heat inside.
4. Check periodically and adjust the angle for maximum sun exposure.
5. Cooking times will vary but expect to wait **twice as long as traditional cooking**.

This solar cooker works best on **clear, sunny days** and can reach temperatures of **200–300°F (93–149°C)**.

Project: Design a Rocket Stove

Materials Needed

- **Metal container** (large coffee can, steel bucket, or metal drum)
- **Metal pipe (3-4 inches in diameter)** (for combustion chamber and chimney)
- **Smaller metal pipe or square tubing** (for fuel intake)
- **Firebricks or clay bricks** (for insulation)
- **Metal cutting tools** (tin snips, angle grinder, or hacksaw)
- **Drill with metal bits**
- **Heavy-duty gloves** (for handling sharp metal edges)
- **Heat-resistant sealant or fireproof mortar**
- **Metal grate or mesh** (to elevate the burning fuel)
- **Sand, clay, or gravel** (for insulation, optional)

Step 1: Choose the Stove's Container

- Select a **sturdy metal container** that will serve as the outer shell of the stove.
- Ensure it is **heat-resistant and free from plastic linings** that could release toxins.
- A **steel bucket or large metal drum** works well for larger stoves, while a coffee can is suitable for smaller designs.

Step 2: Cut the Fuel Intake Hole

- Measure and **mark a hole** near the bottom of the container for the fuel intake pipe.
- The hole should be **slightly larger than the smaller metal pipe** you will insert for the fuel feed.
- Use an **angle grinder, tin snips, or hacksaw** to carefully cut out the marked hole.
- Smooth the edges using a **file or sandpaper** to remove sharp burrs.

Step 3: Prepare the Combustion Chamber

- Cut a **metal pipe (3-4 inches in diameter)** to fit vertically inside the stove's container.
- This pipe will act as the **combustion chamber**, allowing air to flow and heat to rise efficiently.
- At the bottom of this pipe, cut a small opening to align with the **fuel intake pipe from Step 2**.

Step 4: Attach the Fuel Intake Pipe

- Insert the **smaller horizontal pipe** through the hole created in Step 2.
- Position it so that it enters the **vertical combustion chamber at a 90-degree angle**.
- Ensure the pipe **extends a few inches into the combustion chamber** to guide wood into the fire.
- Seal the connection using **fireproof mortar or heat-resistant sealant** to prevent smoke leaks.

Step 5: Insulate the Burn Chamber

- To improve heat retention, surround the combustion chamber with **firebricks, sand, or clay**.
- If using **firebricks**, stack them around the chamber inside the container.
- If using **sand, clay, or gravel**, pour them into the space between the container walls and the combustion chamber.
- Ensure the insulation **does not block airflow** but helps retain heat.

Step 6: Create a Grate for Better Airflow

- Cut a piece of **metal mesh or a small grate** to fit at the bottom of the fuel intake pipe.
- Place it inside the intake pipe to **slightly elevate the wood** as it burns.
- This allows air to circulate underneath the fire, making combustion **hotter and more efficient**.

Step 7: Install the Chimney Pipe

- Attach a **vertical chimney pipe** to the top of the combustion chamber.
- Ensure it is **securely connected** to direct smoke and gases upward.
- The chimney should be at least **12-18 inches (30-45 cm) high** for proper draft and airflow.
- Seal any gaps between the combustion chamber and chimney using **fireproof sealant or mortar**.

Step 8: Build a Stable Cooking Surface

- Place a **sturdy metal plate or cooking grate** on top of the chimney pipe.
- Ensure it has **small gaps to allow heat to escape** while supporting a cooking pot or pan.
- The surface should be **level and stable** to prevent pots from tipping over.

Step 9: Test and Adjust the Rocket Stove

- Place small pieces of **dry wood or twigs** inside the fuel intake pipe.
- Light the fire and **observe the draft** pulling flames into the combustion chamber.
- If the stove is smoking excessively, check for **airflow blockages** or **unsealed gaps**.
- Adjust the **chimney height or intake pipe size** for better efficiency.

Step 10: Perform Safety Checks and Maintenance

- Ensure the **stove is stable and placed on a fireproof surface**.
- Regularly **clean out ash buildup** from the combustion chamber.
- If using the stove long-term, apply **heat-resistant paint** to prevent rust.
- Store the stove in a **dry place** when not in use to extend its lifespan.

Project: Construct a Clay Oven

A clay oven is a highly efficient, wood-fired cooking structure that can bake bread, cook pizzas, and prepare a variety of meals while retaining heat for extended periods. The thick clay walls store heat and release it slowly, making it an excellent off-grid cooking option. This project will guide you through building a durable clay oven using natural materials.

Materials Needed

- **Bricks or concrete blocks** (for the oven base)
- **Firebricks** (for the cooking floor)
- **Sand** (for shaping the dome and mixing with clay)
- **Clay** (natural, untreated clay for making the oven walls)
- **Straw** (for reinforcing the clay mixture)
- **Wooden planks or a board** (for a temporary mold)
- **Shovel and trowel** (for digging and shaping the clay)
- **Bucket** (for mixing clay and water)
- **Water** (for mixing the clay)
- **Metal or clay chimney pipe** (optional, for improved ventilation)

Step 1: Choose and Prepare the Location

1. Select a **flat, dry, and stable area** for the oven to prevent shifting over time.
2. Ensure there is **proper airflow and clearance from flammable materials** such as trees or wooden structures.
3. If constructing on **soil or grass**, lay down **a layer of compacted gravel** to improve stability.

Step 2: Build the Oven Base

1. Lay **bricks, concrete blocks, or large stones** in a sturdy rectangular or circular shape.
2. Stack the materials to form a **solid, stable base** about **24-36 inches (60-90 cm) high** for comfortable access.
3. Ensure the base is **level and secure** before proceeding.

Step 3: Create the Cooking Floor

1. Place **firebricks** on top of the oven base to form the **cooking surface**.
2. Arrange them in a **tight, even pattern**, ensuring no large gaps between the bricks.
3. **Do not use cement or mortar** under the firebricks, as they need to expand with heat.

Step 4: Build a Sand Dome Mold

1. Use **wet sand** to create a **dome-shaped mound** on top of the firebricks.
2. The dome should be about **16-24 inches (40-60 cm) high** and have a **smooth, rounded shape**.
3. This sand mold will define the **oven's interior space**, so ensure it has a **consistent thickness** throughout.
4. Use a **wooden plank or board** to shape and smooth the dome.

Step 5: Prepare the Clay Mixture

1. In a **large bucket or tarp**, mix:
    1.1. **Clay** (50%)
    1.2. **Sand** (50%)
    1.3. **Water** (as needed)

1.4. **Straw** (a handful per batch, for reinforcement)
2. Knead the mixture **thoroughly with your hands or feet** until it forms a **firm, sticky dough-like consistency**.
3. Test a small sample by rolling a **ball of clay** and dropping it from **waist height**—it should hold its shape without crumbling.

Step 6: Apply the First Layer of Clay
1. Begin **pressing handfuls of clay mixture** onto the sand dome, starting at the base.
2. Continue adding clay **until the dome is covered with a layer at least 3 inches (7.5 cm) thick**.
3. Smooth out the surface as you go to **eliminate air pockets and weak spots**.
4. Let the first layer **partially dry** for **24-48 hours**, depending on the climate.

Step 7: Cut the Oven Door Opening
1. Once the clay starts to **firm up but is still slightly soft**, cut an **arched door opening** using a **knife or trowel**.
2. The door should be **wide enough for inserting and removing food** but **small enough to retain heat** (about **10-12 inches (25-30 cm) wide**).
3. Use a **wooden template or a marked line** to ensure a **symmetrical shape**.

Step 8: Remove the Sand Mold
1. Once the clay is **firm enough to hold its shape** (after **48-72 hours**), begin **scooping out the sand** through the oven door.
2. Remove the sand slowly to **avoid collapsing the dome**.
3. Check the **inner walls** and patch any **weak spots** with additional clay mixture if necessary.

Step 9: Add the Final Insulating Layer
1. Prepare another batch of **clay, sand, and straw**, but make it **slightly thicker** than the first layer.
2. Apply **a second layer of clay** over the dome, adding an **extra 2-3 inches (5-7.5 cm) of insulation**.
3. Smooth the surface and **let it dry for several days** before firing the oven.

Step 10: Cure the Oven with Small Fires
1. Start **small fires inside the oven** to **gradually remove moisture** and strengthen the clay.
2. Begin with **small pieces of dry wood** and **low-intensity fires** for the first few burns.
3. Over the next **few days**, increase the **fire size and duration** to allow the oven to fully harden.
4. Once cured, your clay oven is **ready for baking, roasting, and cooking**.

Project: Design and Install a Passive Cooling Ventilation System

A passive cooling ventilation system helps regulate indoor temperatures without relying on electricity. By strategically designing airflow paths and utilizing natural ventilation principles, you can keep your living space comfortable even in warm climates.

Materials Needed
- PVC or metal vent pipes (4-6 inches in diameter)
- Window vents or louvered vents
- Mesh screen (to prevent insects and debris from entering)
- Insulated ducting (if needed for certain areas)
- Adjustable vent dampers (for airflow control)
- Roofing sealant or caulk
- Drill and hole saw (for creating ventilation openings)
- Screws, mounting brackets, and a screwdriver
- Measuring tape and level

Step 1: Identify Ventilation Zones
- Determine the areas where airflow needs improvement (e.g., attic, kitchen, sleeping areas).
- Identify natural cross-ventilation opportunities by locating windows, doors, and existing vents.
- Choose areas for installing vents that will maximize airflow, such as high and low openings to create a natural chimney effect.

Step 2: Design the Airflow Path
- Plan for **low intake vents** to bring in cool air and **high exhaust vents** to expel hot air.
- If possible, align vents with prevailing wind directions to enhance natural airflow.
- Avoid placing vents where obstructions (furniture, walls) might block the air movement.

Step 3: Install Low Intake Vents
- Choose locations near the floor or shaded exterior walls for intake vents.
- Use a hole saw to cut openings in the wall or window frame where the vents will be placed.
- Insert louvered vents or ducting into the openings and secure with screws.
- Attach a mesh screen inside the vent to prevent pests and debris from entering.

Step 4: Install High Exhaust Vents
- Identify the highest possible points in the room, such as near the ceiling or in the attic.
- Cut openings for exhaust vents using a hole saw or jigsaw.
- Insert vent pipes or louvered covers and secure them with screws.
- Apply roofing sealant around the edges to prevent leaks if installing vents through the roof.

Step 5: Adjust Airflow for Maximum Efficiency
- Install adjustable dampers on vents to control the amount of airflow as needed.
- If additional cooling is required, connect ducting from intake vents to areas where airflow is weak.
- Ensure intake and exhaust vents remain unobstructed for continuous ventilation.

Step 6: Test the Ventilation System
- Open intake vents and check if fresh air is entering the space.
- Monitor how well the hot air escapes through the exhaust vents.
- Observe airflow changes during different times of the day to see how ventilation adjusts naturally.

Step 7: Maintain and Optimize the System
- Regularly clean vents and mesh screens to remove dust, debris, and insect buildup.
- Inspect seals and connections to prevent leaks or damage over time.
- If necessary, make seasonal adjustments to vent openings to optimize cooling.

Project: Natural Beeswax or Tallow Candles

Making candles from natural beeswax or tallow provides a reliable off-grid lighting solution without relying on synthetic materials. These candles burn cleanly, produce little smoke, and can be made with simple tools.

Materials Needed
- **Beeswax or tallow** (depending on preference and availability)
- **Cotton wick** (pre-waxed or natural)
- **Glass jars, tin molds, or candle molds**
- **Double boiler or heat-safe container** (for melting wax or tallow)
- **Wooden skewers or pencils** (to hold the wick in place)
- **Essential oils** (optional for scent)
- **Dye chips** (optional for colored candles)
- **Scissors** (for trimming wicks)

Step 1: Prepare the Wick and Mold
- Cut the wick to about **2 inches (5 cm) longer** than the height of the mold or jar.
- Tie one end of the wick to a **wooden skewer, pencil, or wick holder** and lay it across the top of the mold or jar.
- Ensure the wick hangs straight down into the mold and touches the bottom.

Step 2: Melt the Beeswax or Tallow

- If using **beeswax**, break it into small pieces for even melting.
- If using **tallow**, strain it through a cheesecloth to remove impurities before use.
- Place the wax or tallow in a **double boiler** or a heat-safe container inside a pot of simmering water.
- Stir occasionally until fully melted into a **smooth, clear liquid**.
- If desired, add a few drops of **essential oil** for fragrance or dye chips for color, stirring well.

Step 3: Pour the Wax or Tallow into the Mold

- Slowly pour the melted wax or tallow into the prepared mold or jar.
- Hold the wick in place to prevent it from shifting.
- Fill the mold, leaving about ½ **inch (1.25 cm)** of space at the top.
- Let the candle cool for **several hours** until fully solidified.

Step 4: Trim the Wick and Cure the Candle

- Once the candle has hardened, remove the wick holder and trim the wick to about ¼ **inch (0.6 cm)** for optimal burning.
- Allow the candle to **cure for 24 hours** before lighting for the best burn quality.

Step 5: Store and Use Your Candle

- Keep candles in a **cool, dry place** away from direct sunlight.
- Burn in a stable holder on a heat-resistant surface.
- Always monitor a lit candle and trim the wick before each use for a steady flame.

Module I. Essential Skills, Tools & Communication

41. General Repair Works, Basic Carpentry, and Tool Use

Essential Hand Tools and Their Uses

To maintain and repair an off-grid home, you need a selection of reliable hand tools. Unlike power tools that rely on electricity or fuel, these tools work manually and last for generations when properly maintained.

- **Hammer** – Used for driving and removing nails, assembling wooden structures, and light demolition work.
- **Handsaw** – A versatile cutting tool for wood and sometimes plastic. Different types of handsaws exist, such as crosscut saws for cutting across wood grain and rip saws for cutting along the grain.
- **Screwdrivers** – Essential for assembling and repairing household structures, electrical fixtures, and furniture. A set with both flathead and Phillips screwdrivers ensures compatibility with various screws.
- **Chisels** – Used for carving and shaping wood, particularly in joinery and furniture-making.
- **Pliers** – Needed for gripping, twisting, and cutting wires or small metal pieces. Locking pliers (vise grips) provide a strong hold for stubborn objects.
- **Wrenches** – Adjustable and fixed-size wrenches are required for plumbing, mechanical repairs, and tightening or loosening nuts and bolts.
- **Level** – Ensures proper alignment of shelves, doors, and construction elements. A bubble level is the simplest option, while laser levels provide increased accuracy.
- **Tape Measure** – Essential for taking accurate measurements in carpentry, plumbing, and general construction.
- **Hand Drill** – A manually operated drill used to create holes in wood, metal, or plastic when no power is available.
- **Clamps** – Hold pieces of wood or metal together securely during cutting, gluing, or assembly.

Each tool serves a specific purpose, and having a well-maintained set allows you to tackle basic repairs and improvements effectively.

Basic Carpentry for Off-Grid Living

Carpentry is a foundational skill for off-grid living. Learning the basics ensures that you can build, maintain, and repair essential structures without relying on professional services.

Basic Skills You Should Learn

- **Measuring and Marking** – Precision is critical in carpentry. Using a measuring tape, square, and marking tools ensures accurate cuts and joints.
- **Cutting** – Handsaws, chisels, and other cutting tools allow you to shape wood into usable components for furniture, roofing, and storage.
- **Joining** – Basic joinery techniques such as lap joints, mortise and tenon, and dowel joints create strong wooden connections without relying on screws or nails.
- **Finishing** – Sanding, sealing, and painting wood extend its lifespan and improve its appearance.

Understanding these core techniques allows you to build functional and durable structures, from shelves to small storage sheds.

Masonry and Metalwork Basics

Masonry and metalworking skills expand your ability to construct, repair, and maintain off-grid infrastructure.

Mixing and Applying Cement for Small Construction Projects

Cement is a critical material for off-grid construction. Whether reinforcing structures or building a simple foundation, proper mixing and application ensure long-lasting results.

- **Choose the right mix** – A general-purpose concrete mix includes **1 part cement, 2 parts sand, and 3 parts gravel**, with water added until a workable consistency is achieved.
- **Prepare the site** – Clear debris, level the surface, and install wooden forms to contain the cement.

- **Mix thoroughly** – Combine dry materials first, then slowly add water while stirring until the mix reaches a thick, spreadable texture.
- **Apply in layers** – Pour in manageable layers, smoothing each section with a trowel to prevent air pockets.

Once set, properly cured concrete structures can last for decades with minimal maintenance.

Welding and Forging Basics for Tool and Equipment Repairs

Welding and forging allow you to repair broken metal tools, fabricate brackets, and reinforce structures.

- **Basic welding setup** – Requires a welding torch, protective gear, and metal filler rods. A gas or arc welding system is necessary for fusing metal components.
- **Forging tools and techniques** – Heating metal in a forge and shaping it with a hammer and anvil allows you to create custom hardware, hinges, and tool parts.

Both welding and forging require safety precautions, including eye protection, heat-resistant gloves, and proper ventilation.

Repairing and Maintaining Off-Grid Infrastructure

An off-grid home requires constant maintenance to ensure longevity and resilience.

Patching Leaks and Fixing Roofing Issues

- **Inspect for damage** – Look for missing shingles, rusted metal panels, or cracked roofing materials.
- **Clean the surface** – Remove debris and old sealant before applying a patch or new roofing material.
- **Apply waterproofing** – Use roofing cement, tar, or specialized coatings to seal leaks and prevent further damage.

Plumbing Repairs: Fixing Leaks and Installing Basic Pipes

- **Detect leaks** – Check joints and fittings for drips, condensation, or pressure loss.
- **Tighten or replace fittings** – Use a wrench to tighten loose connections or install new gaskets and pipe sealants.
- **Install new pipes** – Cut and connect PVC, PEX, or copper pipes, securing them with fittings, solder, or compression joints.

42. Communication Methods: Staying Connected Beyond Traditional Grid Services

Two-Way Radios and Shortwave Communication

Reliable communication is crucial when living off-grid, especially for emergencies or coordinating with others. Walkie-talkies (also called two-way radios) offer a dependable communication method for distances up to several miles.

- **License-Free Options** – FRS (Family Radio Service) and GMRS (General Mobile Radio Service) radios are commonly used for short-range communication.
- **Durability** – Look for waterproof and shock-resistant models suitable for outdoor use.
- **Recommended Brands and Models** – Popular choices include **Baofeng UV-5R, Midland GXT1000VP4, and Motorola T600 H2O**.

These devices are essential for homesteaders, hikers, and emergency preparedness.

Emergency and Survival Communication Strategies

Creating an alert system ensures that you and your community can respond quickly to threats or emergencies.

- **Audible Signals** – Whistles, bells, or horns can serve as warning signals.
- **Visual Alerts** – Flashing lights, flags, or colored markers communicate different alert levels.
- **Radio Networks** – Establish a scheduled check-in system with local radio users.

Fire, Smoke, and Flag Signaling Techniques for Long-Range Communication

In low-tech situations, traditional signaling methods can convey messages over long distances.

- **Smoke Signals** – Controlled puffs of smoke, using damp grass or leaves, can be used to send coded messages.
- **Flag Codes** – Bright-colored flags or large symbols displayed on high ground provide clear visual communication.

These methods are useful when electronic devices fail or when traditional radio communication is unavailable.

Satellite and Alternative Connectivity Options

Satellite phones provide a direct communication link when cell towers and landlines are unavailable.

- **Global Coverage** – Works anywhere with a clear view of the sky.
- **Recommended Models** – Iridium 9575 Extreme, Inmarsat IsatPhone 2, and Thuraya X5-Touch.
- **Power Considerations** – Satellite phones require battery power but can be charged using solar panels or hand-crank chargers.

They are ideal for off-grid travelers, homesteaders in remote locations, and emergency responders.

43. Basics of Vehicle Maintenance

Conducting a Pre-Trip Inspection for Reliability

Before heading out on any off-grid journey, you need to ensure your vehicle is in top condition. A pre-trip inspection helps you can detect potential issues early on, preventing them from escalating into major problems, reducing the likelihood of breakdowns when you're far from assistance.

Start with a general visual check. Walk around the vehicle, looking for any signs of damage, leaks, or parts that seem out of place. Check under the vehicle for fluid drips, which can indicate oil, coolant, or transmission leaks.

Move to the tires. Look for proper inflation, visible wear, or embedded objects like nails or glass. If your trip involves off-road terrain, ensure you have the correct tire pressure to balance traction and fuel efficiency.

Open the hood and inspect fluid levels. Check the oil, coolant, brake fluid, transmission fluid, and windshield washer fluid. Each should be at the recommended level, and none should look milky, sludgy, or burnt—these are signs of contamination or breakdown.

Ensure all lights work. Test headlights, taillights, brake lights, hazard lights, and turn signals. If you're traveling in low-visibility conditions, proper lighting is critical.

Check the battery. If your battery terminals are corroded, clean them using a wire brush and baking soda solution. A weak battery can leave you stranded, so if it's older than three years, consider replacing it.

Test the brakes. Pump the brake pedal a few times and listen for unusual sounds. If you hear grinding or feel resistance, inspect your brake pads and fluid levels.

Understanding the Importance of Oil, Coolant, and Brake Fluid Maintenance

Fluids are the lifeblood of your vehicle. Neglecting them can lead to overheating, brake failure, and even engine seizure.

Oil: The engine relies on oil to lubricate moving parts and prevent friction. Old or dirty oil thickens over time, losing its ability to protect the engine. Ideally, oil should be changed every 5,000 to 7,500 miles (8,000 to 12,000 km), but if you're operating in extreme conditions, such as dusty roads or cold weather, check it more frequently.

Coolant: Your radiator needs coolant to regulate engine temperature. Low coolant levels or leaks can cause overheating, which can lead to costly engine damage. Always keep a spare jug of coolant and regularly check for leaks in hoses or the radiator itself.

Brake Fluid: A vehicle's brake system relies on hydraulic pressure, which brake fluid provides. If the fluid is low or contaminated, your brakes will feel soft or unresponsive. Unlike oil, brake fluid doesn't need frequent changes but should be checked every few months.

Multi-Tools and Hand Tools for Basic Maintenance

Whether you're performing routine maintenance or fixing an emergency issue, the right tools make all the difference.

• Socket Set: Essential for removing bolts, especially for changing oil, removing batteries, or fixing parts in the engine bay.
• Wrenches: A combination of adjustable, open-end, and box wrenches allows you to tighten or loosen nuts and bolts.
• Screwdrivers: Both flathead and Phillips screwdrivers are needed for removing panels and accessing small components.
• Pliers and Wire Cutters: Handy for electrical repairs and gripping hard-to-reach parts.
• Tire Pressure Gauge: Maintaining proper tire pressure improves gas mileage and prevents uneven wear.
• Jumper Cables: A dead battery is a common issue in off-grid travel. Jumper cables can get your vehicle running again if you have another battery to connect to.

For off-grid settings, having a multi-tool with a knife, small screwdriver, and file can help with quick fixes when carrying a full toolkit isn't practical.

Diagnosing and Fixing Common Mechanical Issues

If your vehicle suddenly won't start or behaves erratically, troubleshooting the issue quickly can prevent a major problem.

Engine Won't Start:
• Check the battery terminals for corrosion.
• Try jump-starting the vehicle if the battery is dead.
• If the engine cranks but doesn't start, check the fuel level and spark plugs.

Transmission Issues:
• Slipping gears often mean low transmission fluid.
• Rough shifting could indicate worn-out transmission components.
• Check fluid color—dark or burnt-smelling fluid needs replacement.

Brake Problems:
• Soft brake pedal: Check brake fluid levels and look for leaks.
• Grinding noise: Worn brake pads require replacement.
• Pulsing pedal: Rotors may be warped and need resurfacing.

Fixing Broken Belts, Hoses, and Electrical Wiring

A broken belt or hose can leave your vehicle stranded, but quick fixes can get you back on the road.

Temporary Hose Repair:
• Identify the leak or crack.
• Cut a rubber patch and secure it over the hole with hose clamps.
• For small leaks, use high-temperature tape as a temporary seal.

Belt Replacement:
• Loosen the tensioner pulley to remove the damaged belt.
• Thread the new belt in place following the engine's diagram.
• Adjust the tension and tighten the pulley.

Fixing Electrical Wiring:
• Strip the damaged section of wire.
• Use electrical tape or crimp connectors to join the ends.
• Secure the wiring to prevent further wear.

44. Essential Skills, Tools & Communication Projects

Project: Build a Simple Hand-Crank Water Pump

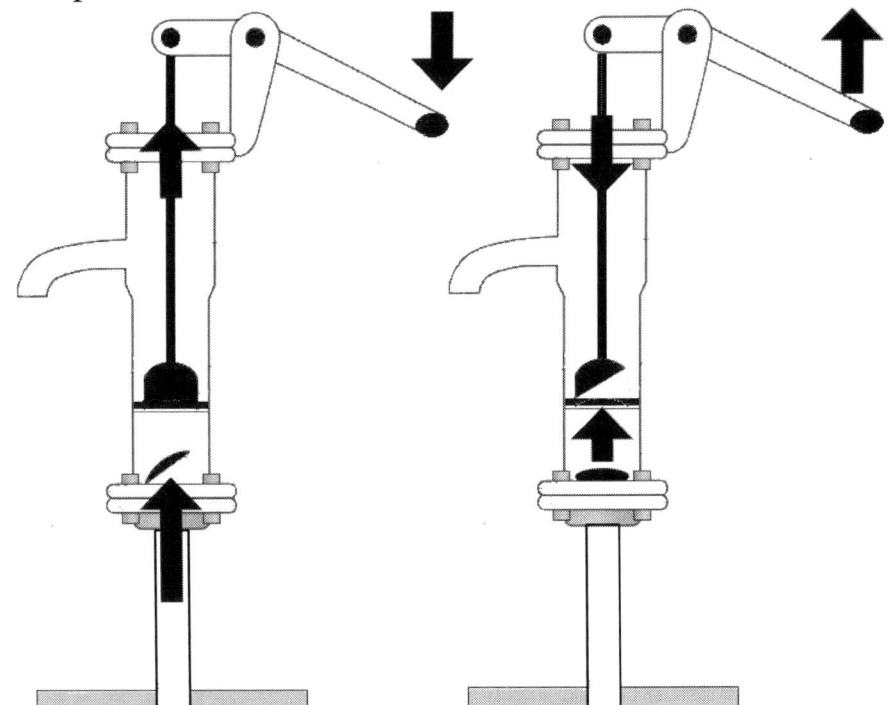

A hand-crank water pump is a reliable, non-electric solution for drawing water from a well, rainwater storage, or other water sources. It is particularly useful in off-grid settings where power may be limited. This project will guide you through constructing a simple pump using affordable materials and basic tools.

Materials Needed

- **PVC pipe** (1-inch diameter, approximately 5 feet long)
- **PVC pipe** (½-inch diameter, approximately 5 feet long)
- **PVC pipe fittings** (elbows and couplers)
- **Check valves** (two, one-way, 1-inch size)
- **Hand crank handle** (can be from an old drill or fabricated from metal or wood)
- **Rubber or silicone washer** (to create a tight seal)
- **PVC cement or strong waterproof adhesive**
- **Hose clamps** (for securing tubing connections)
- **Drill with hole saw attachment**
- **Hacksaw or pipe cutter**
- **Measuring tape**
- **Waterproof grease or lubricant**
- **Optional: Small plastic or metal bucket** (for testing)

Step 1: Cut the PVC Pipes to Size

- **Measure and cut** the 1-inch PVC pipe to approximately 4 feet long. This will serve as the outer casing of the pump.
- **Cut a second piece** of 1-inch PVC pipe to approximately 1 foot. This will serve as the base section.
- **Measure and cut** the ½-inch PVC pipe to approximately 5 feet. This will function as the plunger shaft inside the pump.

Step 2: Prepare the Base of the Pump

- **Attach a check valve** to the bottom of the 1-inch PVC pipe. Ensure the valve allows water to flow in only one direction—upward into the pipe.
- **Seal the check valve** using PVC cement or waterproof adhesive to prevent leaks.
- **Drill small holes** near the bottom of the 1-inch PVC pipe (just above the check valve) to allow water to enter freely.

Step 3: Assemble the Pump Plunger

- **Attach the second check valve** to one end of the ½-inch PVC pipe. This valve should also allow water to flow upward but not downward.
- **Insert a rubber washer** or silicone seal around the check valve to create a snug fit inside the 1-inch PVC pipe. This will help generate suction when the plunger moves.
- **Test the fit** by sliding the ½-inch PVC pipe into the 1-inch pipe. It should move freely but with some resistance to create suction.

Step 4: Install the Crank Handle
- **Drill a hole** through the top of the ½-inch PVC pipe, about 2 inches from the top.
- **Attach the crank handle** using a bolt, metal rod, or wooden dowel. Ensure it rotates freely.
- **Apply waterproof grease** to the rotating parts to reduce friction.

Step 5: Assemble the Full Pump System
- **Insert the plunger (½-inch PVC pipe)** into the main pump body (1-inch PVC pipe).
- **Test the movement** by turning the crank handle up and down. The check valves should allow water to move up with each stroke.
- **Secure all connections** with hose clamps or additional adhesive where necessary.

Step 6: Test the Pump
- **Place the bottom of the pump in a water source** (bucket, well, or rain barrel).
- **Turn the crank handle** and observe whether water is being pulled up through the check valves.
- **Adjust the seals if needed** to improve suction.

Step 7: Final Adjustments and Installation
- **Ensure the pump is easy to operate** and does not require excessive force.
- **Install the pump permanently** near your water source by securing it with a bracket or mounting it to a stable base.
- **Regularly lubricate moving parts** to maintain efficiency.

Project: Build a Simple Emergency Off-Grid Communication System

Having a reliable communication system is essential for staying connected during emergencies, particularly when living off-grid. This project will guide you through setting up a basic, battery-powered emergency communication system using a two-way radio and signal-based backup methods.

Materials Needed
- **Two-way radios (walkie-talkies)** with a minimum range of 2 miles
- **Rechargeable batteries** and a **solar charger**
- **Hand-crank radio** for receiving emergency broadcasts
- **Notebook and waterproof marker** (for written messages if voice communication is unavailable)
- **Signal mirror** (for long-range visual signaling)
- **Whistle or air horn** (for short-range audible alerts)
- **Colored flags or fabric** (for creating visual signals)
- **Flare or signal torch** (optional, for emergency night signaling)
- **Compass** (to provide accurate directions for signaling)

Step 1: Set Up and Test the Two-Way Radios
- **Fully charge** the batteries for the two-way radios using a power source or a solar charger.
- **Turn on and test** the radios by speaking into one and ensuring the signal is received on the other.
- **Adjust the channel settings** to a pre-determined emergency channel that you and your family or group members will use.
- **Test the range** by having one person move a short distance away while maintaining communication. Increase the distance until you identify the maximum effective range in your area.

Step 2: Establish an Emergency Communication Plan
- **Decide on primary and backup communication methods** depending on different situations. Example: Use walkie-talkies for daily communication and signaling devices when out of range.
- **Define a check-in schedule** (e.g., every hour, morning and evening, or at pre-set times during an emergency).
- **Establish a location for message drop-offs** in case direct communication fails. Use a waterproof container to store written messages in a designated location.
- **Assign different signals for different situations** (e.g., three short whistle blasts mean "emergency," a red flag means "danger ahead," etc.).

Step 3: Set Up a Solar Charging System for Your Radios
- **Choose a location** with maximum sunlight exposure for your solar charger.

- **Connect the solar panel** to the battery pack or directly to the walkie-talkies if compatible.
- **Monitor the charging** process and ensure batteries are fully charged before storage.

Step 4: Create a Visual Signaling System
- **Select a high-visibility area** where a flag or signal mirror can be seen from a distance.
- **Attach a colored fabric or flag** to a long pole, tree, or building to serve as a visual signal.
- **Practice using a signal mirror** by angling it towards a light source and reflecting a beam towards a target.
- **Use a compass** to help direct the reflected light towards a specific location.

Step 5: Prepare Audible and Fire-Based Signals for Backup Communication
- **Place a whistle or air horn** in an easily accessible location for emergency use.
- **Determine a signal pattern** that others will recognize as a distress call (e.g., three short blasts repeated).
- **Store flares or signal torches** in a waterproof container and use them only in life-threatening emergencies.

Step 6: Conduct a Communication Drill
- **Test all communication methods** with a friend or family member at various distances.
- **Ensure all members** of your household or group understand the emergency communication plan.
- **Make adjustments as needed** to improve clarity and reliability of signals.

Step 7: Maintain and Store Your Equipment Properly
- **Recharge radio batteries regularly** and store backups in a dry, cool place.
- **Keep all signaling devices in a designated emergency kit** for quick access.
- **Replace batteries or parts** as needed to ensure the system remains functional.

Module J. Community Building & Financial Preparation

Building an off-grid life isn't just about self-sufficiency—it's also about creating resilience through community and financial stability. While living independently can reduce reliance on external systems, no one thrives in complete isolation. A strong off-grid lifestyle is built on sustainable financial planning, diversified income streams, and the ability to exchange skills and goods without dependence on traditional economies. Whether you're starting a self-reliant community, budgeting for long-term survival, or finding ways to monetize your skills, this module will guide you through the essential strategies to ensure security, stability, and prosperity in an off-grid world.

45. Starting an Off-Grid Community

Planning & Legal Considerations

Creating an off-grid community requires careful planning, legal awareness, and foresight. The decision to live independently or build a structured community involves various considerations, from legal requirements to long-term sustainability. The framework for your community should be designed with an understanding of local regulations and what's possible in terms of land use, zoning, and cooperative models.

Choosing Between an Independent Homestead vs. a Structured Community

- An independent homestead offers complete autonomy but might lack shared resources and collaborative opportunities.
- A structured community, on the other hand, creates a support network but requires cooperation and shared decision-making.
- Consider what kind of lifestyle you envision, as well as the practicalities of collaboration versus self-sufficiency.
- Think about the scale of the community and whether you're aiming for a few families or a more extensive group of people.

Zoning Laws, Land Ownership, and Cooperative Models

- Research local zoning laws that regulate land use and check what type of structures or farming practices are allowed.
- Understand the pros and cons of land ownership versus leasing. In some cases, cooperative models, where ownership is shared by the group, may be beneficial for long-term security.
- Establish clear land use agreements and legal responsibilities among the community members.

Water, Energy, and Waste Management Regulations

- Explore local regulations around water rights and collection systems, including rainwater harvesting and well water use.
- Determine the regulations for energy production and how to meet those standards for renewable energy sources, like solar or wind power.
- Learn about waste management laws, such as septic systems, composting toilets, and waste disposal requirements to avoid legal issues.

Finding Like-Minded Individuals

The success of any off-grid community depends on the individuals who choose to live there. Establishing the right environment requires bringing together people who share common values, work ethic, and skill sets.

Defining the Vision and Principles of the Community

- Create a shared vision for the community that includes mutual values such as sustainability, cooperation, and independence.
- Write a clear mission statement or community charter that outlines the goals, guiding principles, and responsibilities for everyone involved.
- Discuss the social and environmental goals of the community to ensure alignment.

Recruiting Members with Complementary Skills

- Identify the skills needed for the community to thrive, such as farming, carpentry, teaching, or medical knowledge.
- Look for people who are not only willing to live off-grid but who bring valuable skills that complement the existing team.
- Create a recruitment strategy—use local networks, online platforms, or word-of-mouth to find individuals who align with your vision.

Establishing Rules, Responsibilities, and Conflict Resolution Systems

- Establish a system of governance or decision-making, whether it's democratic, council-based, or something else.
- Set clear rules for work responsibilities, contributions, and conduct within the community.
- Develop a conflict resolution system, where community members can air grievances and resolve issues without escalating tensions.

Infrastructure & Shared Resources

One of the key benefits of an off-grid community is the ability to share resources. By pooling resources, you can achieve more than individuals working alone.

Designing Shared Spaces: Communal Kitchens, Workshops, and Meeting Areas

- Designate areas where the community can gather to cook, work, and meet. Communal kitchens, shared workshops, and a community hall are important spaces for collaboration.
- Plan these spaces to be efficient, multifunctional, and easily accessible to all members.
- Think about sustainable design—use natural materials, energy-efficient technologies, and shared resources for heating and cooling.

Resource Pooling for Energy, Food, and Water Systems

- Collaborate on the installation of renewable energy sources such as solar panels or wind turbines.
- Build a central food storage and processing system, such as shared greenhouses, communal gardens, and livestock management.
- Create a collective water management system to share rainwater harvesting and well usage.

Creating a Self-Sustaining Ecosystem with Gardening, Animal Husbandry, and Trade

- Design a self-sustaining food system that includes permaculture gardening and animal husbandry.
- Plan how food, fiber, and medicinal plants will be grown and harvested collectively.
- Develop a local trade system to exchange goods and services among community members—whether through barter, skill exchanges, or a local currency.

Security & Governance

Ensuring the safety of all community members is paramount, as is effective governance.

Setting Up Leadership or Decision-Making Models

- Decide how leadership will function—will it be a direct democracy, where everyone votes on issues, or a council-based system with elected representatives?
- Ensure that the decision-making model is transparent, fair, and inclusive of all members.
- Define the roles and responsibilities of leaders or decision-makers.

Developing Safety Protocols and Property Defense Strategies

- Establish clear community safety rules, such as emergency protocols, communication systems, and physical security measures.
- Set up physical barriers, such as fences or gates, and determine patrols or watch shifts to ensure the security of the property.
- Consider emergency exits and hiding places for the protection of all community members during dangerous situations.

Establishing Emergency Preparedness and Crisis Management Plans

- Develop and practice emergency response plans for natural disasters, medical emergencies, and external threats.

- Ensure that community members know how to handle specific crises, whether it involves medical care, fire safety, or dealing with intruders.
- Have a reliable communication system in place to contact external support in case of severe emergencies.

46. Financial Planning: Budgeting for Sustainable Independence

Creating an Off-Grid Budget

Financial planning is the foundation of long-term sustainability. Managing your resources carefully is essential to ensure your off-grid community can thrive.

Estimating Start-Up vs. Long-Term Operational Costs

- Break down the initial costs of land, infrastructure, and equipment.
- Estimate the ongoing costs, including maintenance, food production, energy needs, and healthcare.
- Create a separate savings fund for emergencies or unexpected expenses.

Essential Expenses: Land, Water, Energy, Housing, and Food Production

- Identify and plan for the core expenses: land purchase, water access, food production, shelter, and energy systems.
- Factor in costs for land improvements, plumbing, irrigation systems, and construction materials.
- Calculate the budget for ongoing food production and system maintenance.

Ongoing Maintenance, Repairs, and Healthcare Costs

- Account for the cost of maintaining your off-grid systems, such as solar power, wind turbines, and water pumps.
- Factor in necessary repairs for infrastructure and equipment over time.
- Ensure that healthcare costs (both emergency and preventative) are part of the budget, including medical supplies and professional care when needed.

Reducing Dependence on Traditional Money

Living off-grid often means reducing reliance on cash. You'll need to adopt alternative ways to manage your finances and resources.

Cutting Unnecessary Expenses and Downsizing Lifestyle Costs

- Assess areas where costs can be cut—such as reducing reliance on commercial products, minimizing waste, and living in simpler homes.
- Grow your own food, make your own goods, and focus on self-sufficiency to minimize external costs.
- Implement frugal living habits like reusing materials, recycling, and upcycling.

Utilizing Alternative Currencies and Local Exchange Networks

- Explore the possibility of using barter or local exchange systems to trade goods and services.
- Investigate alternative currencies such as local community currencies or cryptocurrency that support off-grid and sustainable living.
- Develop an internal economy that promotes mutual support and reduces the need for traditional money.

Growing Food and Producing Goods to Offset Costs

- Grow your own food to reduce grocery expenses and increase self-sufficiency.
- Focus on high-yield crops, livestock, and storage methods to ensure food security.
- Produce goods such as homemade cleaning products, clothing, and tools to eliminate the need for purchasing these items externally.

Long-Term Financial Security

Ensuring that your off-grid lifestyle remains financially sustainable in the long run requires planning for the future.

Investing in Durable, Off-Grid-Friendly Assets

- Invest in long-lasting and essential assets such as durable housing, solar energy systems, and tools that can withstand the test of time.
- Focus on acquiring self-sufficient assets that will continue to provide value over the years.
- Prioritize investments that help you live without dependency on external systems.

Understanding Tax Implications for Off-Grid Living

- Research how living off-grid will impact your taxes—both property taxes and income taxes.
- Be aware of any tax credits or incentives available for renewable energy systems or sustainable living practices.
- Plan for any potential changes in tax regulations as governments adapt to the off-grid movement.

Retirement Planning Without Relying on Traditional Pensions

- Establish a retirement plan based on off-grid principles, such as investing in land, growing food, and having a sustainable lifestyle that requires little financial input.
- Look into alternative retirement plans that support off-grid lifestyles, such as self-managed IRAs, land trusts, or other options that provide financial security without relying on traditional pension plans.
- Build a financial cushion that supports long-term independence, ensuring that you won't be reliant on external financial systems.

47. Income Opportunities: Monetizing Skills and Resources Off the Grid

Living off the grid offers a unique opportunity to rely on your skills and resources to generate income without the need for traditional employment or infrastructure. By utilizing your knowledge, creativity, and off-grid resources, you can create a sustainable income stream that aligns with your values and independence. Here's how you can monetize your skills and resources while living off the grid.

Producing & Selling Goods

One of the most direct ways to earn an income while living off-grid is by producing and selling goods. Whether it's food, handmade products, or crafts, your off-grid homestead can become a small business that supports your lifestyle.

Selling Homegrown Produce, Dairy, Eggs, or Preserved Foods

- If you have a garden, a small farm, or livestock, you can sell fresh produce, eggs, dairy, and preserved foods. People who value organic, locally grown food are often willing to pay a premium for high-quality, homegrown goods.
- Consider selling surplus produce at farmers' markets, to local grocery stores, or through a community-supported agriculture (CSA) program.
- You can also preserve fruits, vegetables, meats, and dairy for sale. Jams, pickles, fermented foods, and cured meats are all highly sought after.
- Remember to check local regulations regarding the sale of food, especially preserved or dairy products. Some areas require specific licensing or inspections.

Creating Handmade Products: Soap, Candles, Herbal Remedies, Textiles, etc.

- Off-grid living often allows for the production of handmade goods such as soap, candles, herbal remedies, and textiles. These products can be sold locally, online, or at craft fairs.
- You can learn skills like soap-making, candle-making, or knitting, and create products that people are looking for in their everyday lives.
- Herbal remedies such as tinctures, salves, and teas can be made from plants you grow or forage, adding value to your homestead's natural resources.
- Setting up an online store (e.g., Etsy) allows you to sell your products to a wider market.

Woodworking, Blacksmithing, and Crafting Furniture for Local or Online Sales

- If you have carpentry, woodworking, or blacksmithing skills, you can craft custom furniture, tools, and other useful items. Handmade wooden furniture, crafted with care and skill, is always in demand.
- Selling locally to neighbors, or even shipping items via online platforms, can help you generate a steady stream of income.
- Blacksmithing can also provide you with the skills to make useful tools and metal goods such as knives, hooks, and other practical items.
- Invest in your workshop, tools, and the skills needed to make quality products that people will appreciate and use.

Remote & Digital Work

While living off-grid might seem like a lifestyle that disconnects you from technology, there are many opportunities for earning income remotely through digital work. With the right setup, you can make money without leaving your homestead.

Freelancing, Writing, and Online Consulting with Satellite Internet

- If you have skills in writing, design, or consulting, the internet allows you to offer your services to clients around the world.
- With satellite internet or even low-tech options like using a hotspot for internet access, you can stay connected to clients and work from anywhere.
- Many off-grid workers use freelancing websites, such as Upwork or Fiverr, to find clients for various projects, including writing, web design, and virtual assistance.

Teaching Off-Grid Skills through Courses or YouTube Content

- Sharing your knowledge about off-grid living can become a profitable venture. You can create online courses, write eBooks, or produce YouTube videos teaching people about homesteading, off-grid energy systems, gardening, and other skills.
- Platforms like Teachable, Udemy, and YouTube allow you to monetize content on these topics, attracting people who are interested in off-grid living and sustainable practices.
- Not only does this allow you to share your expertise, but it also establishes you as an authority on off-grid living, creating long-term income opportunities.

Managing an Online Store for Handmade or Surplus Goods

- If you're producing goods on your homestead, setting up an online store can be a great way to sell your products beyond your local area.
- Use platforms like Etsy, eBay, or even create your own website to manage inventory, accept payments, and handle shipping.
- Your online store can be a mix of handmade products like soap, herbal remedies, textiles, or surplus farm produce.

Service-Based Income Streams

As an off-grid homesteader, you can also earn money by providing services to others in your community or beyond. These services might be related to your off-grid expertise or simply offering the skills you've developed through your homesteading journey.

Off-Grid Homesteading Workshops and Eco-Tourism

- If you have expertise in off-grid living or sustainable practices, hosting workshops or eco-tourism experiences can be a great way to share your knowledge and skills with others.
- Offer hands-on experiences such as gardening workshops, permaculture design, building an off-grid cabin, or homesteading basics.
- Eco-tourism opportunities could involve guided tours of your off-grid homestead, teaching visitors about how to live sustainably.

Repair Services, Tool Rentals, or Construction Work within the Community

- If you have experience with tools and construction, you can offer your services to others in the community.

- Renting out tools you own or providing repair services for off-grid systems (solar, wind, water) or general homesteading needs could be another income stream.
- Offering construction services or labor for off-grid building projects can also generate money, especially in remote areas where professional contractors are in short supply.

Hosting Farm Stays, Retreats, or Skill-Exchange Programs

- As an off-grid homesteader, your property can become an income-generating resource by hosting visitors looking for farm stays or retreats.
- Guests can stay on your property in exchange for payment or through a skill-exchange arrangement (e.g., help with farm chores in exchange for accommodation).
- You can also run weekend retreats on topics like mindfulness, homesteading skills, or outdoor survival, attracting guests who wish to learn more about sustainable living.

48. Barter Systems: Trading Skills and Goods

Bartering is an ancient form of exchange and an excellent method for people living off-grid to meet their needs without relying on traditional currency. By trading skills, services, and goods within your community or with other off-grid dwellers, you can create a self-sufficient system that thrives without money.

How Bartering Works in an Off-Grid Setting

Bartering is based on a system of trade where goods or services are exchanged directly for something else of value. Off-grid communities often use barter as a primary method of exchange, and understanding the principles of bartering can help you thrive in such a system.

Understanding Fair Exchange Value

- In a barter system, you need to determine what your goods or services are worth. This can sometimes be tricky, but having a clear idea of the time, effort, and resources involved in producing something helps set a fair exchange value.
- For example, if you offer two hours of carpentry work, you may exchange it for a bushel of produce or a bundle of firewood.
- Barter value depends on the individual agreement and can fluctuate based on scarcity and demand.

Balancing Supply and Demand in a Self-Sufficient Community

- In an off-grid community, it's essential to balance the supply of what you can offer with what others need.
- If your neighbor is growing a large crop of tomatoes and you're harvesting herbs, you can trade your fresh basil for their tomatoes.
- Keep track of the resources you have available and think about how to best manage what you can share or trade to help the entire community meet its needs.

Creating a Local Trade Directory for Skills, Products, and Services

- Create a community directory to keep track of who has what skills and resources available for trade.
- This directory can be a simple bulletin board in a common area, a printed flyer, or an online list if the community has internet access.
- A directory helps people know who to turn to for specific needs, such as carpentry, tutoring, or animal care, and allows them to make more efficient trades.

Setting Up a Local Barter Network

Establishing a successful barter network within your community or among nearby off-grid communities requires effort and coordination. Here's how to get started:

Finding Trusted Barter Partners Within and Outside the Community

- To start a barter network, it's essential to find reliable, trustworthy barter partners.
- You can begin by introducing the idea to your neighbors and others in nearby off-grid communities.

- Engage in trades with people you trust and make sure they are dependable in upholding their end of the agreement.

Establishing Trade Meetups or Barter Markets
- Create regular meetings or barter markets where community members can come together to exchange goods and services.
- These gatherings provide an opportunity to build relationships, discuss needs, and form new barter agreements.
- A barter market can be an informal exchange, or it could be a more structured event, depending on your community's preferences.

Using Barter Credits or Alternative Currencies for Consistency
- To maintain fairness and transparency, consider using barter credits or an alternative currency.
- Barter credits are essentially a system of points that are earned by providing goods or services and can be spent on other goods or services.
- Alternative currencies like local tokens can be a way to ensure that trades are fair and tracked in an organized manner.

Building a Resilient Off-Grid Economy
Creating a resilient off-grid economy requires collaboration with other communities and reducing reliance on external financial systems.

Developing Cooperative Trade Agreements with Other Communities
- Form alliances with neighboring off-grid communities to establish regular trade routes or agreements.
- These partnerships ensure that resources, skills, and services can flow between communities, helping everyone thrive.
- Cooperative agreements can also help reduce the risk of resource shortages in a single community by diversifying the goods and services available.

Reducing Reliance on Cash and Digital Transactions
- The goal of off-grid living is to reduce reliance on traditional currency systems.
- By using bartering as a primary method of exchange, you minimize the need for cash and reduce the impact of inflation or financial crises.
- Bartering helps to create a more self-sufficient community that doesn't need to rely on external economic systems.

Legal Considerations for Barter-Based Businesses
- While bartering can provide many benefits, it's essential to understand the legal implications.
- Some regions require taxes to be paid on barter transactions, so it's crucial to keep accurate records of your exchanges.
- Familiarize yourself with local regulations regarding barter and trade, particularly if you plan to operate a barter-based business.

49. Community Building & Financial Preparation Projects

Project: Set Up a Local Barter Network for Off-Grid Trading

Step 1: Identify the Needs and Available Resources
- Assess the essential goods and services that people in your area may need.
- Identify the surplus resources or skills that community members are willing to trade.
- Consider a broad range of barterable items such as:
- Fresh produce, eggs, dairy, and preserved foods.
- Firewood, water collection, or energy solutions.
- Handmade goods like soap, candles, textiles, or furniture.
- Repair work, medical aid, construction, or childcare services.

Step 2: Connect with Interested Participants
- Start by reaching out to neighbors, local farmers, homesteaders, or off-grid communities.

- Use word-of-mouth, community bulletin boards, or social media groups to spread awareness.
- Schedule an initial meeting with interested participants to discuss expectations and trading ideas.
- Encourage people to list their available goods or skills and what they are seeking in return.

Step 3: Create a Barter Directory
- Develop a written or digital directory that includes:
- A list of available goods and services.
- Contact information for each participant.
- Any specific conditions for trade (e.g., seasonal availability of crops).
- Use a simple spreadsheet or a printed document to keep track of barter offers and requests.
- Ensure that the directory is regularly updated as availability changes.

Step 4: Establish Fair Trade Guidelines
- Define a system for determining the fair value of trades.
- Encourage trading based on necessity rather than direct one-to-one item value.
- Allow flexibility for multiple trades (e.g., one service in exchange for a combination of food and labor).
- Decide if barter credits will be used, allowing participants to "bank" services for future trades.

Step 5: Organize a Barter Meetup or Market
- Choose a recurring time and location where members can meet in person to trade.
- Consider hosting the event in a community space, farm, or market area.
- Set up tables or stations where participants can display their goods.
- Encourage socializing to build trust among traders and foster future barter exchanges.

Step 6: Develop a System for Remote or Ongoing Trades
- If in-person markets are not always feasible, establish a system for ongoing trades:
- A private message board or online group for listing trade offers and requests.
- A designated physical space (e.g., a shed or drop-off point) where trades can be exchanged without face-to-face interaction.
- A contact list that allows participants to schedule direct trades.

Step 7: Encourage Growth and Sustainability
- Continue promoting the barter network by inviting new members.
- Host workshops or training sessions to help participants improve barterable skills.
- Foster a culture of trust by encouraging honest and fair exchanges.
- Monitor challenges such as disagreements or trade imbalances and adjust guidelines accordingly.
- Explore partnerships with other barter networks or nearby communities to expand trade opportunities.

Project: Create a Community Resource Pool for Shared Tools & Equipment

Materials Needed
- Notebook or spreadsheet software (for record-keeping)
- Large storage shed, shipping container, or designated space for tool storage
- Lockable storage cabinets, shelves, or pegboards
- Labels, tags, or inventory markers
- Sign-out sheets or digital tracking system
- Basic maintenance supplies (oil, sharpening tools, cleaning rags)
- Agreement or community guidelines document

Step 1: Identify Community Needs
- Survey community members to determine the most commonly needed tools and equipment.
- List essential categories such as carpentry tools, farming equipment, repair kits, power tools, and emergency supplies.
- Prioritize tools that are expensive, infrequently used, or difficult for individuals to store.

Step 2: Establish a Contribution System
- Decide how tools will be sourced—voluntary donations, collective purchases, or a membership-based system.

- Create a clear agreement on whether tools will be permanently donated or loaned to the pool for shared use.
- Identify any existing equipment that members are willing to contribute temporarily or permanently.

Step 3: Designate a Storage & Maintenance Area

- Select a secure and central location for tool storage, such as a community shed, workshop, or designated space in a communal building.
- Organize the space with shelves, pegboards, or lockable cabinets to keep tools accessible and in good condition.
- Label tools clearly with an inventory number or community mark to prevent loss or mix-ups.

Step 4: Create an Inventory & Tracking System

- Maintain a detailed record of available tools, their condition, and who contributed them.
- Use a simple ledger, spreadsheet, or an online tool-sharing platform to log check-ins and check-outs.
- Assign each tool a unique identifier for easy tracking.
- Develop a system for recording damages, maintenance, and replacements.

Step 5: Set Up Usage Rules & Responsibilities

- Establish a borrowing system with clear guidelines on how long members can keep tools before returning them.
- Set up a simple sign-out and return procedure, either using paper forms or a digital log.
- Define maintenance responsibilities—who cleans, sharpens, or repairs tools before returning them.
- Specify consequences for lost, damaged, or unreturned tools (replacement, repair fees, suspension from borrowing).

Step 6: Maintain the Resource Pool

- Conduct regular maintenance checks to keep tools in working condition.
- Designate a responsible person or rotating volunteers to oversee tool upkeep.
- Hold community meetings to address any issues, review inventory, and make decisions on new tool acquisitions.
- Organize occasional skill-building workshops on proper tool use, safety, and maintenance.

Step 7: Expand & Improve the System

- Assess the effectiveness of the tool-sharing system and make necessary adjustments based on community feedback.
- Look for funding opportunities or bulk purchase discounts for new tools.
- Consider adding specialized equipment, such as solar-powered tools or water filtration devices, based on evolving community needs.
- Encourage knowledge-sharing and mentorship, where experienced members teach others how to use tools safely and efficiently.

Module K. Off-Grid Healthy Living, Mental & Emotional Preparedness & Medical Preparedness

Creating a healthy, balanced life off-grid requires more than just access to water, shelter, and food. It demands a proactive approach to physical health, emotional well-being, and mental resilience. Without the conventional systems that most people rely on, you must create routines that allow you to stay healthy, fit, and emotionally stable in the face of isolation, challenges, and the unpredictability of life outside the grid. This section explores how to build a resilient body, develop physical fitness, and implement mental and emotional strategies for thriving in an off-grid setting.

50. Building a Resilient Body

Living off-grid offers an opportunity to take control of your health in ways that many urban dwellers cannot. One of the most important aspects of long-term health is nutrition. Eating a well-balanced diet is crucial for maintaining energy, supporting immune function, and preventing chronic illnesses. When you're living off the grid, you will likely rely on homegrown or foraged foods, making it important to know how to get the right nutrients from your environment.

Importance of Nutrition in Long-Term Health

Nutrition is the cornerstone of a resilient body. When you grow your own food or forage, you have access to fresh, organic produce, but you also need to ensure that your meals are balanced. A healthy diet helps prevent diseases like heart disease, diabetes, and certain cancers, and it supports your body in times of physical stress, like long work hours or harsh weather conditions.

Key nutrients that are particularly important for maintaining long-term health off-grid include:

- **Protein**: For muscle repair and growth. Sources include beans, nuts, seeds, eggs, and meat from animals you raise or hunt.
- **Carbohydrates**: Your body's primary source of energy, essential for long days of physical labor. Root vegetables, grains, and legumes are excellent sources.
- **Fats**: Healthy fats support brain function and hormone production. Include avocado, nuts, seeds, and animal fats if available.
- **Vitamins and Minerals**: Essential for immune function, bone health, and energy production. You will need a mix of vegetables, fruits, and possibly supplements to ensure you're getting a full range of nutrients.

Maintaining a Balanced Diet Off-Grid with Homegrown or Foraged Foods

The off-grid lifestyle requires some creativity and planning when it comes to food. You will need to grow or forage for much of your diet, which means learning about seasonal cycles, edible plants, and animals. Here are a few key areas to focus on:

- **Vegetable Gardening**: Plant a variety of vegetables that can thrive in your climate. Crucial crops include leafy greens (kale, spinach), root vegetables (carrots, potatoes), and legumes (beans, peas).
- **Herbs and Foraged Foods**: Foraging for herbs and wild edibles can supplement your meals with essential vitamins and minerals. Examples include wild garlic, dandelion greens, and medicinal plants like echinacea.
- **Preservation Methods**: Canning, drying, and freezing will allow you to store food for winter months or when food production is low.
- **Animal Protein**: If raising animals for food, such as chickens or goats, you will have a regular source of protein. Hunting or fishing may also be part of your diet if you're in a location where those resources are available.

You'll want to have a comprehensive understanding of what foods grow well in your region, how to prepare them, and how to preserve them for later use.

Essential Vitamins and Minerals for Self-Sufficiency

In an off-grid setting, it's essential to be aware of the key vitamins and minerals that might be lacking in a diet based on homegrown or foraged foods. While a varied diet should meet most of your nutritional needs, you might need to supplement certain nutrients, especially in the winter months.

- **Vitamin A**: Vital for immune function and vision. It's found in carrots, sweet potatoes, and dark leafy greens.
- **Vitamin C** for immunity
- **Iron**: Important for energy levels and preventing fatigue. Iron-rich foods include beans, lentils, and dark leafy greens.
- **Vitamin D** for bone health
- **Magnesium** for muscle function
- **Calcium**: Crucial for bone health. You can get calcium from dairy products (if you have cows or goats) or plant-based sources like leafy greens and fortified plant milks.
- **Vitamin B12**: Found naturally only in animal products, this vitamin is necessary for brain health. If you don't consume animal products, consider B12 supplements.

Physical Fitness & Injury Prevention

In off-grid living, physical fitness isn't just about looking good—it's about staying strong and capable to handle the daily physical demands of your environment. Whether you're chopping wood, building, or tending to livestock, maintaining strength, flexibility, and endurance will make life off-grid more manageable.

Off-Grid-Friendly Exercises for Strength and Endurance

Exercise in an off-grid setting doesn't need to involve fancy equipment or gym memberships. Many off-grid tasks will naturally build strength and endurance. However, there are simple exercises you can do to target specific areas of fitness that are important for manual labor and survival.

- **Bodyweight Exercises**: Push-ups, squats, lunges, and planks are great exercises that can build strength in your arms, legs, and core.
- **Resistance Training**: If you have access to basic weights or resistance bands, use them to work your upper body, legs, and back muscles.
- **Cardiovascular Exercise**: Walking, hiking, and cycling can help build endurance and stamina, important for long days of work or emergency situations.

A consistent routine of physical exercise will help your body adapt to the physical demands of off-grid living, preventing fatigue and injury.

Stretching, Mobility, and Injury Prevention for Manual Labor

Manual labor can put a strain on your muscles, joints, and ligaments. To avoid injury and improve your efficiency, it's essential to incorporate mobility and stretching exercises into your routine.

- **Stretching**: Focus on stretching areas that will bear the brunt of manual work, such as your back, shoulders, and legs. Hold each stretch for 20–30 seconds to lengthen muscles.
- **Mobility Exercises**: Dynamic stretches and movements like leg swings, arm circles, and hip rotations will help improve your flexibility and range of motion.
- **Rest and Recovery**: Adequate rest is just as important as exercise. After a long day of hard work, be sure to take the time to rest and recover to avoid muscle fatigue and strain.

Ergonomic Work Techniques to Avoid Strain and Long-Term Injuries

The repetitive nature of off-grid labor can lead to chronic injuries over time, such as back pain or joint issues. Implementing ergonomic techniques while working will reduce strain on your body.

- **Posture**: Pay attention to your posture when lifting, bending, or carrying heavy loads. Use your legs instead of your back when lifting.
- **Tools**: Choose tools that are designed for comfort and efficiency. Using tools that reduce strain on your hands, wrists, and back can prevent long-term damage.
- **Pacing Yourself**: Don't overexert yourself during the day. Take regular breaks and rotate between tasks to avoid putting repetitive stress on the same muscles.

51. Mental & Emotional Preparedness for Off-Grid Living

Living off-grid offers many rewards, but it also comes with unique mental and emotional challenges. You will likely experience isolation, stress, and unpredictability. Building mental resilience and emotional stability is essential for thriving in this lifestyle.

Building Mental Resilience

Mental resilience is your ability to cope with difficult situations and adapt to new challenges. When living off-grid, you'll need to rely on your resourcefulness and emotional strength to thrive.

- **Self-Reliance**: Cultivating a mindset of self-sufficiency will help you face the challenges of off-grid living. The ability to solve problems and overcome obstacles independently builds confidence.
- **Positive Thinking**: Maintaining a positive outlook, even during difficult times, helps reduce stress and increases your capacity to handle challenges.
- **Adaptability**: Embrace flexibility and open-mindedness. Being able to adapt to new situations and circumstances is key to long-term survival off-grid.

Coping with Isolation and Stress in Remote Settings

Isolation is one of the most common mental health challenges faced in off-grid living. Being far from neighbors and the social support systems you're accustomed to can create feelings of loneliness or depression.

- **Stay Connected**: Stay in touch with family and friends through alternative communication methods like satellite phones or two-way radios.
- **Create Routines**: Having a daily routine helps bring structure to your day, reducing feelings of aimlessness or anxiety.
- **Focus on the Present**: Take time to appreciate the beauty of your surroundings and focus on tasks at hand. Mindfulness practices can reduce stress and improve your emotional well-being.

Stress Management & Emotional Stability

Living off-grid often means dealing with unpredictable challenges, from bad weather to equipment failures. Stress management is essential for emotional stability.

- **Mindfulness and Breathing Exercises**: Incorporate deep breathing, meditation, or yoga into your daily routine to calm your nervous system and reduce stress.
- **Journaling**: Writing down your thoughts and emotions can help process complex feelings, reduce anxiety, and gain clarity.
- **Creative Outlets**: Engaging in creative activities like painting, woodworking, or crafting offers an opportunity to express emotions and take your mind off stressors.

Adopting these practices will help you maintain a healthy emotional state, allowing you to tackle the challenges of off-grid living with resilience and positivity.

52. Managing Medications & Chronic Illness Off the Grid

Living off-grid requires careful planning, especially when it comes to managing your health and medications. With limited access to pharmacies or medical services, stockpiling medications and learning natural alternatives are essential skills for maintaining your well-being. This section will guide you through creating a sustainable health management plan for chronic illnesses, storing essential medications, and exploring herbal remedies to keep you healthy.

Stockpiling & Storing Medications

Properly managing your medications is crucial when living off-grid. While access to pharmacies might be limited, it's still possible to stockpile the necessary medications and maintain them for long-term use.

Identifying Essential Medications for Off-Grid Survival

The first step in ensuring that you can manage your health while off-grid is to identify the medications that are essential for you or your family members. This includes:

- **Prescription medications**: If you or anyone in your household requires ongoing prescription medications (e.g., for diabetes, heart disease, or mental health conditions), you need to plan for their supply.
- **Over-the-counter medications**: Stock up on common medications for pain, fever, colds, and allergies. Include anti-inflammatory drugs, antihistamines, and stomach relief medications.
- **Emergency medications**: Consider including first-aid supplies, such as antiseptics, bandages, and items like an epinephrine auto-injector if allergies are an issue.

Work with a healthcare professional to create a personalized medication list based on specific needs, and make sure to have enough for at least 6 months to a year.

Proper Storage Techniques to Extend Medication Shelf Life

Medications are typically sensitive to temperature, light, and moisture, so it's crucial to store them properly to preserve their effectiveness. Follow these steps for optimal storage:

- **Cool, dry, and dark places**: Store medications in a place away from heat, humidity, and sunlight. A cool, dry cabinet or a waterproof container in a temperature-controlled area works best.
- **Proper containers**: Keep medications in their original packaging to maintain their integrity. For pills, you can use airtight containers to protect them from moisture.
- **Use desiccants**: Including desiccants in your storage containers can help absorb any residual moisture in the air and prevent degradation of medications.
- **Rotation system**: Regularly check expiration dates and rotate your stock, ensuring that older medications are used first.

Alternatives When Prescriptions Are Unavailable

In some off-grid situations, you may not have immediate access to prescriptions, or it may be difficult to restock medications. Here's what you can do:

- **Herbal substitutes**: Many common herbal remedies can be used as alternatives to pharmaceuticals. For example, peppermint or ginger can aid digestion, while echinacea can help boost immunity.
- **Essential oils**: Certain oils like lavender, peppermint, and tea tree oil can have medicinal properties. They can be used for pain relief, relaxation, and to treat infections.
- **Home remedies**: Invest time in learning home remedies for common ailments. For instance, a mix of honey and lemon can soothe a sore throat, while garlic and turmeric are known for their anti-inflammatory properties.

Natural and Herbal Remedies

Herbal remedies can play a key role in managing health off-grid, offering natural solutions to common ailments. Many plants, herbs, and natural substances can treat and prevent a wide variety of conditions without the need for synthetic medications.

Natural Pain Relief Techniques and Herbal Alternatives

When living off the grid, you may not have quick access to over-the-counter painkillers. Luckily, many herbs have pain-relieving properties. Some options include:

- **Willow bark**: Often called "nature's aspirin," willow bark has pain-relieving properties that can help manage headaches, arthritis, and muscle pain.
- **Turmeric**: Known for its anti-inflammatory properties, turmeric can be used to treat joint pain and inflammation.
- **Ginger**: This herb can help relieve pain and nausea, as well as reduce inflammation in the body.

Best Herbs for Common Ailments

In an off-grid lifestyle, you'll want to cultivate a small herbal medicine garden to help with common health issues. Here are some of the best herbs to grow for daily health:

- **Lavender**: Known for its calming effects, lavender can reduce stress, anxiety, and promote better sleep.
- **Peppermint**: Useful for digestive issues like nausea, bloating, and indigestion.
- **Echinacea**: Known for its immune-boosting properties, it can help prevent colds and flu.
- **Ginger**: Effective for nausea, digestion, and reducing inflammation.

Herbal First Aid for Wounds, Burns, and Infections

Herbs can also be used as a first-aid treatment for cuts, burns, and infections. Some essential plants include:

- **Aloe vera**: Excellent for soothing burns and skin irritations. It can be applied directly to the skin.
- **Calendula**: Known for its healing properties, calendula can be made into a salve for cuts and abrasions.
- **Comfrey**: Often called "knitbone," comfrey can help promote tissue repair for sprains and fractures.

Herbs for Digestion, Sleep, and Mental Well-being

In addition to first aid, herbs can be used for mental well-being, digestion, and sleep support:

- **Chamomile**: Effective for calming the mind, relieving stress, and aiding sleep.
- **Lemon balm**: Known for its ability to reduce anxiety, improve sleep quality, and aid digestion.
- **Fennel**: Useful for bloating, indigestion, and supporting a healthy digestive system.

Natural Antibiotics and Antiviral Herbs

Herbal medicine can also help protect against infections, acting as natural antibiotics and antivirals. Some herbs to consider are:

- **Garlic**: A potent natural antibiotic, garlic can help fight off bacterial infections and boost the immune system.
- **Oregano**: Contains carvacrol, which has antiviral and antibacterial properties.
- **Thyme**: Known for its ability to fight respiratory infections and support overall immune health.

Herbal Medicine Applications

Once you've gathered the right herbs, learning how to prepare and use them is key to managing your health off-grid. Herbal medicine is versatile and can be applied in various forms, such as teas, tinctures, and salves.

Herbal Teas, Tinctures, Decoction, Infusions, Capsules, and Extracts for Health and Healing

- **Teas**: One of the simplest ways to use herbs, making tea can extract beneficial compounds from herbs like chamomile, peppermint, and lemon balm. Steep the herbs in hot water for about 5-10 minutes to allow the beneficial properties to be released.
- **Tinctures**: Alcohol-based extracts of herbs that can be concentrated for stronger medicinal effects. To make a tincture, steep your herbs in high-proof alcohol for several weeks, then strain and bottle.
- **Decoctions and Infusions**: Decoctions are made by boiling tougher herbs, like roots or bark, while infusions are steeping softer herbs in hot water. Both methods help release medicinal properties from different plant parts.
- **Capsules and Extracts**: Capsules and extracts are more concentrated forms of herbal medicine. You can fill empty capsules with powdered herbs for easy consumption or make extracts for specific purposes.

Making Salves, Poultices, and Infusions for External Applications

- **Salves**: Combine your chosen herbs (such as calendula or comfrey) with a base of beeswax and olive oil to create a soothing, healing salve for cuts, burns, and rashes.
- **Poultices**: Crush herbs into a paste and apply them directly to wounds or infections. This method can draw out infection and relieve pain.
- **Infusions**: Herbal infusions made with vinegar or oil can be used for a variety of external applications, like soothing sore muscles or treating skin irritations.

53. Sustainable Hygiene & Natural Sanitation

Living off-grid doesn't mean sacrificing hygiene or sanitation. Instead, you will need to adopt a sustainable and natural approach to maintaining personal cleanliness and waste management.

DIY Handwashing Stations and Alternatives to Soap

While soap is readily available in many places, you may want to explore natural alternatives when living off-grid. Making your own handwashing station can help conserve water and make cleaning more efficient:

- **Handwashing station**: Set up a simple, gravity-fed system using a clean container, spigot, and a basin. Use biodegradable soap or make your own with natural ingredients like olive oil, lye, and herbs.
- **Alternatives to soap**: You can use natural cleaning agents like baking soda, vinegar, or essential oils for personal hygiene. These options are environmentally friendly and can be made easily from ingredients you have at home.

Natural Disinfectants and Cleaning Products

When living off the grid, creating your own natural cleaning solutions can help you avoid harmful chemicals:

- **Vinegar and water**: A simple yet effective disinfectant for surfaces.
- **Essential oils**: Use oils like tea tree, eucalyptus, or lavender for antibacterial properties.
- **Baking soda**: Great for scrubbing surfaces and neutralizing odors.

Natural Toothpaste, Deodorant, and Soap

Personal care items don't have to come from the store. You can make your own natural versions:

- **Toothpaste**: Mix baking soda, coconut oil, and a few drops of peppermint essential oil for a natural toothpaste.
- **Deodorant**: Combine baking soda, arrowroot powder, and coconut oil to create a natural deodorant.
- **Soap**: Making your own soap with oils like olive oil and essential oils is an excellent way to ensure your soap is free of harmful chemicals.

Alternatives to Commercial Shampoos and Skin Care

Commercial skin and hair products often contain synthetic chemicals that may not be available to you off-grid. Here are some natural alternatives:

- **Shampoo**: Use baking soda or apple cider vinegar to wash your hair. For a more luxurious option, you can use a herbal rinse made with chamomile or nettle.
- **Skin care**: Coconut oil, aloe vera, and honey can be used as moisturizers, sunscreens, and skin treatments. You can also make your own face masks using clay and essential oils.

Living off-grid means taking a holistic approach to your health, hygiene, and well-being. By utilizing natural remedies, creating your own cleaning products, and maintaining personal hygiene, you can live sustainably while staying healthy.

54. Off-Grid Healthy Living, Mental & Emotional Preparedness & Medical Preparedness Projects

Project: Create a Personal Off-Grid Fitness Plan

Step 1: Assess Your Current Fitness Level
- Start by evaluating your current fitness level to understand your baseline.
- Perform simple exercises like squats, push-ups, or walking for 10-15 minutes to gauge your endurance and strength.
- Take note of any physical limitations or areas that require improvement.

Step 2: Set Clear Fitness Goals
- Define what you want to achieve with your off-grid fitness plan. Consider goals like increasing strength, improving flexibility, or building stamina.
- Break down these goals into smaller, achievable steps with target dates to track progress.

Step 3: Choose Fitness Activities That Align With Off-Grid Living

Select activities that will be beneficial for off-grid living and do not rely on gym equipment. These might include:

- Bodyweight exercises (push-ups, squats, lunges)
- Walking or hiking (ideal for building endurance)
- Carrying or lifting tasks (building strength for physical work)

Incorporate daily functional activities like gardening, hauling wood, or other homestead chores to mimic real-life off-grid tasks.

Step 4: Design a Weekly Workout Routine

Develop a fitness schedule that works for your lifestyle and the tasks you perform. Consider the following:

- Monday, Wednesday, Friday: Strength training (bodyweight exercises like squats, push-ups, and planks)
- Tuesday, Thursday: Cardio or endurance (walking, hiking, or cycling)
- Saturday: Flexibility and mobility (stretching or yoga)
- Sunday: Rest or light physical activity (gardening, a leisurely walk)

Ensure that each workout is well-rounded to focus on strength, endurance, and flexibility.

Step 5: Incorporate Rest and Recovery

- Rest is crucial for muscle recovery and overall fitness improvement.
- Include at least one rest day per week, and make sure to listen to your body's signals to avoid overexertion.
- Practice active recovery on rest days, such as light stretching or a casual walk.

Step 6: Track Your Progress

- Keep a fitness journal to track your progress, recording your workouts, physical achievements, and any changes in your strength, endurance, or flexibility.
- Regularly reassess your goals and adjust your plan as needed to continue progressing.

Project: Design a Mental Resilience Toolkit

Step 1: Define Your Mental Health Goals

- Think about your mental well-being and what resilience means to you. Do you want to improve emotional stability, reduce stress, or foster a positive mindset?
- Set specific goals for your mental health that can guide your toolkit, like staying calm during a crisis or maintaining a positive outlook in isolation.

Step 2: Incorporate Mindfulness and Relaxation Practices

Add mindfulness exercises to help reduce stress and stay grounded. Consider including:

- Daily meditation for 5-10 minutes
- Deep breathing exercises to help relax during stressful moments
- Journaling to process emotions and reflect on your thoughts

Make sure these practices are simple and can be done without any special equipment or tools.

Step 3: Develop a Routine for Managing Stress

Create a routine that includes activities to handle stress effectively:

- Set aside time each day for relaxation or mindfulness activities.
- Include physical exercise, such as yoga or stretching, to relieve tension.
- Consider activities that reduce mental stress, such as reading, listening to music, or spending time in nature.

Step 4: Cultivate Positive Thinking

Positive thinking is crucial for mental resilience. Integrate the following into your toolkit:

- Affirmations: Write down positive statements that resonate with you, such as "I am strong and capable."
- Gratitude practice: Keep a gratitude journal and list things you are thankful for each day.
- Visualizations: Imagine positive outcomes for challenges or future goals.

Step 5: Build Social Support
- Even in an off-grid lifestyle, human connection is essential. Identify people who can offer emotional support, whether family, neighbors, or online communities.
- Develop systems for maintaining communication, such as using two-way radios or satellite phones when necessary.
- Set up regular check-ins with your social circle to help manage feelings of isolation.

Step 6: Create a Crisis Plan

Think ahead about potential stressors and create an action plan for when stress or negative emotions arise. For example:

- Set aside a space in your home for quiet reflection.
- Know the physical activities you can do to release built-up tension, such as going for a walk or practicing breathing exercises.
- Have a list of coping strategies to rely on during challenging moments, such as calling a friend, journaling, or focusing on positive affirmations.

Step 7: Review and Adapt Your Toolkit
- Mental resilience is an ongoing practice. Periodically evaluate the effectiveness of your mental health toolkit.
- Adjust your strategies as needed, adding new techniques or revising existing practices based on how you are feeling.

Project: Build a Homemade Hygiene Station

Step 1: Choose the Location
- Select a well-ventilated and easily accessible area near your living space or shelter.
- Ensure the location is close to a water source or allows for easy water transport.
- Position the station in an area with proper drainage to avoid standing water and mud buildup.

Step 2: Gather Materials
- **For the frame and base:** Wooden planks, bamboo poles, or sturdy branches.
- **For water storage:** A food-grade plastic container, metal drum, or ceramic water jug.
- **For water flow:** A spigot, faucet, or simple hole with a cork stopper.
- **For handwashing and bathing:** A small basin or bucket to catch used water.
- **For soap and sanitation:** Natural soap bars, liquid soap dispensers, or homemade hand sanitizer.
- **For drying hands and face:** Towels, cloth strips, or a drying rack.

Step 3: Build the Water Supply System
- If using a water container, elevate it on a wooden stand or sturdy platform for gravity-fed water flow.
- Attach a spigot or faucet at the base of the container to control water release.
- Secure the container with ropes or brackets to prevent tipping.
- If installing a foot-operated system, use a simple pedal and tubing mechanism to allow hands-free water dispensing.

Step 4: Set Up the Handwashing Station
- Place a small basin under the water outlet to collect wastewater.
- Install a soap holder or attach a mesh bag containing soap for easy access.
- Keep a separate area for rinsing, ensuring wastewater is directed away from the clean water source.
- If using a foot pump system, connect the foot pedal to a pipe leading to the water container.

Step 5: Create a Dishwashing Area (Optional)
- Designate a second water container for dishwashing with a similar gravity-fed spigot system.
- Place two basins: one for washing with soap and another for rinsing with clean water.
- Position a drying rack nearby to air-dry dishes after washing.

Step 6: Construct a Drainage System
- Dig a small trench leading away from the hygiene station to prevent stagnant water.
- Fill the trench with gravel or small stones to filter water and prevent pooling.
- If possible, direct the wastewater toward plants or a garden area for irrigation.

Step 7: Organize and Store Hygiene Supplies
- Store soap in a covered but well-ventilated area to prevent melting or contamination.
- Keep towels or reusable cloths on hooks or a drying rack for air circulation.
- Place a small waste bin nearby for used tissues or non-recyclable materials.
- Maintain a designated area for restocking hygiene products such as homemade disinfectants and hand sanitizers.

Step 8: Maintain the Hygiene Station
- Refill the water container regularly to ensure a continuous supply.
- Clean the water basin and drainage system frequently to prevent bacterial buildup.
- Rotate and wash towels or drying cloths to keep them fresh and sanitary.
- Inspect the station periodically for any leaks, loose connections, or mold growth.

Module L. Essential Outdoor Survival & Disaster Readiness

Living off-grid includes being prepared for emergencies, whether it's a natural disaster, societal unrest, or unexpected survival situations. This module equips you with essential skills for bug-out planning, shelter building, water sourcing, fire-starting, and navigation without GPS. You'll learn the essentials on how to assess threats, choose a secure location, and sustain yourself off-grid.

Mastering survival knots, alternative power sources, and efficient energy use ensures long-term resilience. Whether in the wild or an urban emergency, these skills empower you to navigate, secure resources, and stay self-sufficient in any situation.

55. Essentials of Bug-Out Planning & Bug-Out Location

As an off-grid living enthusiast, preparing for emergencies requires having a clear plan in place for when it becomes necessary to leave your home and seek safety elsewhere. A bug-out plan involves identifying when the right time is to evacuate and where you should go for the most security and resources. The process also requires thoughtful preparation, including having a bug-out bag packed and ready for quick departure. The following details will help you navigate through this planning, assess risk factors, and gather the necessary resources for a successful bug-out plan.

Understanding When to Bug Out vs. Stay Put

Being able to differentiate between when to stay put and when to evacuate could mean the difference between safety and risk. There are multiple scenarios where evacuation may be required, including natural disasters, civil unrest, or other regional emergencies. Below, we break down these important factors.

Evaluating Threats: Natural Disasters, Societal Collapse, Localized Emergencies

The first step in deciding whether to stay put or leave is to identify and evaluate the threat at hand:
- **Natural Disasters:** Earthquakes, floods, wildfires, hurricanes, and tornadoes often create unpredictable situations. These disasters can leave you vulnerable in your current location, especially if emergency services are overwhelmed, or evacuation routes are blocked.
- **Societal Collapse:** Economic downturns, civil unrest, or potential breakdowns in social order could make staying in a populated area unsafe. Areas affected by looting, violence, or instability often require a retreat to a safer location.
- **Localized Emergencies:** Chemical spills, gas leaks, or other local environmental hazards may force you to evacuate. These emergencies could be temporary, but they still require quick action to avoid exposure to hazardous substances.

Risk Assessment Factors: Mobility, Dependents, Medical Needs, Available Resources

Before deciding to evacuate, it's crucial to assess several factors:
- **Mobility:** Assess your ability to move quickly and safely. If you or your family members have mobility issues, plan for special equipment or transportation needs. Time is critical during an evacuation, and making sure everyone can get out of harm's way quickly is essential.
- **Dependents:** Children, the elderly, or others who are dependent on you require extra considerations. Ensure you have enough supplies for their needs, whether it's food, medicine, or mobility aids.
- **Medical Needs:** Those who have chronic illnesses or require regular medications must consider whether leaving is viable if medical supplies are scarce. Have you accounted for prescription medications, first-aid kits, or the need for regular medical care?
- **Available Resources:** Assess whether your home or current location still offers enough food, water, or shelter to last. If your resources are running low or compromised, moving to a more secure location may be essential.

Signs that Indicate the Necessity to Leave

Look for these signs that it's time to bug out:
- Rising floodwaters or other dangerous natural events.
- Worsening civil unrest or escalating violence.
- Confirmation that local resources (water, food, medical care) are no longer available.
- A sudden, unmanageable local emergency that directly threatens your safety.

Choosing the Right Bug-Out Location

Once the decision to evacuate is made, the next step is determining where to go. A well-chosen bug-out location can make all the difference in your survival.

Proximity vs. Remoteness: Balancing Security and Accessibility

- **Security:** Look for locations that are hidden from view and difficult for outsiders to access. Remote spots are often safer, but you need to ensure that they offer the resources you need.
- **Accessibility:** While remoteness is a key factor for security, accessibility is also critical. You need to be able to reach your bug-out location easily and quickly. Plan for transportation routes and take into consideration how difficult it may be to access during an emergency.

Essential Survival Factors: Water Sources, Food Availability, Defensibility, Terrain

- **Water Sources:** Water is your most crucial resource. Ensure that your bug-out location is close to a clean water source, whether it's a river, stream, well, or lake.
- **Food Availability:** A sustainable food supply is vital for long-term survival. Consider whether your bug-out location allows for hunting, fishing, or growing crops.
- **Defensibility:** Choose a location that is easy to defend. Look for natural barriers like hills or ridges that can keep unwanted visitors at bay.
- **Terrain:** Assess the local terrain and whether it can support farming or shelter-building. It's also important to consider seasonal variations that could affect your ability to survive or grow food.

Evaluating Climate, Seasons, and Sustainability of Long-Term Stay

- **Climate:** The climate of your bug-out location should be manageable, and you should be prepared for temperature extremes, rainfall, or drought. It's essential to evaluate whether you have the proper clothing and shelter to withstand the local weather.
- **Seasons:** Some areas may be hospitable in certain seasons but challenging in others. For example, winter months in colder regions may make it difficult to access water, while summer heat may make the area unbearable. Assess how the seasons affect the sustainability of your location.
- **Sustainability:** Think long-term. Your bug-out location must be able to support you over an extended period. This includes evaluating the natural resources available, including food, water, and shelter materials.

Bug-Out Bag & Supplies

Packing the right supplies is essential for a successful bug-out. A bug-out bag should meet both short-term survival needs (72 hours) and long-term evacuation needs.

Essentials for 72-Hour Survival vs. Long-Term Evacuation

1. **Short-Term Bug-Out Bag (72 hours):**
    1.1. Water: At least 1 gallon (3.8 liters) per person per day.
    1.2. Food: Non-perishable foods like energy bars, freeze-dried meals, or canned goods.
    1.3. First-aid kit: Include basic supplies like bandages, antiseptic wipes, gauze, and any personal medications.
    1.4. Shelter: Lightweight tarp, sleeping bag, or emergency blanket.
    1.5. Fire-starting tools: Matches, lighters, fire starter.
2. **Long-Term Bug-Out Supplies:**
    2.1. Water purification tools: Filters, purification tablets, or a portable distillation system.
    2.2. Cooking supplies: Portable stove, pots, fuel, and utensils.

2.3. Extra clothing: Warm clothing, boots, rain gear.
  2.4. Tools and equipment: Knife, multi-tool, hammer, rope, duct tape.
  2.5. Communication devices: Solar-powered radios or satellite phones.

Packing Considerations: Weight, Durability, Redundancy, and Multipurpose Items

When packing your bag, consider the following:
- **Weight:** Your bag needs to be light enough to carry comfortably over long distances. Keep weight in mind when choosing materials and equipment. Every item should be useful, but don't over-pack.
- **Durability:** Your bag and all equipment should be durable and able to withstand harsh conditions. Prioritize quality over quantity when choosing items.
- **Redundancy:** Bring backup items for key items like fire-starting tools, first aid supplies, and water filtration. You can't afford to risk being caught without a vital piece of equipment.
- **Multipurpose Items:** Choose items that can serve multiple purposes. For example, a large tarp can be used for shelter, a ground cover, or a rain collector. A multi-tool can serve many functions, reducing the need to carry several individual tools.

Tactical Packing Methods for Ease of Transport

- **Packing System:** Organize your gear to maximize access and minimize the time spent searching for items. Pack heavier items closer to your body, and lighter items on the outside or at the top of your bag.
- **Accessing Essentials:** Make sure that items you might need quickly, such as water, snacks, a flashlight, or first aid supplies, are easily accessible. Consider adding external pouches or small bags for these items.
- **Quick Deployment:** Use compression straps or external loops to secure bulky items like a sleeping bag or rolled-up tarp for easy access when needed. You should be able to deploy them without unpacking your entire bag.

56. Shelter Building: Staying Protected in Any Environment

When living off-grid or navigating a survival situation, your shelter is one of the most critical components for survival. Whether you are exposed to extreme weather conditions, wildlife threats, or the risk of being discovered in uncertain situations, building the right type of shelter can mean the difference between comfort and hardship. Understanding the principles of heat retention, wind protection, and insulation will allow you to construct a shelter suitable for any environment, be it the wilderness, urban areas, or a long-term survival situation.

Basic Principles of Survival Shelters

A good shelter serves two primary purposes: protecting you from the elements and conserving energy. Exposure to extreme cold, heat, wind, or precipitation can deplete your energy reserves and lead to life-threatening conditions such as hypothermia or heatstroke.

Understanding Heat Retention, Wind Protection, and Insulation

- **Heat Retention:** In cold environments, your shelter must trap body heat. Use materials like leaves, moss, or snow to add insulation and reduce heat loss.
- **Wind Protection:** Strong winds can make conditions unbearable. Always position your shelter in a way that blocks prevailing winds while allowing some ventilation to prevent condensation buildup.
- **Insulation:** Whether using natural materials or pre-made insulation, your shelter needs to slow down the transfer of heat. The thicker the insulating layer, the better it retains warmth in cold weather or blocks excessive heat in hot climates.

Selecting a Site: Avoiding Flood Zones, Dead Trees, and Exposed Areas

Before building, selecting the right location is critical:

- **Avoid flood zones.** Low-lying areas or dry riverbeds may seem ideal but can flood rapidly.
- **Steer clear of dead trees.** They can fall or drop branches unexpectedly, especially during storms.
- **Stay out of exposed areas.** Hilltops and wide-open spaces leave you vulnerable to strong winds, rain, and unwanted attention.
- **Look for natural windbreaks.** Large rocks, dense foliage, or land formations can help shield your shelter.

Quick Shelter vs. Long-Term Construction Considerations

- **Quick shelters** are designed for immediate use, providing short-term protection in a crisis. These include lean-tos, tarp shelters, and debris huts, which can be constructed in under an hour.
- **Long-term shelters** are more durable and intended for extended stays. These require more planning and resources, such as log cabins, earthbag structures, or underground shelters.

Types of Wilderness Shelters

Wilderness shelters come in many forms, depending on available materials and environmental conditions. The key to building a reliable shelter is using natural or readily available resources efficiently.

Lean-Tos, Debris Huts, A-Frame Shelters, Tarp Shelters

- **Lean-To Shelter:** Uses a basic framework of branches against a support structure, such as a fallen tree or rock wall. This is easy to construct and ideal when time is limited.
- **Debris Hut:** A fully enclosed, insulated shelter made from leaves, branches, and natural debris. Best for colder conditions where heat retention is essential.
- **A-Frame Shelter:** A simple structure shaped like an "A" using poles and natural materials to create a sturdy design that sheds rain effectively.
- **Tarp Shelter:** One of the most versatile shelters, a tarp can be rigged in multiple ways for protection against rain, sun, or wind. It's lightweight and easy to carry.

Constructing a Shelter with Available Materials

- **Branches and Logs:** Can be stacked or leaned together for structural support.
- **Leaves, Grass, and Moss:** Act as insulation, keeping warmth inside the shelter.
- **Rocks:** Can be used for anchoring materials or forming windbreaks.
- **Snow:** Provides excellent insulation in winter environments, with shelters like igloos and snow caves.

Waterproofing Techniques and Insulation for Warmth

- **Layer natural materials.** Leaves and grass should be packed thick for insulation.
- **Use mud or clay.** Applying a mud mixture over a shelter's surface can help with waterproofing.
- **Elevate your bedding.** Sleeping on the cold ground draws heat away. Use dry leaves, branches, or a survival blanket to stay insulated.
- **Angle your roof properly.** A steeper roof will help shed rain and snow more efficiently.

Urban Survival Shelters

In urban survival scenarios, access to natural materials may be limited, but creative use of buildings, furniture, and other resources can provide adequate protection.

Safe Indoor Areas for Natural Disasters and Urban Conflicts

- **Basements and interior rooms** offer protection from extreme weather, explosions, and social unrest.
- **Stairwells, elevator shafts, or underground parking garages** may provide shelter if traditional buildings are unsafe.
- **Office buildings and malls** may offer supplies but should be entered cautiously.

Creating Hidden, Reinforced, or Disguised Sleeping Areas

- **Hiding in plain sight** is often the best defense. Sheltering in abandoned or overlooked areas reduces the risk of being discovered.

- **Reinforce weak walls or doors** with furniture, metal sheets, or additional barricades.
- **Create a camouflaged sleep area** by using dark materials and blending into the surroundings.

Repurposing Existing Structures for Protection

- **Use furniture and rubble** to create barriers or fortified spaces.
- **Turn abandoned cars into temporary shelters,** insulating windows with blankets or sunshades.
- **Create rooftop or attic hideouts** where visibility is low, and access is restricted.

Reinforced & Long-Term Survival Shelters

When long-term shelter is required, more durable construction methods are needed. These shelters provide better insulation, security, and sustainability.

Building Semi-Permanent Shelters from Logs, Clay, and Rock

- **Log Cabins:** A classic survival structure, using logs stacked and sealed with clay or mud for insulation.
- **Cob or Adobe Shelters:** Made from a mixture of clay, sand, and straw, these structures maintain temperature regulation.
- **Stone Shelters:** Using large rocks stacked and reinforced with clay or cement, these shelters offer durability against harsh weather.

Creating Underground or Earthbag Shelters for Extreme Conditions

- **Underground Shelters:** Dug into the earth and reinforced with wood, concrete, or metal. Offers excellent insulation and protection from extreme temperatures.
- **Earthbag Shelters:** Constructed using sandbags or sacks filled with earth, stacked, and sealed with plaster or clay. They provide excellent insulation and security.

Camouflaging Shelters for Security in Uncertain Environments

- **Use natural surroundings** like trees, bushes, or rock formations to break up the visual outline.
- **Paint or cover structures** with mud or vegetation to blend into the landscape.

- **Limit smoke and light emissions** to avoid drawing attention in hostile situations.

Final Considerations

The key to survival shelter success is adaptability—using the resources available in your environment efficiently. Whether you need a short-term lean-to in the woods, a reinforced underground structure, or a hidden urban hideout, these shelter-building techniques will give you the protection needed to survive in any condition.

Wilderness Water Solutions: Finding, Purifying & Storing Water

When you're off-grid, in the wilderness, or in an emergency situation, securing a clean and reliable water source is the first priority. You can go weeks without food, but just a few days without water can put your survival at risk. Understanding how to locate, purify, and store water can be the difference between life and death in remote or disaster-stricken areas.

Locating Water in the Wild

Finding water in nature requires observation, an understanding of terrain, and resourcefulness. While large bodies of water like rivers and lakes are the easiest to spot, sometimes survival depends on extracting water from less obvious sources.

Recognizing Terrain Features That Indicate Water Sources

- **Low-lying areas** naturally collect water. Valleys, depressions, and dried-up riverbeds often hold underground moisture.
- **Vegetation clues** can lead you to water. Lush, green plants such as willows and reeds thrive near water sources.
- **Animal behavior** can provide guidance. Birds often fly low toward water sources at dawn and dusk, and insects like mosquitoes congregate near standing water.

- **Rock formations and soil moisture** suggest underground water. Digging in damp areas, especially in riverbeds, can sometimes yield water.
- **Hearing running water** in quiet areas can alert you to streams, especially in hilly terrain.

Finding Underground Water, Dew Collection, and Tree Tapping

- **Digging in dry riverbeds**: Water may be just below the surface of sandy or clay areas. Dig a hole several feet deep and let the moisture seep in.
- **Dew collection**: In the early morning, grass, leaves, and non-poisonous plants often hold dew. Dragging a cloth through vegetation and wringing it out can yield small amounts of water.
- **Tree tapping**: Some trees, like birch and maple, store drinkable sap inside. Cutting a small notch or inserting a hollow twig can allow sap to drip out for consumption.

Evaluating Water Quality Before Consumption

Even if water looks clear, it can still be contaminated with bacteria, parasites, or chemicals. Before drinking:

- **Look at the water's color and clarity**. Murky water, green algae, or foam on the surface suggests contamination.
- **Smell the water**. A rotten or chemical smell may indicate toxins.
- **Test in small amounts** if no purification is available. Taking a small sip and waiting a few hours for reactions (such as stomach cramps) can be a last-resort method.

Water Purification Methods

Once water is located, it needs to be purified before drinking to prevent illness. There are several effective methods, each with its own advantages and limitations.

Boiling: Effective But Fuel-Intensive

- **Why it works**: Boiling kills bacteria, viruses, and parasites, making it one of the most effective purification methods.
- **How to do it**: Bring water to a rolling boil for **1 minute** at lower elevations and **3 minutes** at high elevations.
- **Limitations**: Requires fuel and a fire source, making it impractical in fuel-scarce environments.

Filtration: Using Natural Materials or Portable Filters

1. **DIY natural filtration**: A homemade water filter can remove debris and large contaminants.
   1.1. Layer a container with **sand, charcoal, and gravel**.
   1.2. Pour water through slowly to trap particles before further purification.
2. **Survival water filters**:
   2.1. **Straw filters** are lightweight and can purify water instantly when drinking.
   2.2. **Gravity-fed filtration systems** work well for larger amounts of water.

Chemical Treatment: Iodine, Chlorine, and Improvised Disinfectants

1. **Iodine tablets or liquid**:
   1.1. Add **5 drops per quart/liter** (double for cloudy water).
   1.2. Let stand for **30 minutes** before drinking.
2. **Chlorine (household bleach, unscented)**:
   2.1. Use **8 drops per gallon (4 liters)** and wait at least **30 minutes**.
3. **Improvised treatment**:
   3.1. Crushed charcoal from a fire can absorb impurities.
   3.2. Pine resin and some plants have antimicrobial properties.

Solar Distillation and UV Purification Methods

- **Solar still**: Dig a hole, place a container in the center, cover with plastic, and let sunlight condense and collect purified water.
- **UV purification**: Fill a **clear plastic bottle** and leave in direct sunlight for **6+ hours** to kill pathogens.

Water Storage & Conservation

Once water is secured and purified, it needs to be stored efficiently to last through emergencies.

Improvised Containers for Transport and Storage

- **Plastic bottles and collapsible water bladders** for portability.
- **Clay or ceramic pots** for keeping water cool and fresh.
- **Rain barrels, cisterns, or underground tanks** for long-term storage.

Maximizing Efficiency with Solar Stills and Rainwater Catchment

- **Solar stills** help extract moisture even in arid conditions.
- **Rainwater collection** using tarps, gutters, or funneling systems maximizes available water.

Techniques to Reduce Dehydration and Conserve Water in Survival Situations

- **Sip small amounts frequently** instead of large gulps.
- **Reduce exertion** to minimize sweating in hot conditions.
- **Use covered or shaded storage** to prevent evaporation.

57. Wilderness Water Solutions: Finding, Purifying & Storing Water

When you're off-grid, in the wilderness, or in an emergency situation, securing a clean and reliable water source is the first priority. You can go weeks without food, but just a few days without water can put your survival at risk. Understanding how to locate, purify, and store water can be the difference between life and death in remote or disaster-stricken areas.

Locating Water in the Wild

Finding water in nature requires observation, an understanding of terrain, and resourcefulness. While large bodies of water like rivers and lakes are the easiest to spot, sometimes survival depends on extracting water from less obvious sources.

Recognizing Terrain Features That Indicate Water Sources

- **Low-lying areas** naturally collect water. Valleys, depressions, and dried-up riverbeds often hold underground moisture.
- **Vegetation clues** can lead you to water. Lush, green plants such as willows and reeds thrive near water sources.
- **Animal behavior** can provide guidance. Birds often fly low toward water sources at dawn and dusk, and insects like mosquitoes congregate near standing water.
- **Rock formations and soil moisture** suggest underground water. Digging in damp areas, especially in riverbeds, can sometimes yield water.
- **Hearing running water** in quiet areas can alert you to streams, especially in hilly terrain.

Finding Underground Water, Dew Collection, and Tree Tapping

- **Digging in dry riverbeds**: Water may be just below the surface of sandy or clay areas. Dig a hole several feet deep and let the moisture seep in.
- **Dew collection**: In the early morning, grass, leaves, and non-poisonous plants often hold dew. Dragging a cloth through vegetation and wringing it out can yield small amounts of water.
- **Tree tapping**: Some trees, like birch and maple, store drinkable sap inside. Cutting a small notch or inserting a hollow twig can allow sap to drip out for consumption.

Evaluating Water Quality Before Consumption

Even if water looks clear, it can still be contaminated with bacteria, parasites, or chemicals. Before drinking:

- **Look at the water's color and clarity**. Murky water, green algae, or foam on the surface suggests contamination.
- **Smell the water**. A rotten or chemical smell may indicate toxins.

- **Test in small amounts** if no purification is available. Taking a small sip and waiting a few hours for reactions (such as stomach cramps) can be a last-resort method.

Water Purification Methods

Once water is located, it needs to be purified before drinking to prevent illness. There are several effective methods, each with its own advantages and limitations.

Boiling: Effective But Fuel-Intensive
- **Why it works**: Boiling kills bacteria, viruses, and parasites, making it one of the most effective purification methods.
- **How to do it**: Bring water to a rolling boil for **1 minute** at lower elevations and **3 minutes** at high elevations.
- **Limitations**: Requires fuel and a fire source, making it impractical in fuel-scarce environments.

Filtration: Using Natural Materials or Portable Filters
3. **DIY natural filtration**: A homemade water filter can remove debris and large contaminants.
   3.1. Layer a container with **sand, charcoal, and gravel**.
   3.2. Pour water through slowly to trap particles before further purification.
4. **Survival water filters**:
   4.1. **Straw filters** are lightweight and can purify water instantly when drinking.
   4.2. **Gravity-fed filtration systems** work well for larger amounts of water.

Chemical Treatment: Iodine, Chlorine, and Improvised Disinfectants
4. **Iodine tablets or liquid**:
   4.1. Add **5 drops per quart/liter** (double for cloudy water).
   4.2. Let stand for **30 minutes** before drinking.
5. **Chlorine (household bleach, unscented)**:
   5.1. Use **8 drops per gallon (4 liters)** and wait at least **30 minutes**.

6. **Improvised treatment**:
   6.1. Crushed charcoal from a fire can absorb impurities.
   6.2. Pine resin and some plants have antimicrobial properties.

Solar Distillation and UV Purification Methods
- **Solar still**: Dig a hole, place a container in the center, cover with plastic, and let sunlight condense and collect purified water.
- **UV purification**: Fill a **clear plastic bottle** and leave in direct sunlight for **6+ hours** to kill pathogens.

Water Storage & Conservation

Once water is secured and purified, it needs to be stored efficiently to last through emergencies.

Improvised Containers for Transport and Storage
- **Plastic bottles and collapsible water bladders** for portability.
- **Clay or ceramic pots** for keeping water cool and fresh.
- **Rain barrels, cisterns, or underground tanks** for long-term storage.

Maximizing Efficiency with Solar Stills and Rainwater Catchment
- **Solar stills** help extract moisture even in arid conditions.
- **Rainwater collection** using tarps, gutters, or funneling systems maximizes available water.

Techniques to Reduce Dehydration and Conserve Water in Survival Situations
- **Sip small amounts frequently** instead of large gulps.
- **Reduce exertion** to minimize sweating in hot conditions.
- **Use covered or shaded storage** to prevent evaporation.

58. Navigation Without GPS: Finding Your Way in Any Terrain

In a world increasingly dependent on technology, the ability to navigate without relying on GPS or digital tools is a vital survival skill. Whether you're venturing into the wilderness, bugging out during an emergency, or simply trying to stay oriented in unfamiliar terrain, knowing how to use traditional navigation techniques is indispensable.

Map & Compass Basics

A map and compass are two of the most reliable tools in wilderness navigation. While GPS has made things simpler, understanding the basics of reading maps and using a compass will allow you to navigate confidently when the grid fails.

Reading Topographic Maps and Understanding Contour Lines

Topographic maps are your detailed guide to understanding the landscape. Unlike regular maps, they show the elevation changes through contour lines. These lines represent different elevations, so they are vital for understanding the terrain.

- **Contour lines** are curved lines drawn at equal intervals. The closer they are to each other, the steeper the slope. If they are farther apart, the terrain is relatively flat.
- **Identifying elevation**: Each line is labeled with its elevation above sea level. Knowing how to read these lines helps you assess how to traverse a hill or valley, anticipate difficulty, and plan your route accordingly.
- **Reading features**: Features like mountains, valleys, ridges, and depressions are shown by the arrangement of contour lines. For example, concentric circles indicate a mountain peak, while lines that form "V" shapes point toward a stream or river.

Adjusting for Magnetic Declination and Orienting a Compass

A compass points to magnetic north, but the Earth's magnetic poles do not align with true north, so you must adjust for magnetic declination.

- **Magnetic declination**: This is the angle between magnetic north and true north and varies by location. You can find the magnetic declination on most maps, and it will tell you how many degrees to adjust your compass to get an accurate heading.

How to orient the compass:
1. Hold your compass level.
2. Align the compass needle with the orienting arrow.
3. Rotate the housing so that the needle points to magnetic north, then set the map's north direction to align with your compass.

By knowing how to adjust for declination, you can confidently read a map and navigate through even the most remote terrains.

Triangulation Techniques to Pinpoint Location

Triangulation is a method used to pinpoint your exact location using three known landmarks. Here's how it works:

- **Step 1**: Identify three prominent landmarks visible on your map and in your environment. These could be mountains, large buildings, or distinct natural formations.
- **Step 2**: Use your compass to take a bearing to each landmark.
- **Step 3**: Transfer those bearings onto your map using a protractor or compass. Where the lines intersect is your location.

Triangulation can be the most effective way to determine your exact position when you're in the wilderness without any reliable markers.

Navigating with Natural Landmarks

Sometimes, navigating the terrain without the use of technology is necessary. In such cases, natural landmarks become your guide.

Using the Sun, Moon, and Stars for Direction

The sun, moon, and stars have been guiding travelers for millennia. These celestial bodies are reliable indicators of direction and can help you find your way even when you don't have a map or compass.

- **The sun**: In the Northern Hemisphere, the sun rises in the east and sets in the west. At noon, the sun will be directly south. You can use this information to approximate directions.
- **The stars**: The North Star (Polaris) is located directly above the North Pole. If you can locate it in the night sky, you'll know you're facing north. In the Southern Hemisphere, the Southern Cross constellation can be used to navigate south.
- **Using the moon**: The moon's position in the sky can also be used to determine direction. At certain times of the month, the moon rises in the east and sets in the west.

Recognizing Patterns in Wind, Plant Growth, and Animal Behavior

The natural world can provide numerous clues to help orient yourself. Animals and plants respond to environmental factors that can indicate direction.

- **Wind**: In some areas, wind patterns can be predictable. In coastal areas, winds often blow from the ocean toward the land during the day and from the land toward the ocean at night. Recognizing these patterns can give you a sense of direction.
- **Plant growth**: In many parts of the world, plants grow toward the sunlight, so if you find a group of trees or plants with clear growth patterns, they may guide you toward or away from the sun.
- **Animal behavior**: Birds, especially waterfowl, often follow specific migratory paths. If you observe their flight patterns, you can approximate which way is north or south.

Estimating Distance Using Pace Counting and Shadow Sticks

- **Pace counting**: Count your steps as you walk to estimate the distance between two points. Typically, an average adult's pace is about **2.5 feet** per step. By counting your steps, you can estimate the distance you've traveled, which is particularly useful in unfamiliar terrain.
- **Shadow sticks**: A simple but effective tool for estimating direction and time. Place a stick upright in the ground and mark the tip of its shadow. As the day progresses, you'll see the shadow move. This method can help you track the time and determine north and south based on the shadow's position.

Improvised Navigation Tools

In a pinch, you can create effective navigation tools using available materials in your environment.

Making a Sun Compass with a Stick and Shadow Method

A sun compass can be created with just a stick and the sun. This tool allows you to estimate direction without any technological help.

- **Step 1**: Place a stick upright in the ground. It should be long enough to cast a shadow that you can track.
- **Step 2**: Mark the tip of the shadow at regular intervals, such as every 15-30 minutes.
- **Step 3**: After several hours, the shadow will have moved, and you can draw a line between the markings. The first mark shows **west**, the last one shows **east**, and the line between the two shows **north-south**.

Crafting a Floating Needle Compass

A floating needle compass can be made with a needle, a leaf, and some water. This can serve as a backup navigation tool.

- **Step 1**: Rub the needle against a magnet or piece of magnetite to charge it with a magnetic field.
- **Step 2**: Place the needle on a floating leaf or piece of cork on a bowl of water. The needle will align itself along the Earth's magnetic field, indicating direction.

Using Water Reflections and Cloud Movement for General Direction

- **Water reflections**: If you're near a body of water, you can often use reflections of the sun or landmarks to help determine direction. The sun's reflection will help identify the position of east and west, based on its movement throughout the day.
- **Cloud movement**: In the Northern Hemisphere, clouds generally move from west to east. Observing the cloud movement can help you identify the general direction and assess your position.

Transportation & Routes

Once you have your bearings, planning routes and understanding terrain is crucial for navigation.

Planning Multiple Escape Routes and Alternatives

When navigating off-grid, it's important to plan for multiple routes, especially if your primary route is blocked. Having alternatives ensures you can always move toward safety or resources.

- **Evaluate the terrain**: Plan routes that avoid known hazards, like steep terrain, river crossings, or areas prone to natural disasters.
- **Prepare backup routes**: Depending on the situation, you might need to take a detour. Know where roads, trails, or passable terrain are located on your map.

Navigating Without GPS: Understanding Roadblocks, Detours, and Environmental Obstacles

Without GPS, it's important to stay flexible and adapt as you navigate.

- **Roadblocks**: Know common obstacles like rivers, cliffs, or urban congestion. Plan alternate routes for navigating around these.
- **Detours**: Sometimes environmental factors, like a wildfire or flood, can force you to change course. Know how to adjust your route accordingly.
- **Environmental obstacles**: Learn to identify and plan for environmental factors like shifting sand dunes, snow-covered terrain, or dense forest that could delay your progress.

Utilizing Off-Road Vehicles, Bikes, Boats, or Walking Routes

- **Off-road vehicles**: When traveling through rough terrain, using ATVs or UTVs can help cover ground faster. However, be aware of fuel limitations and terrain hazards.
- **Bikes**: In less rugged terrain, bicycles are an excellent alternative to off-road vehicles, offering maneuverability and ease of transport.
- **Boats**: If navigating rivers or lakes, boats can provide a quick and effective method of transportation.
- **Walking routes**: For areas where vehicles can't go, walking may be the only option. Understanding how to estimate walking time and distance is essential.

59. .Emergency Power Solutions

When living off-grid or preparing for an emergency, power becomes one of the most critical considerations. Whether you're in a survival scenario or simply trying to maintain comfort in an off-grid setting, understanding how to generate and efficiently use energy is essential.

Portable Power Sources

In an emergency situation, having access to portable power is vital. These power sources are easy to transport and offer flexibility when you're on the move or when your fixed power system is unavailable. Let's take a closer look at selecting the right options.

Selecting the Right Battery Banks and Solar Chargers for Survival

Battery banks and solar chargers are essential components of off-grid energy systems. These devices allow you to store energy and use it when needed, whether for lighting, communication, or powering small appliances.

1. **Choosing the right battery bank**: When selecting a battery bank for survival purposes, you need to consider its capacity (measured in amp-hours, or Ah) and its discharge rate. Look for deep-cycle batteries, as they are designed to handle frequent charging and discharging.
2. **Storage capacity**: The larger the capacity of the battery bank, the more energy you can store. For an emergency scenario, you should aim for a battery bank that can power your essentials for at least 24-48 hours, depending on your needs.

3. **Solar chargers**: Solar chargers are an excellent way to keep your battery bank topped off. These come in various sizes, from small foldable panels to larger, stationary units. The efficiency of the charger depends on the size of the panel and how much sunlight it receives.
   3.1. **Solar panel wattage**: Match the wattage of the solar charger to the capacity of your battery bank. A general rule is to have a panel that can charge your bank in 4-6 hours of good sunlight.
   3.2. **Portability**: For emergencies, a foldable or portable solar panel is a great option. These can be easily packed and carried to different locations, providing flexibility if you need to set up your power source in different areas.

Hand-Crank Generators and Pedal-Powered Devices

When you don't have access to sunlight or if your battery bank runs low, hand-crank generators or pedal-powered devices can serve as backup power sources.

1. **Hand-crank generators**: These generators work by converting manual energy into electricity. They are small and compact, making them an ideal solution for charging small devices such as radios, lights, or phones in an emergency. The downside is that they require physical effort, which may not be sustainable for long periods of time.
   1.1. **Choosing a hand-crank generator**: Look for a model with a built-in USB port for charging your devices, and consider one with a larger capacity if you plan to use it for longer periods.
   1.2. **Maintenance**: Keep the generator in good working condition by ensuring the cranking mechanism is lubricated and that it is stored properly when not in use.
2. **Pedal-powered devices**: Pedal-powered generators offer a more efficient way to generate power over an extended period. They can power larger devices, such as small appliances, or recharge larger battery banks.
   2.1. **Pedal-powered bike generator**: This is essentially a stationary bicycle attached to a generator. By pedaling, you generate electricity that is stored in a battery or directly powers devices. It's a good option for when you have time to generate power and can keep pedaling.
   2.2. **Energy consumption**: Pedal-powered generators require consistent effort. You should be able to generate between 50-150 watts per hour depending on your pedaling speed and the device used.

Alternative Energy Generation

While solar panels and hand-crank generators are great options for small-scale energy production, sometimes you need to think bigger to meet your energy needs. Alternative energy generation methods such as wind, hydro, and biomass power can supplement or replace solar energy in off-grid environments.

Small-Scale Hydro, Wind Turbines, and Biomass Fuel Options

These systems can provide a continuous, renewable source of energy if you are located in the right environment.

1. **Small-scale hydro**: If you are near a stream or river, a small hydroelectric generator can provide constant power. These systems work by harnessing the energy from flowing water to turn a turbine, which generates electricity.
   1.1. **Installation**: The generator is typically installed in a flowing water source, with a turbine connected to the water flow. A battery bank or storage system is necessary to store the energy for use.
   1.2. **Maintenance**: Regular maintenance includes checking for debris in the water that could clog the turbine and ensuring the

system is properly grounded to prevent electrical issues.
2. **Wind turbines**: Wind energy is a great renewable resource if you live in an area with consistent wind. Small-scale wind turbines can be used to generate electricity for lighting, tools, and appliances.
   2.1. **Choosing the right turbine**: Ensure the turbine is sized appropriately for your energy needs. Residential turbines generally range from 400 watts to 10 kilowatts.
   2.2. **Installation**: The turbine should be placed at a height of 30-50 feet above the ground to maximize exposure to the wind.
3. **Biomass fuel**: Biomass energy comes from organic materials, including wood, crop residues, and even animal waste. This material can be burned or converted into biogas or biofuel to generate heat or electricity.
   3.1. **Biogas generators**: A biogas generator uses anaerobic digestion to break down organic waste, producing methane gas that can be burned for heat or electricity.

Converting Vehicle Alternators for Power Generation

A vehicle's alternator is designed to generate power when the engine is running. With a few modifications, you can use this alternator to charge batteries or power devices in an off-grid scenario.

• **How it works**: When a vehicle's engine is running, the alternator converts mechanical energy into electrical energy, which charges the vehicle's battery. By connecting the alternator to a battery bank, you can use the vehicle to generate power.
• **Installation**: You can create a system where the vehicle's alternator is connected to a charging circuit. This allows you to generate electricity when driving and store it for later use.
• **Power output**: The output of an alternator can vary but typically ranges from 40-150 amps, depending on the vehicle. This can be enough to power small appliances or charge batteries for off-grid use.

Efficient Power Use in a Survival Scenario

In a survival scenario, managing energy efficiently is just as important as generating it. You need to prioritize your energy consumption and ensure that you have enough power for critical systems like communication, lighting, and heating.

Prioritizing Energy Needs: Lighting, Heating, Communication

In an emergency or off-grid scenario, the energy demands can quickly outweigh the available supply. Therefore, prioritizing your energy use is crucial.

• **Lighting**: Lighting should be a priority, especially after dark. LED lights are energy-efficient and can provide a steady source of illumination for extended periods. Consider solar-powered or hand-crank lanterns as alternatives when battery power is limited.
• **Heating**: In colder environments, keeping warm is essential. If you have a wood stove or propane heater, they may need to be powered intermittently. During the warmer months, natural cooling methods like ventilation can reduce heating demands.
• **Communication**: Communication devices such as radios and phones should be charged or powered as needed. Solar chargers and hand-crank generators are excellent for this purpose.

Maximizing Battery Efficiency and Extending Power Storage

Battery storage is one of the most important elements of an off-grid power system. To get the most out of your storage, you need to maximize efficiency.

• **Battery type**: Use deep-cycle batteries designed for off-grid applications. These batteries can handle deep discharges and recharges without losing capacity.
• **Energy management**: Keep track of your energy consumption and only use devices when necessary. Solar panels can help charge your batteries during the day, but you may need to regulate your power usage at night.
• **Battery maintenance**: Regularly check the battery's health and maintain it by cleaning

terminals and keeping the system dry to prevent corrosion.

Low-Tech Alternatives to Modern Electrical Appliances

In off-grid scenarios, it's important to adapt and use low-tech solutions when possible. These alternatives can significantly reduce your reliance on electricity.

• **Hand-powered tools**: Instead of using electric-powered tools, opt for manual alternatives like hand drills, saws, or gardening tools. These tools require no electricity and are often more reliable.
• **Cooking alternatives**: Solar cookers, wood stoves, and rocket stoves can help you prepare meals without using electricity. These methods use natural resources like sunlight or wood for cooking.
• **Non-electric lighting**: Use oil lamps, candles, or even firelight for illumination. These methods have been used for centuries and can provide enough light for tasks like reading or cooking.

60. Fire-Starting: Reliable Methods for Any Condition

Fire is one of the most crucial survival tools, and understanding how to reliably start and maintain a fire can mean the difference between life and death in a wilderness or emergency scenario. Whether you're trying to cook food, purify water, or keep warm in cold conditions, fire will be your constant companion. Fire Basics: Understanding the Fire Triangle

Before we delve into fire-starting methods, it's important to understand the fundamental elements that sustain a fire. These are known as the "Fire Triangle," and it consists of three components: oxygen, fuel, and heat.

Oxygen, Fuel, and Heat: How They Interact

• **Oxygen**: Fire needs oxygen to burn. In an open setting, the air around you contains about 21% oxygen, which is sufficient to keep most fires going. However, in a contained space (like a shelter), you'll need to ensure that airflow allows for enough oxygen to keep the fire burning. Lack of oxygen will smother the fire.
• **Fuel**: Fuel is any material that will catch fire and burn. This can be anything from dry wood and grass to paper and even some oils. The key is selecting the right fuel for the fire size you need. For example, small twigs and dry leaves work for kindling, while larger logs are needed for a long-lasting fire.
• **Heat**: Heat is the energy required to ignite the fuel. Without heat, fuel won't combust, no matter how much oxygen is available. When starting a fire, it's important to introduce enough heat to initiate combustion, which is why the initial spark or ember is so critical.

Choosing the Right Fire Location

Once you understand how fire works, the next step is finding the ideal location. A poor fire location can result in inefficiency or even make your fire unsafe.

• **Wind Protection**: Wind can either help or hinder a fire. A strong wind can provide much-needed oxygen, but it can also blow out your fire if it's too strong. When starting a fire, look for a natural windbreak (like a rock outcropping or dense brush), or build a windbreak using available materials.
• **Fuel Access**: Choose a location with easy access to fuel, such as dry wood, grass, or branches. Look for dead and dry wood, as wet or green wood is much harder to ignite and sustain.
• **Safety Measures**: Always clear the area of dry leaves, debris, or anything else that could catch fire unintentionally. Make sure you have enough space around the fire to prevent it from spreading. Use a circle of rocks to contain the fire if necessary.

Traditional Fire-Starting Techniques

Now that you understand the basics, it's time to learn how to start a fire. Let's begin with the methods that have been used by people for centuries.

Using Matches, Lighters, and Fire Strikers Efficiently

1. **Matches**: Standard safety matches are often the easiest way to get a fire started. They're simple, fast, and effective when you have dry tinder.
    1.1. **Keep matches dry**: Store matches in a waterproof container to ensure they are

ready when needed. Consider waterproof matches if you're in a wet environment.
  - 1.2. **Strike the match correctly**: Hold the match at an angle and strike it across the rough side of the matchbox. Hold the match away from your body to avoid burning yourself.
2. **Lighters**: Lighters are an excellent choice for starting fires quickly. They are easy to use, but they are dependent on fuel.
  - 2.1. **Keep lighters dry**: If you're in a wet environment, store your lighter in a waterproof case to prevent it from getting damp.
3. **Fire Strikers**: A fire striker creates sparks when scraped against rough material like steel wool or flint. It's one of the most reliable methods of fire-starting in wet conditions.
  - 3.1. **Use dry tinder**: Fire strikers require dry material to catch the sparks. Always use tinder that is dry and fibrous (like dry grass, cotton balls, or paper).
  - 3.2. **Proper striking**: Hold the striker close to the tinder and scrape it forcefully. Aim for a spark to land directly on the tinder to ignite it.

Primitive & Emergency Fire-Starting Methods

In an emergency, when matches and lighters aren't available, you'll need to resort to more primitive fire-starting methods. Here's how you can make a fire with minimal equipment.

Friction Fire Techniques: Bow Drill, Hand Drill, Fire Plow

Bow Drill: The bow drill is one of the oldest fire-starting methods. It uses a spindle and hearth board to create friction, which generates heat and eventually an ember.

  - 1.1. **Build the bow**: Select a flexible branch and use a strong cord to form a bow.
  - 1.2. **Create the spindle**: Choose a straight piece of wood, about 10-12 inches long, to serve as the spindle.
  - 1.3. **Start drilling**: Place the spindle on the hearth board, and use the bow to rapidly spin the spindle while applying downward pressure. Once an ember forms, transfer it to your tinder bundle.

Hand Drill: The hand drill method is similar to the bow drill but doesn't use a bow. Instead, you use your hands to spin the spindle against the hearth board.

  - 1.4. **Position the spindle**: Place the spindle vertically on the hearth board, and roll it between your palms to create friction. This method requires more effort and practice but is highly effective with dry conditions.

Fire Plow: The fire plow is another friction-based method where you use a piece of wood to scrape against the surface of a dry piece of bark or wood. The friction generates heat that can ignite a small amount of dry tinder.

  - 1.5. **Scrape the surface**: Using a sharp edge of wood, scrape across the surface of the bark or wood to generate heat.

Lens-Based Methods: Magnifying Glass, Ice Lens, Water-Filled Bag

Magnifying Glass: A magnifying glass can be used to concentrate sunlight onto tinder, igniting it. This works best in direct sunlight.

  - 1.1. **Find the sun**: Hold the magnifying glass at an angle so that it concentrates sunlight into a small point on the tinder.
  - 1.2. **Maintain focus**: Hold the glass steady until the tinder begins to smolder.

Ice Lens: In the winter months, you can make a lens from clear ice to focus sunlight. This requires finding clear, transparent ice and shaping it into a convex lens.

1.3. **Shape the ice**: Carve the ice into a lens shape and hold it between the sun and your tinder, focusing the light to ignite the material.

Water-Filled Bag: In a pinch, you can use a clear plastic bag filled with water to focus sunlight. The curvature of the water-filled bag creates a lens that concentrates sunlight.

1.4. **Position the bag**: Hold the bag over dry tinder and adjust until the light is focused into a small, concentrated point.

Chemical Reactions for Emergency Fire-Starting

Potassium Permanganate & Glycerin: These two substances can be combined to create an exothermic reaction that generates heat. By combining a small amount of potassium permanganate with glycerin, you can create a fire-starting chemical reaction.

- **Prepare the chemicals**: Carefully place a small amount of potassium permanganate on a dry surface.
- **Add glycerin**: Drop glycerin onto the potassium permanganate and watch the reaction begin. The mixture will heat up and eventually ignite.

Outdoor Cooking Fires

Once you've mastered fire-starting, you'll want to use your fire for cooking. Understanding the best way to cook outdoors in different conditions is critical for survival.

Building an efficient cooking fire: Once your fire is started, arrange the fuel into a manageable shape for cooking. The type of fire you build will depend on what you're cooking:

- **Cooking with direct flame**: Some foods can be cooked directly over the flame. Fish, meat, and vegetables can be skewered or placed on a grate over the fire.
- **Using embers**: For slow-cooked foods like stews, it's often better to use the embers of the fire rather than the flames. Allow the fire to burn down to coals and then cook your food on or around the embers.
- **Building a reflector oven**: A reflector oven uses the heat from the fire and reflects it onto your cooking pot or food. This method allows for more controlled heat and is perfect for baking bread or roasting meats.

61. Handy Knots Everyone Should Learn

Whether you're setting up camp, building a shelter, or simply securing gear, knowing how to tie the right knot at the right time can make all the difference. In survival situations, knots serve as one of the most versatile tools you can use. They are critical for creating shelter, securing gear, lifting heavy loads, and even saving lives in rescue operations. In this section, you'll learn the essential knots every off-grider should know. These knots are easy to learn and can be applied to a wide range of practical survival tasks.

Essential Survival Knots

When it comes to knots, it's important to know which one to use for the task at hand. Here are the most commonly used survival knots that you should have in your toolkit.

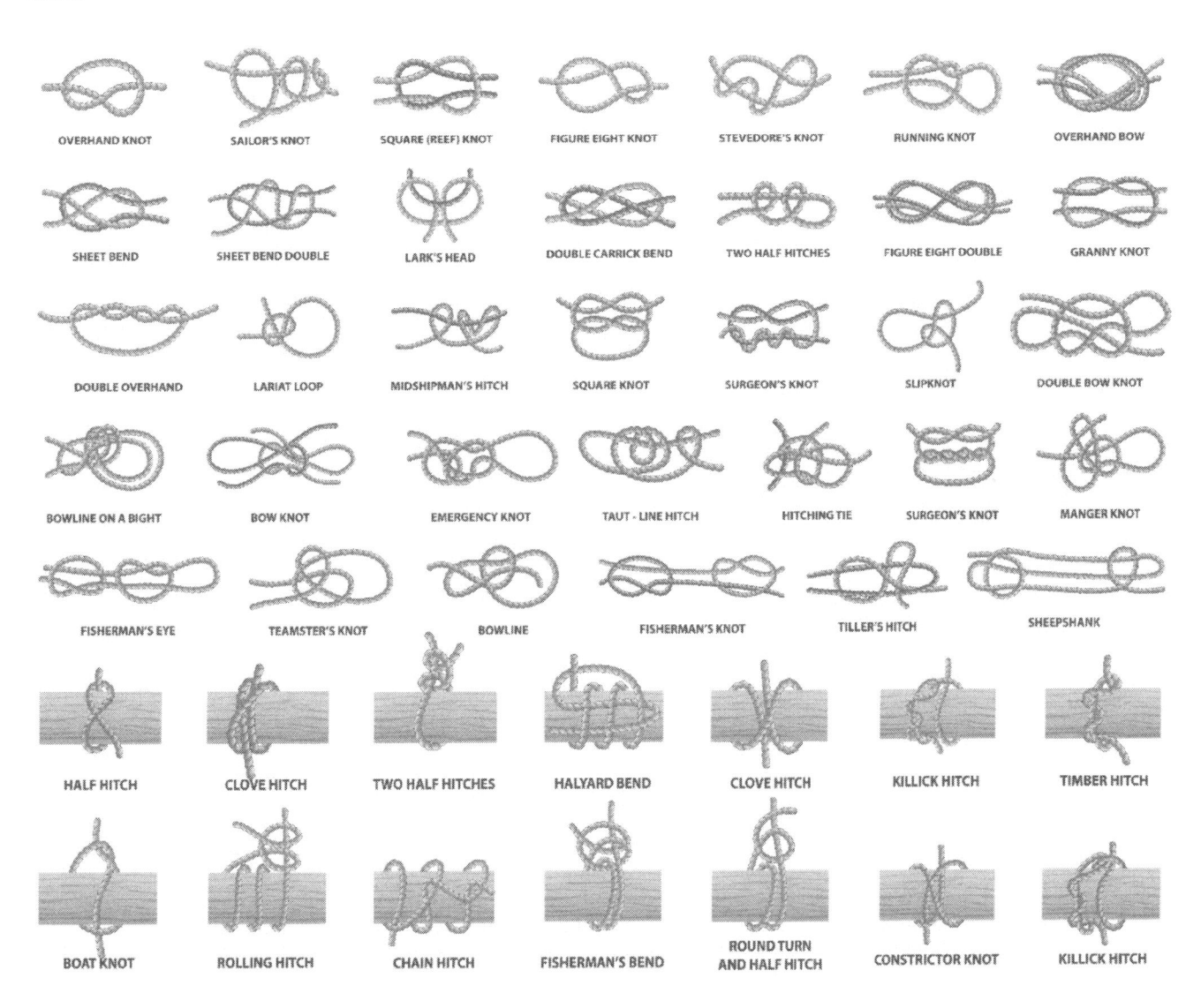

OVERHAND KNOT	SAILOR'S KNOT	SQUARE (REEF) KNOT	FIGURE EIGHT KNOT	STEVEDORE'S KNOT	RUNNING KNOT	OVERHAND BOW	
SHEET BEND	SHEET BEND DOUBLE	LARK'S HEAD	DOUBLE CARRICK BEND	TWO HALF HITCHES	FIGURE EIGHT DOUBLE	GRANNY KNOT	
DOUBLE OVERHAND	LARIAT LOOP	MIDSHIPMAN'S HITCH	SQUARE KNOT	SURGEON'S KNOT	SLIPKNOT	DOUBLE BOW KNOT	
BOWLINE ON A BIGHT	BOW KNOT	EMERGENCY KNOT	TAUT-LINE HITCH	HITCHING TIE	SURGEON'S KNOT	MANGER KNOT	
FISHERMAN'S EYE	TEAMSTER'S KNOT	BOWLINE	FISHERMAN'S KNOT	TILLER'S HITCH	SHEEPSHANK		
HALF HITCH	CLOVE HITCH	TWO HALF HITCHES	HALYARD BEND	CLOVE HITCH	KILLICK HITCH	TIMBER HITCH	
BOAT KNOT	ROLLING HITCH	CHAIN HITCH	FISHERMAN'S BEND	ROUND TURN AND HALF HITCH	CONSTRICTOR KNOT	KILLICK HITCH	

Bowline: Creating a Secure Loop That Doesn't Slip

The bowline is one of the most versatile knots and is known for creating a strong, secure loop at the end of a rope. It's easy to tie and, most importantly, it doesn't slip under pressure, making it ideal for rescue operations, securing items, or setting up tarps.

- **Step 1**: Create a small loop in the rope, leaving enough tail to work with.
- **Step 2**: Pass the working end of the rope through the loop from underneath.
- **Step 3**: Bring the working end of the rope around the standing part of the rope.
- **Step 4**: Pass the working end back through the loop.
- **Step 5**: Tighten the knot by pulling on the standing part and the working end.

Square Knot: Binding Two Ropes of Equal Thickness Together

The square knot, also known as a reef knot, is one of the simplest knots. It is used for tying two ropes of the same thickness together securely. It's great for bundling items or tying up packages.

- **Step 1**: Take the right rope and lay it over the left rope.
- **Step 2**: Bring the right rope under the left and pull it tight.
- **Step 3**: Now, take the left rope and lay it over the right rope.
- **Step 4**: Bring the left rope under the right and pull it tight.
- **Step 5**: Ensure the knot is secure by pulling on both ropes.

Clove Hitch: Securing Objects to Poles and Branches

The clove hitch is one of the simplest knots for securing a rope to a pole, tree, or other cylindrical objects. It's quick and easy to tie, and it is essential for tasks like securing a tarp, tying a boat to a dock, or fixing a rope to a post.

- **Step 1**: Wrap the rope around the object, crossing over itself.
- **Step 2**: Loop the rope around again and pass the working end of the rope under the first loop.
- **Step 3**: Tighten the knot by pulling both ends of the rope.

Figure-Eight Knot: Preventing Rope Slippage in High Tension

The figure-eight knot is often used to create a stopper knot to prevent ropes from slipping through carabiners or other devices under tension. It's essential for climbing and rescue operations but also useful in any high-stress situation where you need to secure a rope and prevent it from untying.

- **Step 1**: Form a loop in the rope.
- **Step 2**: Bring the working end of the rope around the standing part of the rope.
- **Step 3**: Pass the working end back through the loop.
- **Step 4**: Tighten the knot by pulling on both the working end and the standing part.

Knots for Shelter Building & Repairs

Building shelters in off-grid environments often requires you to tie down tarps, secure ropes, and bind materials together. The following knots are essential for shelter construction.

Taut-Line Hitch: Adjustable Tension Knot for Tents and Shelters

The taut-line hitch is a highly useful knot when you need to adjust tension on a rope. It's ideal for securing a tent or tarp, as it allows you to easily adjust the tension, ensuring your shelter stays taut and secure.

- **Step 1**: Wrap the rope around a secure object.
- **Step 2**: Form a loop around the rope.
- **Step 3**: Wrap the working end of the rope around the standing part twice.
- **Step 4**: Pass the working end through the loop and pull tight.
- **Step 5**: Slide the knot to adjust tension.

Sheet Bend: Joining Two Different Thickness Ropes Securely

The sheet bend is useful for tying two ropes of different thicknesses together. It's a strong knot that

is commonly used when combining ropes of various sizes in survival situations.
- **Step 1**: Lay the thicker rope parallel to the thinner rope.
- **Step 2**: Pass the thinner rope through the loop formed by the thicker rope.
- **Step 3**: Wrap the thinner rope around the thicker rope and pass it through the loop.
- **Step 4**: Tighten the knot by pulling both ropes in opposite directions.

Knots for Climbing, Rescue, and Utility Use

Some knots are specifically designed for more technical purposes, such as climbing, rescue operations, or securing heavy loads. Here are some critical knots you should be familiar with in these situations.

Prusik Knot: Ascending Ropes for Climbing or Rescue

The prusik knot is used for climbing or ascending ropes. It allows you to attach a rope to another rope in such a way that it holds under weight but slides when not under pressure. This knot is essential in rescue operations.
- **Step 1**: Tie a loop in a smaller rope (the prusik loop).
- **Step 2**: Wrap the prusik loop around the main rope two or three times.
- **Step 3**: Pull the loop tight and ensure that it slides when not under load, but locks when weight is applied.

Fisherman's Knot: Securing Critical Loads

The fisherman's knot is a reliable and straightforward knot used for joining two ropes of similar diameter. It is particularly useful in fishing, climbing, and other outdoor applications where a secure yet compact knot is needed.

- Step 1: Take the two rope ends and overlap them.
- Step 2: Using one end, tie an overhand knot around the other rope.
- Step 3: Repeat the process with the other rope, tying an overhand knot around the first rope.
- Step 4: Pull both ropes in opposite directions to tighten the knots securely against each other.

62. Essential Outdoor Survival & Disaster Readiness Projects

Project: Design Your Ideal Bug-Out Location

Choosing the right bug-out location is one of the most important decisions you can make for your long-term survival and security. Whether you are anticipating natural disasters, societal unrest, or other emergencies, having a well-thought-out, secure, and sustainable place to retreat to is essential. This project will guide you through the process of designing your ideal bug-out location, considering factors such as accessibility, defensibility, resources, and sustainability.

Step 1: Define the Criteria for Your Ideal Location

- Assess your needs and goals: Consider what you need from a bug-out location. Are you looking for a place that provides immediate shelter and food, or is long-term sustainability the priority?
- Consider accessibility: The location should be relatively easy to access, but not so accessible that it becomes vulnerable to looters or threats.
- Security and defensibility: Choose a location that can be easily defended. Think about natural barriers, such as mountains, rivers, or forests, that provide security.

Step 2: Choose the Location's Terrain and Environment

- Water access: Ensure that the location has access to clean, reliable water sources, whether from natural

springs, rivers, or rainwater catchment. Water is critical for survival.
• Food availability: Check if the land has fertile soil for growing crops, or if there are nearby forests and wildlife for hunting and foraging.
• Climate and seasons: Understand the local climate and seasonal variations. The ideal location should offer comfort year-round, and you must consider challenges like winter cold or summer heat.
• Proximity to dangers: Evaluate potential threats in the surrounding area, including proximity to industrial sites, flood zones, or high-crime areas.

Step 3: Select the Size and Layout of Your Property
• Space requirements: Based on the number of people and animals you plan to accommodate, decide on the size of the property. A few acres might suffice for a small family, while a larger group may need more space for farming, shelter, and livestock.
• Zoning and divisions: Plan for a space that can be divided into different zones, such as living quarters, food production, storage, and defensive areas.
• Buildable land vs. natural resources: Balance between forested or wild areas and the land available for building shelter, growing food, and creating necessary infrastructure.

Step 4: Evaluate the Local Resources for Shelter and Building Materials
• Natural resources: Look for access to building materials such as timber, clay, stone, or sand. These materials will be crucial for building shelters, storage facilities, and other essential structures.
• Long-term sustainability: Make sure that there is a sustainable way to acquire these resources over time without depleting the local environment.
• Consider alternative building methods: Think about alternative construction methods, such as cob, straw bale, or earthbag homes, which require fewer resources and can be built with local materials.

Step 5: Plan for Renewable Energy Sources
• Solar power: Ensure that the location has adequate sunlight to support solar panels for energy. This will be essential for powering lights, small appliances, or other systems.

• Wind or hydro power: If located in an area with consistent wind or access to flowing water, wind turbines or small hydroelectric systems can provide a sustainable power source.
• Backup options: Consider backup power sources, such as hand-crank generators or pedal-powered systems, in case the primary energy source fails.

Step 6: Design for Self-Sufficiency
• Water management: Design systems for collecting and storing rainwater or accessing natural water sources. Consider using solar stills, rainwater catchment systems, and filtration methods.
• Food production: Plan for gardening, small-scale farming, and livestock. Consider crop rotation, polyculture, and perennial crops for year-round food production.
• Waste management: Set up systems for composting human waste, greywater, and food scraps to maintain sanitation and improve soil health.

Step 7: Consider Safety and Security Features
• Defensible perimeter: Create natural or man-made barriers, such as fencing, walls, or thorn hedges, to secure the perimeter of your bug-out location.
• Emergency escape routes: Plan multiple escape routes in case of emergency. These routes should be safe, hidden, and easily accessible from multiple locations on your property.
• Lookout points and surveillance: Plan for lookout points or tree platforms to keep an eye on any movement around your location. Surveillance is key for spotting potential threats early.

Step 8: Build Shelter and Establish Living Spaces
• Shelter types: Depending on your needs and resources, plan for temporary or permanent shelters. Options include tents, cabin-style buildings, or earth homes like underground bunkers or earthbag homes.
• Insulation and heating: Ensure your shelter is insulated to protect against extreme temperatures. Plan for wood stoves or other off-grid heating solutions.
• Living quarters layout: Design your living spaces with functionality in mind. Include areas for

sleeping, eating, storage, and communal activities. Ensure that all spaces are multi-functional and efficient.

Step 9: Set Up Communication and Warning Systems

- Emergency communication tools: Set up basic communication tools such as two-way radios, satellite phones, or signal mirrors to stay in touch with others.
- Warning systems: Consider setting up smoke signals, flare guns, or alarms to alert others of danger. Make sure these systems are easy to use and can cover a wide area.
- Signal fires or flags: In case of isolation, a signal fire or a colored flag system can help others locate you in an emergency.

Step 10: Test and Update the Plan Regularly

- Conduct regular checks: Periodically assess the bug-out location for any changes in the environment, security concerns, or resource availability.
- Update your supplies: As your bug-out location develops, update your supplies and storage systems. Ensure that all equipment, tools, and provisions are functional and in good condition.
- Simulate evacuation scenarios: Practice evacuating to your location in different weather conditions, at night, or under stress, to ensure your plan is robust and well-tested.

Project: Assemble a Bug-Out Bag

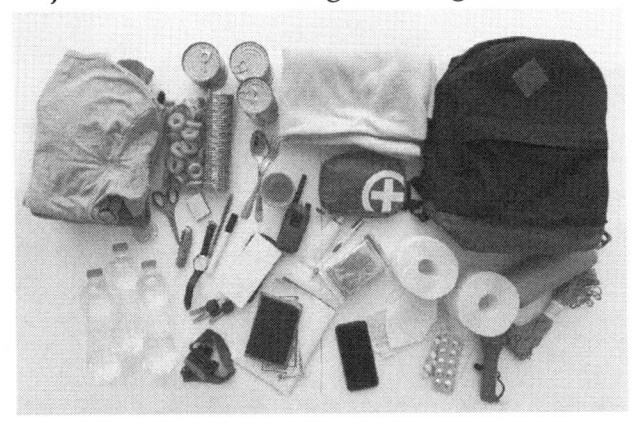

Assembling a bug-out bag (BOB) is a crucial step in preparing for emergencies. A well-packed bag ensures that you are ready to leave at a moment's notice, equipped with all the essentials to survive for at least 72 hours. The contents should be versatile, durable, and tailored to your specific needs, climate, and the types of emergencies you might face. This project will guide you through the process of assembling a bug-out bag, focusing on the essentials for survival, mobility, and security.

Step 1: Choose the Right Bag

- Select a durable, lightweight bag: Your bug-out bag should be easy to carry, especially over long distances. Look for a backpack or rucksack made from heavy-duty materials like nylon or polyester that can withstand the weight of your gear.
- Size considerations: The bag should be large enough to fit all your essentials but not so big that it becomes cumbersome. A 40-60 liter bag is ideal for a 72-hour kit.
- Multiple compartments: Choose a bag with multiple compartments to organize your gear, making it easier to access items quickly when needed.

Step 2: Pack Water and Hydration Essentials

1. **Water bottles:** Include at least two durable water bottles or hydration bladders. Opt for bottles that are BPA-free and durable enough to handle rugged conditions.
2. **Water purification methods:**

2.1. **Water purification tablets:** Pack iodine or chlorine tablets to purify water if you can't access clean water sources.
2.2. **Portable water filter:** A lightweight, compact water filter like a LifeStraw or Sawyer Mini can provide safe drinking water from natural sources.
2.3. **Collapsible water container:** In case you need to carry water for longer distances, include a collapsible water container for extra capacity.

Step 3: Select Shelter and Sleeping Gear
• Tent or tarp: A lightweight, weather-resistant shelter is crucial for protecting yourself from the elements. Choose a small, compact tent or a multi-use tarp that can be set up quickly.
• Sleeping bag: Choose a sleeping bag suited for the climate in your area. Make sure it's lightweight and can pack down to fit in your bag.
• Sleeping pad or mat: A compact sleeping pad provides insulation from the ground, making it more comfortable to rest and sleep.
• Emergency blanket: Include a thermal, space-saving emergency blanket to help retain body heat if needed.

Step 4: Pack Food and Cooking Supplies
• Non-perishable food: Pack high-energy, lightweight foods like granola bars, freeze-dried meals, jerky, and nuts. Make sure the food is easy to prepare and doesn't require refrigeration.
• Portable stove: A small, lightweight stove such as a pocket rocket or camping stove will allow you to cook food or boil water. Don't forget the necessary fuel.
• Cooking utensils: Include a lightweight cooking pot, utensils, and a small knife for food preparation.
• Waterproof matches/lighter: Keep a reliable lighter or waterproof matches for lighting your stove or starting a fire.

Step 5: Prepare First Aid and Medical Supplies
• Basic first aid kit: Pack a compact first aid kit that includes bandages, antiseptic wipes, gauze, medical tape, tweezers, and pain relievers.
• Prescription medications: If you take any medications, be sure to include a 72-hour supply in your bag.
• Personal hygiene items: Include items like hand sanitizer, a toothbrush, toothpaste, and soap, preferably biodegradable to minimize environmental impact.
• Insect repellent: Pack a small bottle of insect repellent to prevent bites from mosquitoes and other pests.

Step 6: Include Clothing and Weather Protection
• Weather-appropriate clothing: Choose lightweight, moisture-wicking clothes that can be layered. Pack extra socks, underwear, and a hat to protect from the sun.
• Rain gear: Include a lightweight rain poncho or jacket to stay dry in wet conditions.
• Sturdy footwear: A pair of well-worn, durable boots or shoes suitable for hiking and long-distance walking is essential. Make sure they are comfortable and broken in.
• Gloves and scarf: For colder environments, pack gloves, a scarf, and a beanie to protect against the cold.

Step 7: Include Tools and Self-Defense Items
• Multi-tool or knife: A compact multi-tool with pliers, a knife, and various attachments is a versatile and essential piece of gear for survival.
• Fire-starting tools: Pack waterproof matches, a lighter, and a fire starter to help you start a fire in any condition.
• Duct tape and paracord: Duct tape is great for quick repairs, and paracord is a versatile tool for shelter building, securing items, or as a makeshift rope.
• Signal mirror and whistle: These can be used for signaling for help in an emergency, especially if you're lost or need to alert rescuers.

Step 8: Pack Navigation and Communication Tools
• Map and compass: A reliable map of your area and a compass will help you navigate without GPS.

Learn basic map reading and compass skills beforehand.
- Two-way radio or whistle: A two-way radio can help communicate with others in your group, while a whistle is useful for attracting attention in case of an emergency.
- Chargers and power bank: If you plan on using electronics, pack a portable power bank to charge your phone or other devices. Consider solar-powered chargers as a sustainable alternative.

Step 9: Secure Your Personal Documents
- Copies of important documents: Store photocopies of your ID, passport, insurance, and any medical records in a waterproof bag.
- Cash or alternative currency: In case digital payments aren't available, pack some cash in small denominations and consider carrying gold or silver as a backup.
- Emergency contacts: Write down the names, phone numbers, and addresses of your emergency contacts in case your phone is unavailable.

Step 10: Review, Pack, and Practice
- Review your gear: Before packing, make sure all your items are functional and that your bag isn't too heavy to carry. Remove any non-essential items that might add unnecessary weight.
- Test your bag: Pack your bag with all the items listed above and practice carrying it. Make sure it fits comfortably and that you can access critical items quickly.
- Prepare for evacuation: Practice scenarios where you might need to evacuate quickly, and plan the quickest route from your home to your bug-out location. Familiarize yourself with the contents of your bug-out bag so you can respond quickly in an emergency.

Project: Create a Personal Bug-Out Plan

Creating a personal bug-out plan is a crucial step in preparing for any emergency where you may need to leave your home quickly. The plan ensures that you and your family have the necessary resources, information, and strategies to stay safe when evacuation becomes necessary. Follow these steps to create an organized and effective bug-out plan.

Step 1: Assess Potential Risks
- Identify local risks and emergencies: Consider natural disasters like floods, wildfires, earthquakes, or hurricanes. Also, factor in man-made emergencies like civil unrest or industrial accidents.
- Evaluate the likelihood of each risk: Understand the probability of each event occurring in your area to prioritize your responses.
- Understand your immediate needs: Consider mobility, medical needs, and the presence of dependents (children, elderly family members, pets) that will influence your plan.

Step 2: Establish Bug-Out Triggers
- Define when to evacuate: Clearly outline the signs that indicate it's time to leave. This could include extreme weather conditions, loss of utilities, or escalations in local conflicts.
- Use real-time information: Plan to stay updated on current events via local news, radio broadcasts, or trusted online sources. This will help you react quickly when necessary.
- Set decision-making criteria: Decide who in the household will make the final call to bug out based on the situation.

Step 3: Choose Your Bug-Out Locations
- Primary location: Choose a safe location to evacuate to first. This should be a place that's accessible, secure, and has necessary resources like water and shelter. It can be a relative's house, a cabin, or a pre-selected shelter.
- Secondary location: Identify an alternative destination in case your primary location becomes unreachable or compromised. This should be located at a reasonable distance to avoid potential roadblocks or hazards.
- Consider terrain and climate: Choose a location with manageable terrain, an adequate climate for survival, and natural resources like water or wood.

Step 4: Assemble Your Bug-Out Bag
1. **Essentials for 72 hours:** Pack enough supplies for at least 72 hours, focusing on portability, efficiency, and durability.
    1.1. **Food and water:** Non-perishable food items, high-calorie energy bars, and a

portable water filter or purification tablets.
  - 1.2. **Clothing:** Pack durable, weather-appropriate clothing, including extra socks, jackets, and thermal wear if necessary.
  - 1.3. **First aid kit:** Include basic medical supplies such as bandages, antiseptics, pain relievers, and any prescription medications.
  - 1.4. **Tools and gear:** Flashlight, multi-tool, fire-starting materials, and a first-aid kit.
2. **Tactical considerations:** Choose bags that are lightweight and easy to carry, such as a tactical backpack or duffel bag, with multiple compartments for easy access.
3. **Add personal documents:** Include copies of important documents such as IDs, insurance papers, and any other legal documents that might be necessary in an emergency.

Step 5: Plan Escape Routes

1. **Primary and secondary routes:** Identify your main route to your bug-out location, considering factors such as roadblocks, traffic congestion, and natural obstacles.
   - 1.1. **Use topographic maps and apps:** Plan your route using reliable maps, and note alternate routes that avoid known hazards like highways or major intersections.
   - 1.2. **Travel at different times:** If you can, practice traveling during different times of the day to test the ease of access and potential for bottlenecks.
2. **Vehicle readiness:** Ensure your vehicle is capable of taking you to your bug-out location. Keep the fuel tank full and check that the vehicle is in good working condition (tires, oil, brakes, etc.).
   - 2.1. **Off-road capabilities:** If you plan to use off-road vehicles or ATV/UTVs, ensure they are in working order and stocked with appropriate fuel and tools.

Step 6: Communicate with Your Family or Group

- Establish communication plans: Make sure everyone in your household or group knows when and how to communicate if you get separated. Consider using walkie-talkies or satellite phones if traditional communication methods are unavailable.
- Create a meeting point: If you become separated, have a designated meeting point at the bug-out location or along the route.
- Emergency contacts: Set up an emergency contact list of people outside your immediate area who can help relay information or assist with communications during a crisis.

Step 7: Prepare for Pets and Livestock

- Pet supplies: Pack food, water, leashes, and medical supplies for pets. Also, include carriers for small animals or plans for transporting larger animals.
- Livestock evacuation: Consider how you'll transport and care for larger animals such as chickens, goats, or horses. If evacuation isn't possible, consider arranging for someone to care for them.
- Vaccinations and documentation: Ensure that your pets and livestock have up-to-date vaccinations, and carry necessary documentation.

Step 8: Secure Your Home Before You Leave

- Lock and protect valuables: Ensure that your home is secure, valuables are stored in a safe place, and doors/windows are locked.
- Turn off utilities: Disconnect gas, water, and electricity if necessary to prevent hazards like gas leaks or electrical fires.
- Leave instructions for neighbors (if possible): If you have a trusted neighbor, inform them about your departure and give instructions for handling any emergencies that may arise in your absence.

Step 9: Practice the Plan

- Run evacuation drills: Practice bugging out with your family or group to ensure that everyone knows

their role and understands how to execute the plan efficiently.
- Review and update the plan regularly: Regularly review your bug-out plan and supplies to ensure that everything is up-to-date.
- Evaluate challenges: During practice runs, identify any weak points in your plan, whether it's the route, the bug-out bag, or communications, and address them.

Step 10: Monitor and Adapt
- Stay informed: Keep track of changing conditions in your area, such as weather patterns, political instability, or natural disasters, that might affect your plan.
- Adapt the plan: As conditions change, so should your bug-out plan. Be prepared to adjust the location, routes, or supplies depending on new information.
- Review your resources: Check the status of your vehicle, supplies, and communication methods regularly. Replace expired items, refill fuel, and test equipment.

Project: Build a Portable Emergency Shelter

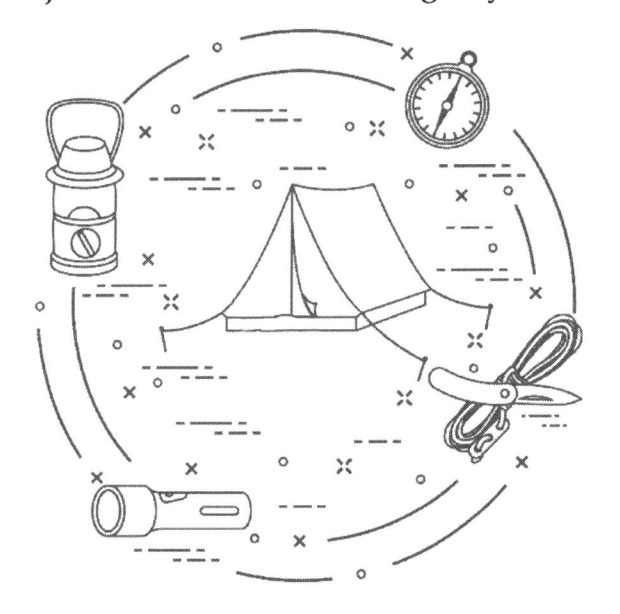

Building a portable emergency shelter can be crucial in a survival situation where you need protection from the elements. This project will guide you through constructing a shelter that is lightweight, compact, and easy to set up in various environments.

Step 1: Gather Materials
- Tarp or waterproof fabric: Choose a durable material that will keep you dry in rain or snow.
- Rope or paracord: Strong enough to tie and secure your shelter.
- Stakes or heavy-duty pins: To anchor your shelter to the ground.
- Branches or poles: For frame support (can be gathered from the environment).
- Survival blanket or mylar blanket: To help with insulation and reflect heat.
- Small tarp or groundsheet (optional): To provide additional protection from the ground moisture.

Step 2: Select the Location
- Flat ground: Find a spot that is level to avoid discomfort while resting and to ensure stability.
- Avoid low areas: Do not set up your shelter in valleys, depressions, or flood-prone areas to prevent water runoff.
- Wind protection: If possible, set up near natural windbreaks like trees or large rocks.
- Proximity to resources: Make sure you are near a water source but not too close to avoid flooding risks.

Step 3: Construct the Frame
- Find sturdy branches or poles: Look for branches that are about 4-6 feet long and sturdy enough to hold the weight of the tarp or fabric.
- Create a basic frame: Lay out two longer poles parallel to each other. These will serve as the sides of the shelter.
- Secure the frame: Use shorter poles or branches across the longer poles to create a triangular frame. Make sure everything is secured with rope or cord.
- Tie the frame to a solid anchor: Ensure the frame is stable by anchoring one end of each long pole to the ground with stakes or large rocks.

Step 4: Attach the Tarp or Waterproof Cover
- Place the tarp over the frame: Drape your tarp or waterproof fabric over the frame to provide shelter from the elements.
- Secure the tarp: Use rope or paracord to tie the tarp tightly to the frame, making sure it is taut. If the weather is expected to be windy, make sure to pull the tarp tight to prevent it from flapping.

• Ensure coverage: Ensure the tarp covers all sides of the frame, providing full protection from the elements.

Step 5: Secure the Shelter to the Ground
• Use stakes to anchor the corners: Drive stakes or heavy-duty pins into the ground at the four corners of the shelter, pulling the tarp tight.
• Tie the rope: If the shelter is large, consider adding additional anchor points along the sides for extra stability. Tie the rope around the stakes and the shelter frame.
• Check for stability: Make sure the shelter is firmly anchored and will not be blown away by the wind. Adjust the ropes or reposition the stakes if needed.

Step 6: Add Insulation and Comfort
• Lay a survival blanket or mylar blanket on the ground: This will act as a barrier between you and the cold ground, improving comfort and insulation.
• Fill gaps with natural materials (optional): Gather dry leaves, pine needles, or grass to stuff around the edges of the shelter to help with insulation.
• Create a small fire pit (if safe): If the situation allows, consider building a small fire in front of your shelter to provide warmth.

Step 7: Test and Adjust
• Check for leaks: After setting up the shelter, test it by applying water to see if it holds up against rain. If any leaks appear, seal them with waterproof tape or reposition the tarp.
• Test wind resistance: If possible, simulate wind by testing the shelter with a fan or in a breezy area to check how well it stays secure.
• Evaluate comfort: Lie inside the shelter for a few minutes to see if it's comfortable and protects from the ground and elements. Adjust the shelter accordingly for optimal comfort and protection.

Step 8: Pack It Back Up
• Disassemble the shelter: Once you're done testing or using the shelter, take it down carefully.
• Clean the site: Leave no trace by collecting any materials you used and making sure the area is clear of debris.
• Store the shelter materials: Roll up the tarp, fold the poles, and pack the ropes in a way that they can be quickly reassembled next time.

Project: Emergency Surface Water Collection and Filtration Setup

Surface water, such as water from rivers, lakes, or streams, can be an invaluable resource for an off-grid home, but it requires proper filtration and purification before it can be used safely. The water may contain contaminants, bacteria, or other harmful substances that need to be removed. This project will guide you through the steps of collecting surface water from a nearby stream, river, or pond and setting up a filtration system to ensure the water is clean and safe for use in your off-grid home.

Materials Needed
• A water container or barrel for collecting surface water
• A large mesh filter or cloth for pre-filtration
• A fine filtration system (such as activated carbon, sand, or ceramic filters)
• A clean bucket or container for storing filtered water
• A portable pump or siphon (if needed)
• A UV sterilizer or water purification tablets (optional for final disinfection)
• A water testing kit (optional but recommended for assessing water quality)
• A tarp or large container for creating a temporary collection basin (if necessary)

Step 1: Choose the Right Surface Water Collection Site

The first step in collecting surface water is selecting an appropriate collection site. Ideally, you want to choose a clean, flowing water source that is as free from contamination as possible.

• **Rivers or streams** are typically better than lakes or ponds because the moving water is less likely to harbor stagnant contaminants.

- Look for **clear water** that shows minimal visible debris, algae, or pollutants. Avoid areas that are near agricultural runoff, industrial waste sites, or human settlements.
- When collecting water, always take water from the **center of the water body** where it is less likely to be contaminated by debris or animal waste near the shoreline.
- If the water source is in a **remote area**, consider whether it is accessible year-round and if you can easily transport the water back to your off-grid location.

Step 2: Set Up the Water Collection System

Once you have selected the collection site, the next step is to gather the water. You may need to use a bucket, container, or a portable pump to collect the water depending on the proximity and the amount of water you need.

1. Gather water with a container: If the source is shallow and easily accessible, you can simply use a bucket or large container to scoop water directly from the source.
2. Use a siphon or pump system: If the source is a deep river or lake, a portable pump or siphon system can be used to extract water without needing to dip containers into the water directly.
3. Pre-filter the water: Before storing water, pass it through a mesh cloth or a large mesh filter to remove visible debris such as leaves, sticks, or other large particles. This step ensures that your filters don't clog quickly and increases the efficiency of your filtration system.

Step 3: Set Up a Filtration System

Once you have collected the surface water, the next step is filtration. Filtration removes particles and contaminants from the water, making it safer for drinking, cooking, and other uses. There are several types of filtration systems that can be set up depending on the available resources.

Option 1: Sand and Gravel Filter

A sand and gravel filter is one of the most cost-effective and simple ways to filter water. This filter uses different layers of materials to trap debris and particles.

1. Prepare a container: Use a large bucket or barrel as the main filtration container.
2. Layering materials: Start by adding a layer of gravel at the bottom of the container. This acts as a coarse filter and will catch larger debris.
3. Add a layer of sand on top of the gravel to filter smaller particles. The finer the sand, the better it will filter out contaminants.
4. If available, add a layer of activated charcoal above the sand to help remove odors, chemicals, and some contaminants.
5. Slowly pour the collected surface water through the filter and into a clean container below.

This filter works well for basic filtration but may need to be followed by additional treatment for bacterial contamination.

Option 2: Ceramic or Bio-Filters

A ceramic filter or a bio-filter can be used as a more efficient filtration system, especially for drinking water. These filters are capable of removing both physical contaminants and microorganisms.

1. Purchase or make a ceramic filter: Ceramic filters are available for purchase, or you can create one by inserting a ceramic filtration device into a clean container.
2. If creating a DIY bio-filter, fill a large container with alternating layers of charcoal, sand, and gravel.
3. Place the filter directly over a clean container to catch the filtered water.
4. Let the water pass through slowly for best results.

Ceramic filters typically remove bacteria, protozoa, and larger pathogens, making them suitable for drinking and cooking water.

Step 4: Final Disinfection of Water

While filtration removes many contaminants, it does not eliminate harmful bacteria and viruses. For water to be fully drinkable, it must be treated to kill any remaining microorganisms. There are two main methods to do this: UV light treatment and chemical disinfection.

UV Light Treatment

A portable UV sterilizer is a great option for disinfecting water without chemicals. UV light kills bacteria and viruses by disrupting their DNA, rendering them harmless.

1. Fill the UV device with filtered water.

2. Activate the UV light for the required amount of time (usually 2-5 minutes, depending on the device).
3. After treatment, the water is safe to drink and can be stored in a clean container.

Chemical Disinfection

If UV sterilization is not an option, water purification tablets or liquid chlorine can be used for disinfection.

1. Follow the manufacturer's instructions on the water purification tablets or chlorine drops.
2. Add the required amount of disinfectant to your water and let it sit for 30 minutes to an hour.
3. Once the waiting period is over, the water is safe for consumption.

Step 5: Storing Filtered and Treated Water

After collecting, filtering, and disinfecting the water, the next step is safe storage. Storing water properly ensures that it remains clean and ready for use when needed.

1. Store the treated water in a clean, covered container to prevent contamination.
2. Use food-grade plastic containers or stainless steel tanks for storing water to avoid leaching chemicals.
3. Label the containers and use them within a reasonable time frame to ensure that the water remains fresh. If you store large amounts of water, rotate your supply to avoid stagnation.

Step 6: Regular Maintenance

A surface water collection and filtration system requires regular cleaning and maintenance to ensure it remains functional.

1. Clean the filters regularly to prevent clogging. Depending on the filter type, you may need to replace or scrub the media (sand, charcoal, etc.) every few months.
2. Inspect and clean the water collection basin (e.g., tarp or collection container) to prevent the buildup of algae, mold, or debris.
3. Check the water storage tanks for signs of contamination and clean them periodically.

Thank You

Creating this guide has been a rewarding journey—driven by dedication and a deep passion for helping people like you take back control of their lives. In today's uncertain world, the ability to be self-reliant and prepared is more important than ever. This guide is here to equip you with proven strategies and practical solutions to create a secure, self-sustaining future for yourself and your family.

Your support is what keeps this mission alive. It's people like you—ready to take action and embrace self-sufficiency—who inspire us to continue sharing these tools and systems with others. Thank you for being a vital part of this growing movement toward independence and resilience.

We'd love to hear your thoughts! Your feedback is incredibly valuable—not just to us, but to others who are searching for reliable, time-tested ways to fortify their homes, secure essential resources, and protect their families. Sharing your experience only takes a moment, and it can help inspire someone else to start their own off-grid journey.

Here's how you can share:

Option A: Record a quick video and tell us what excites you most about using this guide. Whether you're setting up solar panels, collecting rainwater, or growing your own food, we'd love to hear how you're putting these strategies into action.

Option B: Prefer to stay behind the scenes? No problem! Snap a few photos of the book in use—maybe next to one of your DIY project or just write a few words about your experience.

Your input not only acknowledges the effort and care that went into this guide, but also helps others discover a resource that can truly change lives.

While it's completely optional, your voice makes a real difference—and we're genuinely grateful for your support.

Scan the QR code below to share your excitement!

Access Your Extra Content

To access your extra content

Off Grid Video Tutorials	Printable Prepper Checklists
Prepping & Homesteading Budget Template	First Aid Kit Essentials Cheat Sheet
Printable Emergency Communication Cards	Examples of Off-Grid Home Layouts

Scan the QR code below:

Or go to → rebrand.ly/off-grid-extra

References

- agrolearner.com
- Aishwarya Naik, Bhagyashree Lanje, Mrunal Chimote, Swapnil Khubalkar. "Design and Implementation of Automated Solar-Based Irrigation System with Sensor-Based Pump Activation", 2023 World Conference on Communication & Computing (WCONF), 2023
- Ali Çelik, Emre Mandev, Orhan Ersan, Burak Muratçobanoğlu, Mehmet Akif Ceviz, Yusuf AliKara. "Thermal performance evaluation of PCM-integrated interior shading devices inbuilding glass facades", Journal of Energy Storage, 2025
- Amit Kumar Tyagi, Shrikant Tiwari. "AI and Blockchain in Smart Grids - Fundamentals, Methods, and Applications", CRC Press, 2025
- betterbythelake.com
- botanikks.com
- canadagrowsupplies.com
- cellsaviors.com
- chinabrewtec.com
- D. Yogi Goswami, Frank Kreith. "Handbook of Energy Efficiency and Renewable Energy", CRCPress, 201909
- diyhappy.com
- doorwinwindows.com
- Doug Oughton, Ant Wilson. "Faber & Kell's Heating and Air-Conditioning of Buildings", Routledge, 2014
- durablevintage.com
- electricsaver1200.com
- energymatters.com.au
- energyscaperenewables.com
- everydayshooter.com
- expressivecandles.com
- fastercapital.com
- "Food Security, Nutrition and Sustainability Through Aquaculture Technologies", Springer Science and Business Media LLC, 2025
- friendlyaquaponics.com
- futuristarchitecture.com
- global-batteries.com
- groundwatergovernance.org
- growagoodlife.com
- hayleyssolar.com
- hydroponichorizons.com
- iamhomesteader.com
- Ibrahim Dincer, Dogan Erdemir. "Chapter 1Solar Energy Systems", Springer Science and Business Media LLC, 2024
- Issa Zaiter, Mohamad Ramadan, Ali Bouabid, Ahmad Mayyas, Mutasem El-Fadel, Toufic Mezher. "Enabling industrial decarbonization: Framework for hydrogen integration in the industrial energy systems", Renewable and Sustainable Energy Reviews, 2024211
- Jeffrey M. Farber, Ewen C. Todd. "Safe Handling of Foods", CRC Press, 2019
- John Ashurst, Francis G Dimes. "Conservation of Building and Decorative Stone", Routledge,2007244
- kaizenaire.com
- Michael A. Richards. "Regreening the Built Environment - Nature, Green Space, and Sustainability", Routledge, 2024
- mindcull.com
- Moncef Krarti. "Energy Audit of Building Systems - An Engineering Approach", CRCPress, 2020
- mrswheelbarrow.com
- mtu-solutions.com
- naturallivingideas.com
- oasyswater.com
- openmindplace.com
- outdoorapothecary.com
- Peter F. Varadi. "Sun Above the Horizon -Meteoric Rise of the Solar Industry", Pan Stanford, 2014
- pluginhighway.ca
- portlandiaelectric.supply
- powerequiphub.com
- Pragnaleena Debroy, Priyanka Majumder, Amrit Das, Lalu Seban. "Model-based predictive greenhouse parameter control of aquaponic system", Environmental Science and Pollution Research, 2024
- prepperswill.com
- rainwaterharvesting.tamu.edu
- Robert Ehrlich, Harold A. Geller. "Renewable Energy - A First Course", Routledge, 2017
- Saffa Riffat, Mardiana Idayu Ahmad, AliffShakir. "Chapter 6 Eco-Cities: Sustainable Urban Living", Springer Science and Business Media LLC, 2025
- security-zone.info
- Shivam Gupta, Sushil Kumar Himanshu, Pankaj Kumar Gupta. "Agri-Tech Approaches for Nutrients and Irrigation Water Management", CRC Press, 2024
- shuncy.com
- skyryedesign.com
- slimhome.eu
- solarlesson.com

- solartown.com
- southernliving.com
- Sven Ruin, Göran Sidén. "Small-Scale Renewable Energy Systems – Independent Electricity for Community, Business and Home", CRC Press, 2019
- thekitchencommunity.org
- vancouversun.com
- verticalfarmingplanet.com
- walkingsolar.com
- waterbergbiosphere.org
- Weihao Huang, Qifan Xu, Bolun Zhao, Qian Lv, Guangyuan Wang. "Urban multi-scale building energy modeling (BEM) and computationalfluid dynamics (CFD) integration: Tools, strategies and accuracy for energy microclimate analysis", Journal of Building Engineering, 2025
- worldradiohistory.com

Made in United States
Orlando, FL
09 November 2025